I0748833

36TH CONGRESS, 1st Session. | SENATE. | REP. COM. No. 278.

IN THE SENATE OF THE UNITED STATES.

JUNE 15, 1860.—Ordered to be printed.

JUNE 21, 1860.—*Resolved*, That twenty thousand extra copies of the reports of the majority and minority of the Select Committee on the Harper's Ferry Invasion, with the testimony accompanying, and other papers, be printed for the use of the Senate.

Mr. MASON submitted the following

REPORT.

The Select Committee of the Senate appointed to inquire into the late invasion and seizure of the public property at Harper's Ferry, beg leave to submit their report:

On the 14th of December, 1859, the resolutions annexed were adopted by the Senate of the United States:

"*Resolved*, That a committee be appointed to inquire into the facts attending the late invasion and seizure of the armory and arsenal of the United States at Harper's Ferry, in Virginia, by a band of armed men, and report—

"Whether the same was attended by armed resistance to the authorities and public force of the United States, and by the murder of any of the citizens of Virginia, or of any troops sent there to protect the public property;

"Whether such invasion and seizure was made under color of any organization intended to subvert the government of any of the States of the Union; what was the character and extent of such organization; and whether any citizens of the United States not present were implicated therein, or accessory thereto, by contributions of money, arms, munitions, or otherwise;

"What was the character and extent of the military equipment in the hands or under the control of said armed band; and where and how and when the same was obtained and transported to the place so invaded.

"That said committee report whether any and what legislation may, in their opinion, be necessary on the part of the United States for the future preservation of the peace of the country, or for the safety of the public property; and that said committee have power to send for persons and papers."

In conducting this inquiry the committee examined a number of witnesses, who were summoned before them from different States of the Union. Their testimony in full will be found annexed to this report.

Upon the first subject of inquiry to which their attention was directed by the resolutions, to wit: Whether "the invasion and seizure of the armory and the arsenal of the United States at Harper's Ferry, in Virginia, by a band of armed men, was attended by armed resistance to the authorities and public force of the United States, and by the murder of any of the citizens of Virginia, or of any troops sent there to protect the public property."

The committee find, from the testimony, that this so-called invasion originated with a man named John Brown, who conducted it in person. It appears that Brown had been for some previous years involved in the late difficulties in the Territory of Kansas. He went there at an early day after the settlement of that Territory began, and either took with him or was joined by several sons, and, perhaps, sons-in-law, and, as shown by the proofs, was extensively connected with many of the lawless military expeditions belonging to the history of those times. It would appear, from the testimony of more than one of the witnesses, that, before leaving the Territory, he fully admitted that he had not gone there with any view to permanent settlement, but that, finding all the elements of strife and intestine war there in full operation, created by the division of sentiment between those constituting what were called the free-State and slave-State parties, his purpose was, by participating in it, to keep the public mind inflamed on the subject of slavery in the country, with a view to effect such organizations as might enable him to bring about servile insurrection in the slave States.

To carry these plans into execution, it appears that, in the winter of 1857–58, he collected a number of young men in the Territory of Kansas, most of whom afterwards appeared with him at Harper's Ferry, and placed them under military instruction at a place called Springdale, in the State of Iowa, their instructor being one of the party thus collected, and who, it was said, had some military training. These men were maintained by Brown; and in the spring of 1858 he took them with him to the town of Chatham, in Canada, where he claimed to have summoned a convention for the purpose of organizing a provisional government, as preliminary to his descent upon some one of the slave States. The proceedings of this convention, with the form of the provisional government adopted there, were taken amongst the papers found with Brown's effects after his capture, and were before the committee, and will be found in the appendix to this report. So far as the committee have been able to learn from the testimony, the convention was composed chiefly of negroes who were residents in and about this town of Chatham, in Canada. The only white persons present were Brown and those who accompanied him. The presiding officer of the convention was a negro, and a preacher. At the close of the convention Brown returned with the party he had taken there back to Ohio, and permitted most of them to disperse, upon the agreement that they would be at his command whenever called for. Two of them, however, to wit: John E. Cook, afterwards executed in Virginia, and Richard Realf were sent on the following missions: Cook was sent to Harper's Ferry, in Virginia, with directions to remain there and thereabout subject to the future call of his chief. Realf was sent to the city of New York, as shown by his testimony, for the following purposes: it would seem from the testimony that a man named Hugh Forbes, an Englishman, who, it was said, had the reputation of military experience in some of the revolutions in Southern Europe, had been engaged by Brown to take charge of his military school in Iowa. Differences, however, arising between them, Forbes, who had gone to the West with that view, abandoned the project and returned to New York. Whilst the convention was sitting at Chatham, Brown received information which led him to believe that Forbes had betrayed his counsels, and Realf was dispatched to New York with instructions, if practicable, to get possession of such correspondence with Brown as might prove the facts of his intended descent upon some one of the slave States should his plans be divulged—a mission which, for the reasons stated in the testimony of Realf, altogether failed.

In conducting the inquiry, the committee deemed it a matter of importance to have the testimony of Forbes. It appeared, however, that not long after the explosion at Harper's Ferry, Forbes left the country, and the committee were not able to procure his attendance before them.

As to the attack itself at Harper's Ferry, the committee find that Brown first appeared in that neighborhood early in July, 1859. He came there under the assumed name of Isaac Smith, attended by two of his sons and a son-in-law. He gave out in the neighborhood that he was a farmer from New York, who desired to rent or purchase land in that vicinity, with a view to agricultural pursuits, and soon afterwards rented a small farm on the Maryland side of the river, and some four or five miles from Harper's Ferry, having on it convenient houses, and began farming operations in a very small way. He had little or no intercourse with the people of the country; and when questioned through the curiosity of his neighbors, stated further that he was accustomed to mining operations, and expected to find deposits of metal in the adjacent mountains. He lived in an obscure manner, and attracted but little attention, and certainly no suspicion whatever as to his ulterior objects. Whilst there, he kept some two or three of his party, under assumed names, at Chambersburg, Pennsylvania, who there received, and from time to time forwarded to him, the arms of different kinds of which he was subsequently found in possession. Cook, one of his men spoken of above, it appears, had resided at Harper's Ferry and its neighborhood for some twelve months before Brown appeared, pursuing various occupations. He left the Ferry a few days before the attack was made, and joined Brown at his country place. The whole number assembled with Brown at the time of the invasion were twenty-one men, making with himself in all twenty-two.

On Sunday night, the 16th of October, 1859, between 11 and 12 o'clock at night, Brown, attended by probably eighteen of his company, crossed the bridge connecting the village of Harper's Ferry with the Maryland shore, and, on reaching the Virginia side, proceeded immediately to take possession of the buildings of the armory and arsenal of the United States. These men were armed, each, with a Sharp's rifled carbine, and with revolving pistols. The inhabitants of the village asleep, the presence of this party was not known until they appeared and demanded admittance at the gate leading to the public works, which was locked. The watchman in charge states that on his refusal to admit them, the gate was opened by violence and the party entered, made him prisoner, and established themselves immediately in a strong brick building used as an engine-house, with a room for the watchmen adjoining it. They brought with them a wagon, with one horse, containing arms and some prepared torches.

The invasion thus silently commenced, was as silently conducted, none of the inhabitants having been aroused. Armed parties were then stationed at corners of the streets. Their next movement was to take possession, by detached parties of three or four, of the arsenal of the United States, where the public arms were chiefly deposited, a building not far from the engine-house; and by another party, of the workshops and other buildings of the armory, about half a mile off, on the Shenandoah river, called "Hall's rifle works." These dispositions made, an armed party was sent into the adjoining country, with a view to the seizure of two or three of the principal inhabitants, with such of their slaves as might be found, and to bring them to Harper's Ferry (in the language of Brown) as "hostages;" Cook, who had become well acquainted with the country around Harper's Ferry, acting as their guide. They thus seized Colonel Lewis W. Washington, with several of his slaves, (negro men,) at his residence, some five or six miles distant; and in like manner a gentleman named Allstadt, who lived near the road leading from Colonel Washington's to the Ferry, two or three miles distant from the latter, with some five or six of his slaves, (also negro men.) They brought off also from Colonel Washington's such arms as they found in his house, with a wagon and four horses, for subsequent use, as will be shown. This party with their prisoners arrived at the Ferry a little before day, and the latter were carried at once to the room adjoining the engine-house, where they were kept in custody.

Having thus far apparently perfected his plans, a party was sent, taking Washington's wagon and horses, and five or six of the captured slaves, into Maryland to bring the arms deposited at Brown's house there to a point nearer the Ferry and more accessible. On their way, they seized a gentleman named Byrne, who lived in Maryland, three or four miles from Harper's Ferry, and whom they afterwards sent to the Ferry and placed amongst the other prisoners at the engine-house. It is shown that their design was to have taken at the same time as many of the slaves of Byrne as might be found, but in this they did not succeed. During Monday, a large portion of the arms, consisting of carbines, pistols, in boxes, and pikes, were brought off in the wagon and deposited in a school-house about a mile from the village of Harper's Ferry, on the Maryland side.

The first alarm that was given, indicating the presence of the hostile party, appears to have been on the arrival there of the mail train of cars on the Baltimore and Ohio railroad, on its way from Wheeling to Baltimore, and which arrived at Harper's Ferry at its usual hour, about half past one o'clock in the morning. On the arrival of Brown's party, he had stationed two men, well armed, on the bridge, with directions to permit none to pass. This bridge is a viaduct for the railroad to cross the river, having connected with it a bridge for ordinary travel. When the train arrived, it was arrested by this guard, and very soon afterwards a negro named Hayward, a free man who lived at Harper's Ferry and was in the service of the railroad company as a porter, was shot by this guard and died in a few hours. His statement was, as shown in the testimony of John D. Starry, one of the witnesses, "that he had been out on the railroad bridge, looking after a watchman who was missing, and he had been ordered to halt by some men who were there; and instead of doing that, he turned to go back to the office, and as he turned they shot him in the back." The alarm, however, did not extend to the inhabitants of the town, the scene of operations, so far, being near the river, at points occupied by the railroad structures and the public works; the principal part of the town being somewhat remote from that quarter. The train of cars, after being detained some hours, was permitted to proceed on its way to Baltimore.

When daylight came, as the inhabitants left their houses, consisting chiefly of workmen and others employed in the public works, on their way to their usual occupations, and unconscious of what had occurred during the night, they were seized in the streets by Brown's men and carried as prisoners to the engine-house, until, with those previously there, they amounted to some thirty or forty in number. Pikes were put in the hands of such of the slaves as they had taken, and they were kept under the eyes of their captors, as sentinels, near the buildings they occupied. But their movements being conducted at night, it was not until the morning was well advanced that the presence and character of the party was generally known in the village.

The nearest towns to Harper's Ferry were Charlestown, distant some ten miles, and Martinsburg, about 20. As soon as information could reach those points, the citizens assembled, hurriedly enrolled themselves into military bands, and with such arms as they could find, proceeded to the Ferry. Before their arrival, however, it would seem that some four or five of the marauders, who were stationed at "Hall's rifle works," were driven out by the citizens of the village, and either killed or captured. In the course of the day, an attack was made on the engine and watch-house by those of the armed citizens of the adjoining country who had thus hurriedly arrived, and the prisoners in the watch-house, adjoining the engine-house, were liberated. The attacking parties were fired on by the marauders in the engine-house, and some were severely wounded. It should have been stated that during the night Brown selected ten of those whom he considered the principal men of his prisoners, and carried them into the engine-house, where they were detained. The rest thus left in the watch-house were those who were liberated during the attack spoken of. The engine-house is a strong building, and was occupied by Brown, with seven or eight of his men.

During the day it appears that all of Brown's party, who were not with him in the engine-house, were either killed or captured, except those who were on the Maryland side engaged in removing the arms, as above stated. Before, however, they were thus captured or destroyed, they shot and killed two persons, citizens of Virginia, in the streets. One of them, a man named Boerley, who lived in the village, was killed by a rifle shot near his own house. He had taken no part in any of the attacks, and does not appear even to have been armed. The other, Mr. George W. Turner, was a gentleman who lived in the country some ten miles distant, and who, it appears, had gone to the village upon information that his neighbor, Mr. Washington, had been seized in his house and carried off during the night. It would seem that, for his safety, he had taken a gun offered to him by some one in the village, and was proceeding along the street unattended, with it in his hand, when he also was killed by a rifle ball.

The party immediately under Brown remained barricaded in the engine-house during the whole of that day, (Monday.) They had confined with them ten most respectable and valuable citizens, kept, as stated by Brown, in the nature of "hostages," for the security of his own party, he assuming that a regard for the safety of the "hostages" would deter their friends and neighbors from attempting their rescue by force.

During the day an irregular fire was kept up against the engine-house by the people who assembled, and which was returned by the party within through loop-holes made in the wall, or through the doorway, partially opened.

In this manner two of Brown's party were killed at the doorway; and in the afternoon a gentleman of the village, Mr. Beckham, was killed by a shot from the engine-house. It was clearly shown that he was entirely unarmed, and had exposed his person only for an instant on the railroad bridge opposite to the house.

To conclude this narrative, it appears that as soon as intelligence could be conveyed to Washington of the state of things at Harper's Ferry, the marines on duty at the navy-yard were ordered to the scene of action, under the command of Colonel Robert E. Lee, of the army.

The official report of Colonel Lee, found in the appendix to this report, will show in what manner the affair was ended by the capture of Brown and his remaining party, and the rescue in safety of those he detained as prisoners.

Colonel Lee, it will be seen, found it necessary to carry the house by storm, the party within refusing to surrender except on terms properly held inadmissible. In this affair one marine was killed, and another slightly wounded.

Such, it is believed, are succinctly the facts attending this great outrage; and the committee find in response to so much of the resolutions of the Senate, that the armory and other public works of the United States were in the possession and under the control of this hostile party more than thirty hours; that besides the resistance offered by them to the military force of Virginia, they resisted by force the lawful authority of the United States sent there to dispossess them, killing one, and wounding another of the troops of the United States, and as shown that, before they were thus overpowered, they killed in the streets three of the citizens of Virginia who were alone and not even in military array, beside the negro who was killed by them on their first arrival.

It does not appear that any of the public property was stolen or carried away, although a large sum of money was in the paymaster's office near to the engine-house, and doubless would have been seized had they known where it was. There was nothing to protect it but the ordinary safety of an iron door.

In answer to the inquiry contained in the third resolution of the series, "Whether such invasion and seizure was made under color of any organization, intended to subvert the government of any of the States of the Union, what was the character and extent of such organization, and whether any citizens of the United States, not present, were implicated therein, or accessory thereto, by contributions of money, arms, munitions, or otherwise," the committee report:

There will be found in the Appendix, a copy of the proceedings of a convention held at Chatham, in Canada, before referred to, of the provisional form of government there pretended to have been instituted, the object of which clearly was to subvert the government of one or more of the States, and of course to that extent the government of the United States. The character of the military organization is shown by the commissions issued to certain of the armed party as captains, lieutenants, &c., a specimen of which will be found in the Appendix. It clearly appeared that the scheme of Brown was take with him comparatively but few men, but those had been carefully trained by military instruction previously, and were to act as officers. For his military force he relied, very clearly, on inciting insurrection amongst the slaves, who he supposed would flock to him as soon as it became known that he had entered the State and had been able to retain his position—an expectation to no extent realized, though it was owing alone to the loyalty and well-affected disposition of the slaves that he did not succeed in inciting a servile war, with its necessary attendants of rapine and murder of all sexes, ages, and conditions. It is very certain from the proofs before the committee, that not one of the captured slaves, although arms were placed in their hands, attempted to use them; but on the contrary, as soon as their safety would admit, in the absence of their captors, their arms were thrown away and they hastened back to their homes.

It is shown that Brown brought with him for this expedition arms sufficient to have placed an effective weapon in the hands of not less than 1,500 men; besides which, had he succeeded in obtaining the aid he looked to from the slaves, he had entirely under his control all the arms of the United States deposited in the arsenal at Harper's Ferry. After his capture, beside the arms he brought in the wagon to the Ferry, there were found on the Maryland side, where he had left them, 200 Sharp's rifled carbines, and 200 revolver pistols, packed in the boxes of the manufacturers, with 900 or 1,000 pikes, carefully and strongly made, the blade of steel being securely riveted to a handle about five feet in length; many thousand percussion caps in boxes, and ample stores of fixed ammunition, besides a large supply of powder in kegs, and a chest that contained hospital and other military stores, beside a quantity of extra clothing for troops.

For an answer to the inquiry, how far "any citizens of the United States, not present, were implicated therein or accessory thereto by contributions of money, arms, munitions, or otherwise," the committee deem it best to refer to the evidence which accompanies this report. It does not appear that such contributions were made with actual knowledge of the use for which they were designed by Brown, although it does appear that money was freely contributed by those styling themselves friends of this man Brown, and friends alike of what they styled "the cause of freedom," (of which they claimed him to be an especial apostle,) without inquiry as to the way in which the money would be used by him to advance such pretended cause. The evidence fully shows that he had the pikes manufactured in Connecticut especially for this expedition, and certainly they would appear to have been the most formidable weapon which could have been placed in the unskillful hands for which they were intended. For a description of this weapon, and the story told by Brown to the manufacturer when he ordered them, the committee refer to the evidence of the latter. They were sent directly from Connecticut to Brown, under his assumed name of Isaac Smith, first to Chambersburg, in Pennsylvania, there received by some of Brown's men, who were placed there also under assumed names, and by whom they were transported to his abode near Harper's Ferry.

The history of the rifles and pistols is most interesting to this inquiry. It appears from the evidence that, in 1856, these 200 Sharp's carbines had been forwarded by an association in Massachusetts called the "*Massachusetts State Kansas Committee*," at first to Chicago, on their way to Kansas. At Chicago they were placed under the control of another association, called the "*National Kansas Aid Committee.*" There being some difficulty, from the disordered condition of the country at that time, in getting them to Kansas, they were sent by this last named association into Iowa, where they remained. In January, 1857, it seems there was a meeting of this National Kansas Committee in the city of New York. That committee was constituted of one member from most of the non-slaveholding States. At that meeting John Brown appeared, and made application to have these arms placed in his possession. It would seem that he wanted them, as he expressed it, "for purposes of defense in Kansas;" but as the troubles there were nearly ended, such pretension seems to have been discredited by those to whom it was addressed.

At page 245 of the testimony, a full account of this application for the arms will be found, as given by H. B. Hurd, who was the secretary of the association. He states that, "When Mr. Brown was pressing his claim for the aid desired, I asked him this question: 'If you get the arms and money you desire, will you invade Missouri or any slave Territory?' to which he replied, 'I am no adventurer; you all know me; you are acquainted with my history; you know what I have done in Kansas; I do not expose my plans; no one knows them but myself, except, perhaps, one; I do not wish to be interrogated; if you wish to give me anything, I want you to give it freely; I have no other purpose but to serve the cause of liberty.'" And he also adds: "Although it had been understood by the members of the committee that Mr. Brown intended to arm one hundred men, to be scattered about in the Territory and to be actual settlers, and engaged in their several pursuits, only to be called out to repel invasion or defend the Kansas free-State settlers, yet this reply was not satisfactory to all, and the arms were voted back to your committee" (meaning the Massachusetts State Kansas Committee) "to be disposed of as you thought best."

How and why these arms (the two hundred Sharp's rifles) were originally purchased by this Massachusetts State Kansas Committee, will appear from the testimony of George L. Stearns, who was its president or chairman, at page 227 of the testimony. It is shown by Hurd that, after the national committee, for the reason stated, had refused to intrust them to Brown, on his application, they "were voted back," as Hurd calls it, to the Massachusetts State Kansas Committee; and, on page 229 of the testimony, will be found a letter from Stearns to Brown, dated at Boston, on the 8th of January, 1857, advising him that he was directed by his committee to send him an order on Edward Clark, of Lawrence, in Kansas Territory, for the two hundred rifles, "with four thousand ball cartridges, thirty-one military caps," (afterwards corrected as thirty-one thousand percussion caps,) which he states were then stored at Tabor, in Iowa, with directions to hold the same as agent of the society, subject to their order, and, at the same time, authorizing him to draw on their treasurer, at Boston, for a sum of money not to exceed five hundred dollars. At page 228 of the testimony will be found the following question, put to Stearns, with his answer:

"Question. Was it at Brown's request that you put him in possession of these arms in January, 1857?

"Answer. No, sir; but because we needed an agent to secure them," &c.

And again, at page 230, he was asked: "Did I understand you to say that this was voluntarily proffered to him, and not at his request?" (Meaning the arms.)

"Answer. Yes, sir.

"Question. Why did you desire to place these arms in his possession?

"Answer. For safe-keeping.

"Question. Were they not in safe-keeping where they were?

"Answer. They were not substantially in our hands. We had passed them into the hands of the National Kansas Committee, to be transported to Kansas," &c.

The committee are not disposed to draw harsh, or perhaps uncharitable conclusions; yet they cannot fail to remark that these arms, which had been refused to Brown by the national committee, for the very satisfactory reason that he gave evasive answers to their inquiry how they were to be used, were proffered to him, and without request on his part, by the Massachusetts committee; and this proffer is found attended by the fact, not a little to be remarked, that contemporaneous with it—that is to say, in January, 1857—this Mr. Stearns gave authority to Brown to purchase from the Massachusetts Arms Company two hundred revolver pistols, which Stearns alleges he paid for out of his own funds, (page 227 of the testimony,) giving to Brown at the same time authority to draw on him at sight for $7,000, "in sums as it might be wanted, for the subsistence of one hundred men, provided that it should be necessary at any time to call that number into the field for active service in the defense of Kansas, in 1857."

Considering the comparative tranquil condition of Kansas at the period referred to, it is not easy to reconcile this act of the "Massachusetts State Kansas Committee" and its chairman with a reasonable regard to the peace of the country, or the lives of their fellow-citizens. These arms, however, with the two hundred rifles, were left from that time in Brown's possession, although as stated by the witness Stearns, at page 228 of the testimony, "the exigency contemplated did not occur," and therefore no part of the $7,000 was drawn by Brown.

At what time Brown procured the pistols, or transported them to the West, appears only from the testimony of Stearns, who says he paid for them, and the freight on them to Iowa, on production to him of the railroad receipt afterwards, in 1858, but it does appear that they were sent along with the Sharp's rifles from Ohio to him, in the neighborhood of Harper's Ferry. In 1858, Brown it appears told Stearns that both the rifles and the pistols were then "stored in Ohio." (Page 232 of the testimony.) From the correspondence of John Brown, jr., signing himself "John Smith," with his father, and with J. H. Kagi, (under the name of "J. Henrie,") shortly before the invasion at Harper's Ferry, printed in the Appendix, it will be seen that they were sent by him from Ashtabula county, Ohio, to his father at Harper's Ferry, *via* Chambersburg.

The testimony of the witnesses, Hurd and Stearns, would show that the arms refused to Brown by the national committee, had been *afterwards* voted to him by the Massachusetts committee—reference to Hurd's statement (page 250) and to the order given by Stearns to Brown (page 234) for the arms, would from their dates seem to contradict this, but only as to the order of time. The facts interesting to this inquiry are only, were the arms placed under control of Brown; by whom; and when? and this is clearly shown.

It is shown fully, from the testimony, that, although Brown when he first went to Kansas was accompanied by two of his sons, with their families, yet that he never removed his family from New York, and that he subsequently freely and fully avowed that he never had an idea of settling in Kansas, but was attracted to remain there only in the hope that by keeping alive the irritation and excited feeling of the settlers on the subject of slavery, and stimulating and accustoming them to war and bloodshed, he would be enabled in some way to lead them across the borders to incite a servile war in Missouri, from whence he might be able to extend it to other slaveholding States. Ultimately disappointed in this, and so early as the fall of 1857, he seems to have conceived the plan of a distinct invasion of one of the slaveholding States, under the organization and in the manner in which it was afterwards carried into execution in Virginia. This, of course, required the command of large sums of money; and he seems to have so successfully impressed himself and his capacity for conducting what he and his associates styled "the cause of freedom," upon the sickly, if not depraved, sensibilities of his allies in such "cause," as to command their confidence, if he did not altogether lull their suspicions. Letters to and from Brown and others, in the Appendix, give much insight into the manner and the sources whence his funds were derived.

The testimony shows generally how these contributions were made—occasionally in large sums paid directly to Brown, but more usually by collections made in the villages and towns throughout the country by itinerant lecturers. These lectures appear to have been patronized by the principal men in the States where they were delivered. Their topics were various, but all directed in some manner to what was called "the general cause of freedom;" sometimes for the creation of a fund to aid fugitive slaves in their escape; at other times with no definite character ascribed to them, except that the funds collected were to be used in promoting human freedom; and at other times, as would seem, for the personal expenses or to reïmburse supposed losses of Brown. See the evidence of J. R. Giddings, pages 150, 151, and 152, of the testimony. He was a lecturer through the Northwestern States, one class of his lectures devoted, as he states, to "an exposition of the doctrines of the higher law," and which he expounds, at page 151 of the testimony, thus:

"What I mean by the higher law is, that power which for the last two centuries has been proclaimed by the philosophers and jurists and statesmen of Germany, Europe, and the United States, called, in other words, the law of nature; by which we suppose that God, in giving man his existence, gave him the right to exist; the right to breathe vital air; the right to enjoy the light of the sun; to drink the waters of the earth; to unfold his moral nature; to learn the laws that control his moral and physical being; to bring himself into harmony with those laws, and enjoy that happiness which is consequent on such obedience."

To the question, "In your lectures, was the theory of that law applied to the condition of African slavery in the United States," he answered:

"Unquestionably, to all. Wherever a human soul exists, that law applies. I mean by the term '*soul,*' that immortal principle in man that exists hereafter, which is called the human soul; and wherever such soul exists there is the right to live; the right to attain knowledge; the right to sustain life, obey the laws of his Creator, and enjoy heaven or happiness.

"Question. Was that theory or doctrine of a higher law, in your lectures, applied specially to the condition of African slavery in this country?

"Answer. To all human beings, wherever they are."

And further, he states:

"I will say that the meanest slave who treads the footstool of God holds from his Creator the same right to live and attain knowledge and to liberty, that you and I possess."

And in answer to a further question, he states:

"The views given in my lectures go to this extent, that whenever, without going into any other State, we have the opportunity to sustain the right of a fellow-being, it is our duty to do it. I have never felt myself called upon to advocate nor to encourage the entering into other States to speak thus to slaves; but wherever, in my own State, where I can do it without violation of law, or enactments erroneously called *law*, I uniformly arm the slave; I uniformly tell him to defend his life and his liberty; I uniformly teach him his rights, so far as I can."

As a further exposition of the views entertained by those devotees to the so-styled "cause of freedom," the committee refer to the evidence of George L. Stearns, at page 240. This gentleman, although not a lecturer, was, as shown by his testimony, one of the most active and successful workers in that "cause." For his views as to the legitimate use of money contributed to this "cause," see page 242, where he states:

"From first to last, I understood John Brown to be a man who was opposed to slavery, and, as such, that he would take every opportunity to free slaves where he could; I did not know in what way; I only knew that from the fact of his having done it in Missouri in the instance referred to; I furnished him with money because I considered him as one who would be of use in case such troubles arose as had arisen previously in Kansas; that was my object in furnishing the money; I did not ask him what he was to do with it, nor did I suppose he would do anything that I should disapprove."

To the question "Do you disapprove of such a transaction as that at Harper's Ferry," he answered:

"I should have disapproved of it if I had known of it; but I have since changed my opinion; I believe John Brown to be the representative man of this century, as Washington was of the last—the Harper's Ferry affair, and the capacity shown by the Italians for self-government, the great events of this age. One will free Europe, and the other America."

And so in the testimony of Samuel G. Howe, a physician of Boston. At page 166, speaking of Brown, he says:

"I contributed to his aid at various times.

"Question. His aid in what way?

"Answer. In the same way that I contributed to the aid of other anti-slavery men; men who give up their occupations, their industry, to write papers or to deliver lectures, or otherwise to propagate anti-slavery sentiments. I give as much money every year as I can possibly afford. I am in the habit of contributing in that way."

And at page 167:

"Question. Will you state what you mean by that phrase 'contributing for the promotion of anti-slavery sentiments?' What is the meaning of that idea?

"Answer. In the same way that I would promote the Gospel among the heathens; I cannot precisely say what. The means are various—lectures, writing, talking, discussing the matter.

"Question. What ends are to be attained by promoting that anti-slavery sentiment? What is the object in view?

"Answer. The promotion of freedom among men; the same object as the fathers in the revolution.

"Question. Was one of its objects the means of attaining the freedom of the African slaves held in this country?

"Answer. That would be the natural and desired result.

"Question. Was that one of the ends to be attained by promoting this anti-slavery sentiment by lecturing and otherwise?

"Answer. It was. I answer these questions out of courtesy to the Chairman, but I must think they are rather wide."

Of these three witnesses, one, Giddings, represented a district in the House of Representatives from Ohio for a long series of years, and is known to the country as an intelligent man; another, Dr. Howe, holds the highest professional and social position in the city of Boston. The other, Mr. Stearns, is a merchant in the same city, of wealth and with all the influence usually attending it. With such elements at work, unchecked by law and not rebuked but encouraged by public opinion, with money freely contributed and placed in irresponsible hands, it may easily be seen how this expedition to excite servile war in one of the States of the Union was got up, and it may equally be seen how like expeditions may certainly be anticipated in future whenever desperadoes offer themselves to carry them into execution. In regard to the one here inquired into, it appears that Brown, after the dispersal of his convention at Chatham, proceeded to the eastern States to provide materials both of arms and money; and in reference to the ease with which the latter was obtained without scrutiny as to the uses to which it was to be put, it will stand upon the record as a remarkable fact, that a check for one hundred dollars given by Gerritt Smith to Brown was handed by him directly, in part payment, to the manufacturer of the pikes with which the slaves were to have been armed. This gentleman, Mr. Smith, is known to the country as a man of large wealth and a liberal contributor to this pretended "cause." By reason of his very infirm health he was not summoned as a witness before the committee; and the use of this particular check is not referred to as proof in any manner that its contributor knew definitely what was to be done with it, but it is referred to as a most persuasive proof of the utter insecurity of the peace and safety of some of the States of this Union, in the existing condition of the public mind and its purposes in the non-slaveholding States. It may not become the committee to suggest a duty in those States to provide by proper legislation against machinations by their citizens or within their borders destructive of the peace of their confederate republics; but it does become them fully to expose the consequences resulting from the present license there existing, because the peace and integrity of the Union is necessarily involved in its continuance.

It has been already stated in this report that Brown, learning, during or just after the adjournment of the convention at Chatham, that Forbes had betrayed his plot, made an effort through his emissary, Realf, to recover the correspondence between himself and Forbes, which, if exposed, would establish it. And it would appear that Forbes considered, by his revelations at Washington, in May, 1858, that he had done what Brown feared he would do. This is referred to in the testimony of Realf, at page 100, where the committee were endeavoring to trace the arms of the Massachusetts Kansas Committee to Brown's possession. The witness states:

"Within a day or two following the convention at Chatham, John Brown said to me that he had received a copy of a letter written by Senator Henry Wilson, of Massachusetts, from Washington city, to Dr. Howe, of Boston," &c.

And, on page 101. he continues:

"On the occasion of which I have just spoken, at Chatham, Brown said to me that Colonel Forbes, maddened by the non-receipt of moneys which he had expected to receive, had threatened to divulge Brown's plans, and had done so by coming to Washington and stating to Senator Henry Wilson, of Massachusetts, that Brown had a purpose in view of effecting an insurrection in the Southern States."

The committee at once apprized the Hon. Henry Wilson, senator of the United States from Massachusetts, of the testimony of this witness, and invited him to attend the committee, as well to put any questions he might think proper to the witness, as to give his own testimony, if any he had, in relation to this matter, The testimony of Mr. Wilson will be found commencing at page 140. It shows that the communication made to him by Forbes induced him to write a letter at once to Dr. Howe, at Boston, the substance of which, from recollection, he gave as follows:

"I wrote to him for the purpose of saying it was rumored that some of the arms that had been contributed by gentlemen in the East for the defence of Kansas had passed into the hands of John Brown, and were held somewhere in his hands, and that they ought to get them out of his hands and put them in the hands of some reliable man in Kansas who would use them only for the purposes of defense, for which they were contributed; that if these arms should be used for any illegal purpose, they would involve the men who contributed for the other purpose in difficulties. That was the substance of the letter; that if they should be used for any illegal purpose whatever, they would be involved in difficulty, and they should get them out of his hands at once."

Mr. Wilson continued:

"I received a letter, three or four days after I wrote mine, from Dr. Howe, to this effect: that they had sent to Brown to deliver the arms into the hands of somebody in Kansas; at any rate, they had sent to him to take the arms into Kansas, or deliver them up in some way; and I supposed at the time the arms were those referred to as being in Iowa, which were sent out there and stationed on the way. I received this letter a day or two after I wrote. That was the substance of it. The whole matter, I supposed then, was a quarrel between Brown and Forbes, and I paid but little attention to it; and never, until the outbreak took place, dreamed or heard from any quarter whatever anything in regard to it. I heard nothing from Forbes or Brown or any other source."

At page 158, in the testimony of Dr. Howe, he says, in answer to a question:

"In the year 1858 I received a communication from a Mr. Forbes, then in Washington, and information from other quarters, that Captain Brown had in his possession arms belonging to the committee which he would probably use for purposes not intended by the committee. A meeting was called. The committee had then been virtually dissolved; it had nothing more to do; but the members were called together. A vote was passed instructing the chairman to write to Captain Brown and direct him, if he held any property, arms or otherwise, belonging to the committee, to take them into Kansas, there to be used only for the defense of freedom in Kansas. Such a vote was passed, such a letter was written, and, I have no doubt, received by him."

This letter, it seems, however, was not written by Dr. Howe himself, but by the chairman of the "Massachusetts Aid Committee." When asked the question, "Who was the chairman who wrote the letter you refer to?" he answered, "I should prefer not to answer that question," adding, "I am here to answer all I have done myself, freely and frankly, but I would respectfully ask to be excused from answering any questions touching the actions of anybody else. I can only answer for my view as one of the committee." He subsequently added, however, "Perhaps I am over sensitive about it, and inasmuch as the gentleman's name is perfectly well known as chairman of the committee, and is in print, I give it—Mr. George L. Stearns."

At page 160, this witness also stated that about the same time with the letter from Forbes he received one from Mr. Wilson of the Senate; that he preserved a copy of Mr. Wilson's letter "until recently, when, in the general destruction of my [his] papers of no consequence, at the beginning of the year, I destroyed it among others, but I have a distinct recollection of its contents.

"Question. Will you state the contents?

"Answer. It was that he had reason to believe that Captain Brown had in his possession arms belonging to the Massachusetts Aid Committee, which he would be likely to use for purposes not contemplated by the committee; that he, Wilson, considered the original movement for procuring anything of the kind to have been a very mistaken and unfortunate one, and he advised by all means that measures be taken to prevent Captain Brown using those arms for any purpose not contemplated in their original purchase. It was a short letter, and that was the amount of it; but I recollect distinctly he expressed his disapprobation of the fact of such arms being in existence, and his disapprobation of John Brown's general career."

This witness having promised, on his return to Boston, to make search for all documents connected with this subject which could be found, replied by letter to the chairman, which will be found at page 172, and in which he states that the letter from Mr. Wilson could not be found. He sent, however, copies of two letters to Mr. Wilson, dated respectively on the 12th and 15th of May, 1858, which will be found at page 176. The latter is brief, and in the following words:

"DEAR SIR: When I last wrote to you, I was not aware fully of the true state of the case with regard to certain arms belonging to the late Kansas committee. Prompt measures have been taken, and will be resolutely followed up, to prevent any such monstrous perversion of a trust as would be the application of means raised for the defense of Kansas to a purpose which the subscribers of the fund would disapprove and vehemently condemn."

And on page 177 will be found two letters of George L. Stearns, as chairman of the Massachusetts State Kansas Committee, dated the 14th and 15th of May, 1858, referred to by Dr. Howe as the measures taken by the committee to divest Brown of these arms. Howe's letter to Wilson of the 15th of May, cited above, shows very clearly that he was then strongly impressed with the necessity of arresting certain measures projected by Brown, whatever they were, and of which it would appear, at the date of his previous letter to Mr. Wilson, he had not been fully aware, but which he then characterizes as "a monstrous perversion of a trust" in "the application of means raised for the defense of Kansas, to a purpose which the subscribers of the fund would disapprove and vehemently condemn." Stearns, however, as shown by his letters to Brown, in executing the orders of the committee to prevent the misuse of the arms by Brown, contented himself with reminding Brown that those arms were "to be used for the defense of Kansas," and warns him only "not to use them for any other purpose," but to hold them subject to his order as chairman; adding that a member of the committee would go to Chatham to confer with Brown as to the best mode of disposing of them. The following day, the 15th of May, he again wrote to Brown, telling him that he could find no member of the committee who could spare the time to go to Chatham, and requesting Brown to meet him in New York city sometime the following week, and that the committee would pay his expenses.

The committee cannot but remark on the feeble, and, as it resulted, the abortive effort of the chairman of the Massachusetts committee to prevent a murderous use of these arms by Brown; certainly in striking contrast with the assurance given by Dr. Howe to Mr. Wilson, that prompt measures had been taken, and would be resolutely followed, to prevent such a "monstrous perversion of the trust" connected with them. But a perusal of the testimony at large of Mr. Stearns may show that he had at best but vague and undefined opinions as to what would be a perversion of the trust spoken of by Dr. Howe.

The history of the large armament collected by Brown at Harper's Ferry is thus clearly traced. The rifled carbines, manufactured in Connecticut, intended, as would appear, to be originally used in intestine strife in Kansas, and sent there for that purpose, were voluntarily, by the Massachusetts Kansas Committee, through its chairman, placed in the hands of Brown, with vague and inexplicit instructions as to their use, about the time when it would appear that he finally conceived the purpose of exciting servile war in some of the slaveholding States. They were allowed to remain in his possession, notwithstanding his failure or refusal to give them up after that committee and its chairman had been warned of his purpose to put them to some use not warranted by those who owned them. The revolver pistols, as shown by the testimony of Stearns, chairman of that committee, was a volunteer gift from him to Brown, at about the same time the carbines were handed over to him, and whether thus beyond his control or not, were not recalled from his possession. The expedition, so atrocious in its character, would have been arrested, had even ordinary care been taken on the part of the Massachusetts committee to ascertain whether Brown was truthful in his professions. Even the modest inquiry made of him by the National Kansas committee, as stated by their secretary,

Hurd, resulted in such equivocation and evasion on his part as led them peremptorily to refuse these arms to him, as their act.

The facts exposed in this part of the testimony speak for themselves. It will be remembered that the period referred to, when Mr. Wilson communicated his suspicions to Dr. Howe, and through him to the chairman of the Massachusetts committee, was so late as May, 1858. Order had then been restored in Kansas. The troops of the United States had been long previously withdrawn, and the only contests remaining in the Territory were conducted through the ballot-box. Notwithstanding all which, it would seem Brown was to be kept afoot, intrusted with arms for military organization, and amply supplied with money. The testimony shows that after his treasonable proceedings at Chatham he went back to New England, traveled through its several villages, collecting money, which was freely contributed under the auspices both of Dr. Howe and Mr. Chairman Stearns and others, with a knowledge that he retained the large supply of arms of which they had failed to dispossess him.

Upon the whole testimony, there can be no doubt that Brown's plan was to commence a servile war on the borders of Virginia, which he expected to extend, and which he believed his means and resources were sufficient to extend through that State and throughout the entire South. Upon being questioned, soon after his capture, by the Governor of Virginia, as to his plans, he rather indignantly repelled the idea that it was to be limited to collecting and protecting the slaves until they could be sent out of the State as fugitives. On the contrary, he vehemently insisted that his purpose was to retain them on the soil, to put arms in their hands, with which he came provided for the purpose, and to use them as his soldiery. (Pages 61, 62.)

This man (Brown) was uniformly spoken of, by those who seemed best to have known him, as of remarkable reticence in his habits, or, as they expressed it, "secretive." It does not appear that he intrusted even his immediate followers with his plans, fully, even after they were ripe for execution. Nor have the committee been enabled clearly to trace knowledge of them to any. The only exception would seem to be in the instance of the anonymous letter received by the Secretary of War in the summer preceding the attack, referred to in his testimony. The Secretary shows that he could get no clue to the writer; nor were the committee enabled in any way to trace him. Considering that the letter was anonymous, as well as vague and apparently incoherent in its statements, it was not at all remarkable, in the opinion of the committee, that it did not arrest the attention of the officer to whom it was addressed.

The point chosen for the attack seems to have been selected from the two-fold inducement of the security afforded the invaders by a mountain country, and the large deposit of arms in the arsenal of the United States there situated. It resulted in the murder of three most respectable citizens of the State of Virginia without cause, and in the like murder of an unoffending free negro. Of the military force brought against them, one marine was killed and one wounded; whilst eight of the militia and other forces of the neighborhood were wounded, with more or less severity, in the several assaults made by them.

Of the list of "insurgents" given in Colonel Lee's report, (fourteen whites and five negroes,) Brown, Stevens, and Coppic, of the whites, with Shields Green and Copeland, of the negroes, captured at the storming of the engine-house, were subsequently executed in Virginia, after judicial trial; as were also John E. Cook and Albert Hazlett, who at first escaped, but were captured in Pennsylvania and delivered up for trial to the authorities of Virginia—making in all seven thus executed. It does not seem to have been very clearly ascertained how many of the party escaped. Brown stated that his party consisted of twenty-two in number. Seven were executed, ten were killed at the Ferry; thus leaving five to be accounted for. Four of these five, it is believed, were left on the Maryland side in charge of the arms when Brown crossed the river, and who could not afterwards join him; leaving but one, who, as it would appear, is the only survivor of the party who accompanied Brown across the bridge, and whose escape is not accounted for.

The committee, after much consideration, are not prepared to suggest any legislation, which, in their opinion, would be adequate to prevent like occurrences in the future. The only provisions in the Constitution of the United States which would seem to import any authority in the government of the United States to interfere on occasions affecting the peace or safety of the States, are found in the eighth section of the first article, amongst the powers of Congress, "to provide for calling for the militia to execute the laws of the Union, suppress insurrections, and repel invasions;" and in the fourth section of the fourth article, in the following words: "The United States shall guaranty to every State in this Union a republican form of government, and shall protect each of them against invasion, and, on the application of the legislature or of the executive, (when the legislature cannot be convened,) against domestic violence." The "invasion" here spoken of would seem to import an invasion by the public force of a foreign power, or (if not so limited and equally referable to an invasion by one State of another) still it would seem that public force, or force exercised under the sanction of acknowledged political power, is there meant. The invasion (to call it so) by Brown and his followers at Harper's Ferry, was in no sense of that character. It was simply the act of lawless ruffians, under the sanction of no public or political authority—distinguishable only from ordinary felonies by the ulterior ends in contemplation by them, and by the fact that the money to maintain the expedition, and the large armament they brought with them, had been contributed and furnished by the citizens of other States of the Union, under circumstances that must continue to jeopard the safety and peace of the Southern States, and against which Congress has no power to legislate.

If the several States, whether from motives of policy or a desire to preserve the peace of the Union, if not from fraternal feeling, do not hold it incumbent on them, after the experience of the country, to guard in future by appropriate legislation against occurrences similar to the one here inquired into, the committee can find no guarantee elsewhere for the security of peace between the States of the Union.

So far, however, as the safety of the public property is involved, the committee would earnestly recommend that provision should be made by the executive, or, if necessary, by law, to keep under adequate military guard the public armories and arsenals of the United States, in some way after the manner now practised at the navy-yards and forts.

Before closing their report, the committee deem it proper to state that four persons summoned as witnesses, to wit: John Brown, jr., of Ohio, James Redpath, of Massachusetts, Frank B. Sanborn, of Massachusetts, and Thaddeus Hyatt, of New York, failing or refusing to appear before the committee, warrants were issued by order of the Senate for their arrest. Of these, Thaddeus Hyatt only was arrested; and on his appearance before the Senate, still refusing obedience to the summons of the committee, he was by order of the Senate committed to the jail of the District of Columbia. In regard to the others, it appeared by the return of the marshal of the northern district of Ohio, as deputy of the Sergeant-at-Arms, that John Brown, jr., at first evaded the process of the Senate, and afterwards, with a number of other persons, armed themselves to prevent his arrest. The marshal further reported in his return that Brown could not be arrested unless he was authorized in like manner to employ force. Sanborn was arrested by a deputy of the Sergeant-at-Arms, and afterwards released from custody by the judges of the supreme court of Massachusetts on *habeas corpus*. Redpath, by leaving his State, or otherwise concealing himself, successfully evaded the process of the Senate.

And the committee ask to be discharged from the further consideration of the subject.

J. M. MASON,
Chairman.
JEFF'N DAVIS.
G. N. FITCH.

IN THE SENATE OF THE UNITED STATES.

June 15, 1860.—Ordered to be printed.

Mr. Doolittle, in the absence of Mr. Collamer, submitted the following

VIEWS OF THE MINORITY,

[Prepared by Mr. Collamer.]

The treasonable conspiracy of John Brown and his associates, and its fatal development at Harper's Ferry, in Virginia, in October last, has become matter of history, and all its details are too well known to require recapitulation.

While the excitement, alarm, and suspicion were rife in the public mind, the Senate adopted the resolutions raising the committee of inquiry in relation thereto. Their only *legitimate* purpose was to inquire whether anything had transpired which required further legislation by Congress for future security. Though drawn in very general and undefined terms, in some part almost implying the exercise of judicial inquisition, yet from an unwillingness to incur the imputation of embarrassing full investigation, no one objected to their adoption. In the exercise of the same feeling we have made no objection to the great latitude of inquiry taken by the committee. We, however, distinctly understand, that if the resolutions and their peculiar phraseology were drawn or are used for any other purpose than that of furnishing to the Senate information for its own legislative action, it is a perversion and departure from the only justifiable purpose of their adoption.

The objects of inquiry, as stated in the resolutions, are the following, which are stated, not in the order of the resolutions, but in the order of their consecutive relation, for the purpose of their more orderly answer, to wit:

First. The facts in relation to the invasion and seizure of the armory and arsenal at Harper's Ferry.

Second. Whether it was in pursuance of an organization, and the nature and purpose thereof.

Third. The arms and munitions there possessed by the insurgents, and where and how obtained.

Fourth. Were any citizens, not present, implicated in, or accessory thereto, by contributions of arms, money, ammunition, or otherwise.

In relation to the first inquiry, the testimony taken before the committee discloses no material facts but such as appeared on the trial of the conspirators, and have been long since published and are fully known. They are briefly as follows:

On the night of the 16th day of October, 1859, John Brown, together with sixteen white men and five negroes as conspirators, took armed possession of the United States armory at Harper's Ferry, in Virginia, killed four of the inhabitants, and were dislodged by armed force, which they resisted, and in the action seven of the white conspirators were killed, and three of the negroes. John Brown was wounded and taken prisoner, and he, together with four others of the white conspirators, and two of the negroes, were tried, convicted, and executed, and five escaped.

2d. This took place in pursuance of a conspiracy commenced in Kansas by John Brown and most of these conspirators, in the last part of 1857 or beginning of 1858. They were young men and entirely under the influence of Brown, and had been, as well as Brown, deeply engaged in the conflicts in Kansas in 1855, 1856, and 1857. From Kansas they passed into Iowa, and from thence they were led by Brown to Chatham, in Canada West. There they, together with a number of negroes, formed a secret organization, with written articles of association, drawn up by Brown, having for its object the raising of slave insurrection in the slaveholding States and subverting the government thereof.

3d. They had two hundred Sharp's carbines and two hundred revolver pistols and about one thousand pikes, together with a quantity of clothing and ammunition. The carbines and revolvers had been procured by contributions in Massachusetts, in 1856, and forwarded to Iowa to be sent into Kansas for the aid and in the defense of the free-State people in the struggle then existing there, and they had been intrusted to John Brown for that purpose, together with the ammunition. The clothing, which had been contributed for the suffering people of Kansas, had been intrusted to him there for that purpose. In 1857 these troubles in Kansas in a great degree subsided. The associations and committees, who had made contributions, ceased operations, and these arms and munitions, in the hands of Brown, came to be almost overlooked and disregarded, until the summer of 1858, when a suggestion came to the persons having control of them, at Boston, that John Brown was about to make some improper use of them, and thereupon he was particularly charged to make no use of them but in Kansas, and for the defense of the free-State people there, the purpose for which they had been furnished. It seems that this, together with being unable to procure money, and an apprehension of being exposed, prevented him from executing the purpose of his conspiracy for that year.

In 1859, he procured to be completed in Connecticut one thousand pikes, for which he had contracted and partly paid in 1856 or 1857, for like service in Kansas, and then in 1859, he procured those pikes, and also those carbines and revolvers, and the ammunition and clothing, to be privately conveyed and secreted at or near Harper's Ferry, without the knowledge or consent of those who had contributed them for use in Kansas, and contrary to the order so given him by those in control.

4th. There is no evidence that any other citizens than those there with Brown were accessory to this outbreak or invasion, by contributions thereto or otherwise, nor any proof that any others had any knowledge of the conspiracy or its purposes in the year 1859, though Realf, Forbes, and some very few may have understood it in 1858, when it failed of execution.

Although some of the testimony tends to show that some abolitionists have at times contributed money to what is occasionally called practical abolitionism—that is, in aiding the escape of slaves—and may have placed too implicit confidence in John Brown, yet there is no evidence to show, or cause to believe, they had any complicity with this conspiracy, or any suspicion of its existence or design, before its explosion.

There was no evidence tending to show that there ever was any conspiracy or design, by any one, to rescue John Brown or his associates from prison in Virginia.

The place and the boldness of this outbreak, the purpose it entertained, the deaths it involved, and the amount of arms and munitions with which it was supplied, combined to produce not only great alarm, but also a strong suspicion of extensive complicity. Time and investigation has happily dissipated much of such alarm and suspicion, and shown that this was but an offshoot from the extensive outrages and lawlessness in Kansas, commenced and continued there, by armed invasions of that Territory to control its own people, the elections, and the government, for the introduction and perpetuity of slavery in that Territory, on the one hand, and resistance or defense on the other. This invited there many men of desperation, and others became so by the irritations and excitements of those collisions. When comparative peace was restored there many, trained by such a school, were ready for new fields of lawless enterprise. It was from such elements that John Brown concocted his conspiracy, consisting of young men and boys, over whom he had entire control, many of them foreigners, and none of substance or position in the country.

By perverting the arms, ammunition, and clothing with which he had been intrusted, from the purpose for which he had received them, he secured his supplies.

It is almost astonishing that in a country like ours, laden with the rich experience of the blessings of security under the protection of law, there should still be found large bodies of men laboring under the infatuation that any good object can be effected by lawlessness and violence. It is the prostration of law, which is the only bond of security. It can, in its nature, beget nothing but resistance, retaliation, insecurity, and disaster. And yet, with all our intelligence and experience, we have most unfortunate and deplorable manifestations of such infatuations. They are dangerous in direct proportion to the extent of public countenance they receive. No object, however desirable, can justify them or prevent their disastrous example and consequences. The unpunished lawless invasions of our weak neighboring nations; the flagrant and merciless breaches of our laws against the African slave trade, "unwhipt of justice;" the lawless armed invasions of our own people in our own weak Territory of Kansas, not only unpunished, but justified, sustained, and even rewarded, all, it is believed, to extend and sustain slavery, tended strongly to suggest acts of lawless violence to destroy it, especially in those who had witnessed and suffered by these collisions. They are, however, all without excuse, and they but add to the experience that no public peace or private security can be found but where every disregard of law meets with the most prompt public rebuke and effective punishment or correction.

While this act of violence and treason, and the alarm, suspicion, suffering, and death it involved are so deplorable, we cannot but see that the lessons which it teaches furnish many considerations of security against its repetition. Ages might not produce another John Brown, or so fortuitously supply him with such materials. The fatal termination of the enterprise in the death and execution of so large a part of the number engaged; the dispersion of the small remainder as fugitives in the land; the entire disinclination of the slaves to insurrection, or to receive aid for that purpose, which was there exhibited; the very limited number and peculiar character of the conspirators, all combine to furnish assurance against the most distant probability of its repetition.

The extent and freedom with which this investigation has been conducted has resulted in showing that the people of the free States have had no complicity with this atrocity; and, if viewed with candor, the evidence will remove the suspicion of extensive complicity which the possession of such a quantity of arms, unexplained, was likely to create, it now fully appearing they were never furnished for such a purpose. This investigation has its value, if its record be examined and treated with candor, as it fully shows that there is no such ground of suspicion and distrust as has been indulged amongst our people, and that lawless violence as to slavery, by efforts from beyond its border, has culminated in this disastrous and abortive experiment.

We have very succinctly stated the origin, agents, instruments, purposes, and result of this deplorable outrage, and briefly stated the reflections, we think, it suggests. The facts disclosed, viewed in the light in which they appear to us, and in which we have presented them, however much calling for reprobation and regret, may be, and we think should be, used and improved to allay excitement, quiet suspicion, and restore tranquillity.

The committee having come to the conclusion that no such facts have been disclosed as call for any congressional legislation, we should regard this as the termination of its duty; but, by its majority, the committee seems to have entertained a different view of the object of the resolutions and purpose of the inquiry. They give, as we suppose, a different construction from our understanding of those words of the resolutions which direct an inquiry "whether any citizens of the United States not present were implicated therein, or accessory thereto, by contributions of money, arms, munitions, or otherwise." We consider that no man can be properly said to be "implicated" in any transaction, or accessory thereto, who had no knowledge of its purpose, character, or existence; and the whole committee consider that there is no evidence that any citizen, not present, had any such knowledge of this. Yet the committee, by its majority, seem to regard it as their duty to inquire whether there are any citizens who, though not "implicated" in this affair, yet hold such opinions and pursue such courses on the subject of slavery as are dangerous to the national tranquillity, even although Congress has no power to take any action in relation thereto. This we regard as a departure from the duty and proper power of the committee. Upon this view of the committee, by its majority, great latitude and range of inquiry has been taken in the examination, and equal latitude of remark indulged in the report. Witnesses, and especially those known or suspected of ultra abolition sentiments, have been freely examined as to their personal sentiments, theories, purposes, conduct, charities, contributions, lectures, and speeches on the subject of slavery. They have even called a witness to prove that he and others had conspired to be guilty of the charity of providing for a poor, wounded prisoner, in a land of strangers, the necessary counsel able to secure him a fair trial, as if that was evidence of their complicity with his guilt. We feel bound to protest against all the conclusions which the same spirit of suspicion which could call such testimony will seek to deduce from it.

So long as Congress, in the exercise of its power over the Territories, is invoked to exert it to extend, perpetuate, or protect the institution of slavery therein ; so long as the policy of the government is sought to be so shaped as to aid to extend its existence or enlarge its power, in any way, beyond its present limits, so long must its moral, political, and social character and effects be unavoidably involved in congressional discussion. Hence, it is equally unavoidable that the people in all parts of the Union will discuss this subject, as they are to select those who are to represent them and their sentiments in congressional action. So long as slavery is claimed before the world as a highly benignant, elevating, and humanizing institution, and as having Divine approbation, it will receive at the hands of the moralist, civilian, and theologian the most free and unflinching discussion ; nor should its vindicators wince in the combat which their claims invite. In this discussion, it is true, as in other topics of exciting debate, wide latitude and license are, at times, indulged, but it seldom or never exceeds in severity the terms of reprehension on this subject which were long since indulged by Washington, Madison, Jefferson, Mason, and, in later times, by McDowell, Faulkner, and their worthy compeers, all of Virginia, whose information and opinions, on this as well as other subjects, the people of the free States have not yet learned to disrespect. We insist, however, that there is no such matter presented in the testimony or existing in fact, as is more than intimated in the report, that even the abolitionists in the free States take courses intended, covertly, to produce forcible violations of the laws and peace of the slaveholding States, much less that any such course is countenanced by the body of the people in the free States. We cannot join in any report tending to promulgate such a view, as we regard it unfounded in fact and ill calculated to promote peace, confidence, or tranquillity, and a departure from the legitimate purpose for which the committee was appointed.

J. COLLAMER.
J. R. DOOLITTLE.

JOURNAL

OF

THE SELECT COMMITTEE

APPOINTED

To inquire into the facts attending the late invasion and seizure of the United States Armory at Harper's Ferry, Virginia.

FRIDAY, *December* 16, 1859.

The Select Committee, appointed in pursuance of the following resolution of the Senate of the United States, adopted on the 14th instant, viz:

"*Resolved*, That a committee be appointed to inquire into the facts attending the late invasion and seizure of the Armory and Arsenal of the United States at Harper's Ferry, in Virginia, by a band of armed men, and

"REPORT

"Whether the same was attended by armed resistance to the authorities and public force of the United States, and by the murder of any of the citizens of Virginia, or of any troops sent there to protect the public property;

"Whether such invasion and seizure was made under color of any organization intended to subvert the government of any of the States of the Union;

"What was the character and extent of such organization, and whether any citizens of the United States not present were implicated therein, or accessory thereto, by contributions of money, arms, munitions, or otherwise;

"What was the character and extent of the military equipment in the hands or under the control of said armed band, and where and how and when the same was obtained and transported to the place so invaded;

"That said committee report whether any, and what, legislation may, in their opinion, be necessary on the part of the United States for the future preservation of the peace of the country, or for the safety of the public property;

"And that said committee have power to send for persons and papers"—

Met this day.

Present—All the members, viz: Mr. Mason, (chairman,) Mr. Davis, Mr. Collamer, Mr. Fitch, and Mr. Doolittle.

The Chairman was directed, by order of the committee, to ask the Senate to grant authority to employ a clerk to the committee.

On motion, the committee adjourned to meet at the call of the Chairman.

TUESDAY, *December* 20, 1859.

The committee met pursuant to a call of the Chairman.

Present—Messrs. Mason, Davis, Collamer, Fitch, and Doolittle.

The Senate having, by a resolution adopted yesterday, authorized the committee to employ a clerk,

The Chairman proposed the appointment of D. F. Murphy; which was agreed to.

The Chairman was authorized, by order of the committee, to summon the following witnesses to appear on the 3d day of January, 1860, viz:

From Jefferson county, Virginia: Andrew Hunter, Lewis W. Washington, John H. Allstadt.

At Harper's Ferry: A. M. Kitzmiller, J. E. P. Dangerfield, A. M. Ball.

And such others in Jefferson county, Virginia, and its vicinity as he may determine.

It was further ordered, that Richard Realf be summoned from Austin, in Texas, by telegraph or otherwise, as the Chairman may direct.

It was also ordered, that all the proceedings of this committee be considered confidential.

On motion, the committee adjourned, to meet at the call of the Chairman.

WEDNESDAY, *January* 4, 1860.

Committee met pursuant to the call of the Chairman.

Present—Messrs. Mason, Davis, Collamer, and Doolittle.

No witnesses being present, the committee adjourned to meet to-morrow morning, at 12 o'clock.

THURSDAY, *January* 5, 1860.

Committee met pursuant to adjournment.

Present—Messrs. Mason, Davis, Collamer, and Doolittle.

On motion of Mr. DAVIS, it was

Ordered, That, during the examination of witnesses, no person shall be present, besides the members of the committee and its clerk, except the particular witness who is giving his testimony.

The committee proceeded to examine witnesses, and examined—

1. John C. Unseld.

On motion, the committee adjourned until to-morrow, at 11, a. m.

FRIDAY, *January* 6, 1860.

The committee met pursuant to adjournment.

Present—Messrs. Mason, Davis, Collamer, and Doolittle.

The committee examined the following witnesses:
2. Terence Byrne.
3. Daniel Whelan.
4. John D. Starry.
5. George W. Chambers.
6. Lewis W. Washington.
7. John H. Allstadt.

On motion, the committee adjourned until Tuesday morning next, the 10th instant, at 10 o'clock.

TUESDAY, *January* 10, 1860.

The committee met pursuant to adjournment.

Present—Messrs. Mason, Davis, Collamer, Fitch, and Doolittle.

The committee examined the following witnesses:
8. Colonel Robert E. Lee, United States army.
9. Theodore Rynders.
10. Archibald M. Kitzmiller.
11. Armistead M. Ball.

On motion of Mr. DAVIS, it was

Ordered, That the CHAIRMAN and Mr. COLLAMER be authorized to open and examine the contents of the trunk brought from New York city by Theodore Rynders, and seized in that city by the United States marshal, under an order and summons from this committee, as the trunk of one Hugh Forbes.

On motion, the committee adjourned until to-morrow, at 11 o'clock, a. m.

WEDNESDAY, *January* 11, 1860.

The committee met pursuant to adjournment.

Present—Messrs. Mason, Davis, Collamer, Fitch, and Doolittle.

The committee examined the following witness:
12. Lind F. Currie.

The committee discharged J. E. P. Dangerfield from attendance, without examining him.

The CHAIRMAN and Mr. COLLAMER reported that, in pursuance of the order of the committee, they had examined the contents of the trunk brought from New York by Theodore Rynders, deputy marshal, purporting to be the trunk of Hugh Forbes, and finding nothing in it pertinent to the inquiry submitted to this committee, had returned it to the United States marshal for the southern district of New York, with directions to replace it in the custody from which it had been taken, the contents undisturbed. The sub-committee further reported that this trunk came under seal from New York, and was returned by them, under seal, to the custody of the deputy marshal.

At Mr. COLLAMER's suggestion, a summons was directed to be issued for B. B. Newton, of St. Alban's, Vermont.

The CHAIRMAN was also authorized to summon the following persons to attend as witnesses before the committee, viz:

John Brown, jr., of West Andover, Ashtabula county, Ohio.
Joshua R. Giddings, of Jefferson, Ohio.
R. Plumb, of Oberlin, Ohio.

Charles Blair, of Collinsville, Connecticut.
F. B. Sanborn, of Concord, Massachusetts.
George De Bapt, of Detroit, Michigan,
James Redpath, of New York or Boston.
E. Morton, of Rochester, New York.
Dr. S. G. Howe, of Boston.
Lewis Hayden, of Boston.
Gerritt Smith, of Peterboro', New York.
W. H. D. Calender, of Hartford, Connecticut.

On motion, the committee adjourned to meet on Friday next, at 11 o'clock, a. m.

FRIDAY, *January* 13, 1860.

The committee met pursuant to adjournment.

Present—Messrs. Mason, Davis, Fitch, and Doolittle.

The committee examined, as a witness,

13. Andrew Hunter.

On motion, the committee adjourned until Monday next, at 11 o'clock, a. m.

MONDAY, *January* 16, 1860.

The committee met pursuant to adjournment.

Present—Messrs. Mason, Davis, Collamer, Fitch, and Doolittle.

The committee examined, as a witness,

14. William F. M. Arny.

Without concluding his examination, the committee adjourned until to-morrow morning, at 10 o'clock.

TUESDAY, *January* 17, 1860.

The committee met pursuant to adjournment.

Present—Messrs. Mason, Davis, Collamer, Fitch, and Doolittle.

The committee resumed and concluded the examination of William F. M. Arny.

On motion, the CHAIRMAN was authorized to summon, from the Territory of Kansas, such witnesses as he may deem necessary and proper.

On motion, the committee adjourned to meet at the call of the Chairman.

SATURDAY, *January* 21, 1860.

The committee met pursuant to the Chairman's call, at 11 o'clock, a. m.

Present—Messrs. Mason, Davis, Collamer, Fitch, and Doolittle.

The committee examined, as a witness,

14. Richard Realf.

At the suggestion of the CHAIRMAN, he was authorized by the committee to address a note to the Hon. Henry Wilson, and inform him of reference that was made by the testimony of the witness, to-day, to him, and say to him that, if he desired it, the testimony of the witness as to that point should be submitted to him; and if he wished it, he should have the opportunity of putting any interrogatories that he thought proper to the witness; and further, to invite him to attend the committee, with a view to obtain his own testimony in relation thereto.

On motion, the committee adjourned until Monday morning, at 11 o'clock.

MONDAY, *January* 23, 1860.

The committee met pursuant to adjournment.

Present—Messrs. Mason, Davis, Collamer, Fitch, and Doolittle.

The Hon. Henry Wilson appeared before the committee, and that part of the testimony of Richard Realf referred to in the note of the Chairman to him having been read to him by the clerk, Mr. WILSON stated that he desired to put a few questions to the witness, Realf, and would give his own testimony with reference to the matter; but as he desired to obtain a certain letter from Boston, the subject was postponed until he would be able conveniently to attend to it.

The committee proceeded to examine the following witnesses, who were summoned to appear this day:

15. William H. D. Callender.
16. Benjamin B. Newton.
17. Charles Blair.

On motion, the committee adjourned to meet on Thursday next, the 26th instant, at 11 o'clock, a. m.

THURSDAY, *January* 26, 1860.

The committee met pursuant to adjournment.

Present—Messrs. Mason, Davis, Collamer, Fitch, and Doolittle.

S. G. Howe and James Redpath, witnesses summoned for the 24th instant, (and service of the summons made by leaving a copy at their houses on the 23d instant,) and F. B. Sanborn, (served personally on the 16th instant,) summoned for the 24th, failed to appear.

The committee examined, as a witness,

18. James Jackson.

The CHAIRMAN was authorized to summon, as witnesses, for Monday next, or any subsequent day next week, Samuel Chilton and E. K. Schaeffer.

On motion, the committee adjourned to meet on Monday next, at 11 o'clock, a. m.

MONDAY, *January* 30, 1860.

The committee met pursuant to adjournment.

Present—Messrs. Mason, Davis, Collamer, Fitch, and Doolittle.

The committee examined, as a witness,

19. Samuel Chilton.

Joshua R. Giddings, Ralph Plumb, and John Brown, jr., who were summoned as witnesses for this day, did not appear.

On motion, the committee adjourned to meet on Wednesday next, at 11 o'clock, a. m.

WEDNESDAY, *February* 1, 1860.

The committee met pursuant to adjournment.

Present—Messrs. Mason, Collamer, Fitch, and Doolittle.

Hon. Henry Wilson appeared before the committee, and expressed his willingness to testify with reference to the matter contained in Richard Realf's testimony with regard to a letter of his, and declined to ask any questions of Realf.

The committee thereupon examined, as a witness,

20. Hon. Henry Wilson.

Thaddeus Hyatt, who was summoned to appear as a witness this day, having sent word to the committee that he was rather indisposed, it was agreed to defer his examination until to-morrow.

On motion, the committee adjourned until to-morrow morning, at 11 o'clock.

THURSDAY, *February* 2, 1860.

The committee met pursuant to adjournment.

Present—Messrs. Mason, Davis, Collamer, Fitch, and Doolittle.

S. G. Howe reported himself to be in attendance for the purpose of giving his testimony.

The committee examined, as a witness,

21. Edward K. Schaeffer, who had been summoned for to-day.

At the suggestion of Mr. FITCH, it was

Ordered, That the Chairman issue a summons for the attendance, as a witness, of John A. Andrew, of Boston.

On motion, the committee adjourned until to-morrow morning, at 11 o'clock.

FRIDAY, *February* 3, 1860.

The committee met pursuant to adjournment.

Present—Messrs. Mason, Davis, Collamer, and Doolittle.

Joshua R. Giddings and Ralph Plumb reported themselves as in attendance for the purpose of giving testimony, according to the summons of the committee.

Samuel G. Howe presented his reasons in writing for desiring not to testify before the committee, which the committee declined to notice further than to receive the paper.

The committee proceeded to examine the following witnesses:

22. Joshua R. Giddings.

23. Samuel G. Howe.

On motion, the committee adjourned to meet on Tuesday morning next, at 10 o'clock.

TUESDAY, *February* 7, 1860.

The committee met pursuant to adjournment.

Present—Messrs. Mason, Davis, Collamer, and Doolittle.

The committee examined, as a witness,

24. Ralph Plumb.

On motion, the committee adjourned until Thursday morning next, at 11 o'clock.

THURSDAY, *February* 9, 1860.

The committee met pursuant to adjournment.

Present—Messrs. Mason, Davis, Collamer, Fitch, and Doolittle.

The committee examined, as a witness,

25. John A. Andrew, who had been summoned to appear this day.

Charles Robinson, who had been summoned for to-day, reported himself to be in attendance.

It appearing to the committee that certain witnesses duly summoned before this committee, pursuant to the order of the Senate, have made default in appearing, it is

Ordered, That the Chairman report the fact to the Senate for such further proceedings therein as may be directed.

On motion, the committee adjourned until to-morrow, at 11 o'clock, a. m.

FRIDAY, *February* 10, 1860.

The committee met pursuant to adjournment.

Present—Messrs. Mason, Davis, Collamer, and Doolittle.

Martin F. Conway, a witness summoned to appear this day, reported himself to be in attendance.

Augustus Wattles, summoned for to-day, did not make his appearance.

Mr. COLLAMER proposed that a summons be issued for his excellency Henry A. Wise, late Governor of Virginia, for the purpose of ascertaining from him what information he had as to any plans for the rescue of John Brown and his associates, while they were confined in the jail at Charlestown, Virginia.

The consideration of the proposition was deferred until Monday next.

The committee examined, as a witness,

26. Charles Robinson.

On motion, the committee adjourned until Monday, the 13th instant, at 11 o'clock, a. m.

MONDAY, *February* 13, 1860.

The committee met pursuant to adjournment.

Present—Messrs. Mason, Davis, Collamer, Fitch, and Doolittle.

The committee examined, as a witness,

27. Martin F. Conway.

The committee proceeded to consider the proposition of Mr. COLLAMER, presented on Friday last, and, after consultation,

Ordered, That it be postponed for the present.

On motion, the committee adjourned to meet on Friday morning, the 17th instant, at 11 o'clock.

FRIDAY, *February* 17, 1860.

The committee met pursuant to adjournment.

Present—Messrs. Mason, Collamer, and Doolittle.

Augustus Wattles, who was summoned for the 10th instant, reported himself to be in attendance.

The CHAIRMAN notified the committee that he had reported to the Senate the names of the defaulting witnesses, and that warrants had been issued for the arrest of F. B. Sanborn, John Brown, jr., and James Redpath.

He also informed the committee that, for the convenience of the Senate, he had instructed the Sergeant-at-Arms that he should be at liberty to have the warrants for these arrests executed by deputy, as had been done theretofore, in the case of the summons issued by the committee, advising that the marshals, or their deputies, in the different States should be so deputized.

The committee examined, as a witness,

28. Augustus Wattles.

On motion, the committee adjourned until Monday morning next, at 11 o'clock.

MONDAY, *February* 20, 1860.

The committee met pursuant to adjournment.

Present—Messrs. Mason, Davis, and Collamer.

On motion, it was

Ordered, That the Chairman ask the Senate to order its warrant to issue for the arrest of Thaddeus Hyatt, for having failed and refused to appear pursuant to the summons of this committee.

On motion, the committee adjourned to meet at the call of the Chairman.

FRIDAY, *February* 24, 1860.

The committee met pursuant to the call of the Chairman.

Present—Messrs. Mason, Davis, Collamer, and Fitch.

The committee examined, as a witness,

29. George L. Stearns.

The CHAIRMAN notified the committee that the Senate had issued its warrant for the arrest of Thaddeus Hyatt.

On motion, the committee adjourned to meet on Monday next, at 11 o'clock, a. m.

MONDAY, *February* 27, 1860.

The committee met pursuant to adjournment.

Present—Messrs. Mason, Collamer, Doolittle, and Fitch.

The committee examined, as a witness,

30. Horace White.

On motion, the Chairman was authorized to request the attendance of the Hon. John B. Floyd, Secretary of War.

On motion, the committee adjourned to meet at the call of the Chairman.

MONDAY, *March* 5, 1860.

The committee met pursuant to the call of the Chairman.

Present—Messrs. Mason, Davis, Collamer, Fitch, and Doolittle.

The committee examined, as a witness,

31. Hon. John B. Floyd.

The CHAIRMAN submitted a letter from J. H. Lane and William A.

Phillips, of Kansas, relative to the testimony said to have been given before this committee by Charles Robinson.

Ordered, That its consideration be postponed.

On motion, the committee adjourned to meet at the call of the Chairman.

MONDAY, *March* 26, 1860.

The committee met pursuant to the call of the Chairman.

Present—Messrs. Mason, Davis, Collamer, Fitch, and Doolittle.

The CHAIRMAN laid before the committee a letter from M. Johnson, United States marshal for the northern district of Ohio, relative to the difficulties which would probably be encountered in any attempt to serve the warrant deputed to him, for the arrest of John Brown, jr., by order of the Senate.

After consultation,

Ordered, That the Chairman request Marshal Johnson to return the warrant, with a statement, under oath, of the facts rendering its service impracticable; and that the same, when received, be laid before the Senate for its action.

On motion, the committee adjourned until Friday morning next, at 11 o'clock.

FRIDAY, *March* 30, 1860.

The committee met pursuant to adjournment.

Present—Messrs. Mason, Davis, Collamer, Fitch, and Doolittle.

The CHAIRMAN submitted a letter from George L. Stearns, dated Boston, March 22, 1860, requesting that it and a letter which it inclosed, from H. B. Hurd, be added to his testimony; which was authorized to be done.

The committee resumed the consideration of Mr. COLLAMER'S proposition to summon Henry A. Wise, which was now presented in writing by Mr. COLLAMER, the mover, in the words following, viz:

"I request that Henry A. Wise be called as a witness, to furnish to the committee all reliable information within his knowledge as to any citizen of the United States, not present at the invasion at Harper's Ferry, being implicated therein or accessory thereto, by contributions of money, arms, ammunition, or otherwise, and, so far as he knows, where said information can be obtained.

"And that said Wise furnish to the committee all reliable information within his knowledge as to any combination or conspiracy in any part of the country for the purpose of rescuing John Brown or his associates from prison, in Viginia.

"I desire not his suspicions, apprehensions, belief, or opinion, or any account or vindication of his course, as Governor of Virginia, on the occasion, but only the facts and information above stated—in short, that knowledge and information in relation to the transaction which he is credibly reported to have said 'rubies could not obtain from him.'"

The CHAIRMAN presented his objections to the proposition, as follows:

"Objected to, because the witness named was, at the time when the affair at Harper's Ferry occurred, and until after the execution of Brown and his followers, Governor of the State of Virginia, and, of consequence, all the information he may have touching that subject was obtained by him in his official character as governor of that State, and acquired the better to enable him to protect the public interests confided to him as such.

"It is not considered competent to any authority of the United States in any manner thus to trench upon, or directly or indirectly to question the separate action of a State administration."

The correspondence annexed, between the Chairman of this committee and the Governor of Virginia, shows that the latter has no documents or correspondence of the character called for by the letter of the Chairman, connected with the inquiries of the committee, not already given to the public:

"WASHINGTON, *December* 15, 1859.

"DEAR SIR: I inclose a copy of a resolution of the Senate of the United States, which was adopted yesterday. It is framed, as you will see, to effect a searching inquiry into everything connected with the late transaction at Harper's Ferry.

"In conducting the investigation, it will be important that the committee should be in possession of the original of every paper or document found with Brown or his confederates which will aid in the investigation, or implicate others at a distance.

"I have already advised Mr. Andrew Hunter, of Charlestown, that I would call for him to bring here all the documents which were used at the trial, and I have to beg the favor of you to cause to be collected together such as may have been taken to Richmond, and put them in sealed packages, certified in such manner as will show here that they were amongst the papers taken from the effects of Brown.

"I will send for them specially to Richmond, and will be responsible for their safety.

"Very respectfully and truly, yours,

"J. M. MASON.

"His Excellency HENRY A. WISE,
"*Governor of Virginia.*"

"EXECUTIVE DEPARTMENT,
"*Richmond, December* 17, 1859.

"SIR: I am directed by the governor to inform you that your letter of the 15th instant, inclosing a copy of a resolution of the Senate of the United States for investigating everything connected with the late treasonable invasion at Harper's Ferry, has been received. He regrets that these papers cannot now be furnished. Most of the originals are are filed, and have become parts of the records of the court in Jefferson. Some of them were sent by Colonel Lee to Washington. Some were carried by privates of military companies to Baltimore, and some maps were sent to Southern States. He has caused copies of the most important, which were in evidence at Charlestown, to be submitted to the general assembly. These have been ordered to be printed, and are now in the hands of the public printer. When printed, they have been referred to a committee of the legislature, and will be under investigation by it. The originals are in the custody of Andrew Hunter, Esq., or of the clerk of the circuit court, and will be wanting in the trial of Stevens and Hazlett, two prisoners, who are yet to be tried.

"Very respectfully, &c.,

"GEORGE W. MUMFORD,

"*Secretary of the Commonwealth of Virginia.*

"Hon. JAMES M. MASON,

"*United States Senate.*"

The papers in the custody of Andrew Hunter, Esq., and the printed report to the legislature of Virginia above referred to, were all subsequently before the committee.

The objections taken by the CHAIRMAN to the motion of Mr. COLLAMER, that Governor Wise be summoned as a witness, were sustained by the committee, and the motion was overruled.

On motion, the Chairman was authorized to request the attendance of the Hon. William H. Seward, as a witness.

On motion, the committee adjourned to meet at the call of the Chairman.

SATURDAY, *April* 7, 1860.

The committee met pursuant to the call of the Chairman.

Present—Messrs. Mason, Collamer, Fitch, and Doolittle.

The CHAIRMAN submitted letters which he had received from the attorney and marshal of the United States for the district of Massachusetts, relative to the circumstances attending the arrest of F. B. Sanborn under the authority of the Senate.

On motion, the committee adjourned to meet at the call of the Chairman.

FRIDAY, *April* 13, 1860.

The committee met pursuant to the Chairman's call.

Present—Messrs. Mason, Collamer, Fitch, and Doolittle.

The CHAIRMAN submitted the return of Silas Carlton, the deputy of the Sergeant-at-Arms, to the warrant for the arrest of F. B. Sanborn.

Ordered, That the return be presented to the Senate by the Chairman, with a request that the same be referred to the Committee on the Judiciary.

It was also

Ordered, That the same course be pursued in reference to the return of the marshal of the northern district of Ohio, to the warrant for the arrest of John Brown, jr.

On motion by Mr. DOOLITTLE, it was

Ordered, That the testimony taken before the committe be printed for its use.

On motion, the committee adjourned to meet at the call of the Chairman.

WEDNESDAY, *May* 2, 1860.

The committee met pursuant to the call of the Chairman.

Present—Messrs. Mason, Davis, Collamer, Fitch, and Doolittle.

The committee examined, as a witness,

32. Hon. William H. Seward.

The committee proceeded to consider the application of J. H. Lane and W. A. Phillips, of Kansas, relative to testimony alleged to have been given by Charles Robinson; and pending its consideration,

On motion, the committee adjourned to meet at the call of the Chairman.

THURSDAY, *May* 24, 1860.

The committee met pursuant to the call of the Chairman.

Present—Messrs. Mason, Davis, Collamer, and Doolittle.

The CHAIRMAN notified the committee that summons had been issued, pursuant to order, for E. Morton, Gerritt Smith, George De Bapt, and Lewis Hayden, which had not been served, because it appeared by the returns of the officers intrusted with them that the said Morton could not be found, that Mr. Smith's health was such as to render it improper to bring him here, and that said De Bapt and Hayden were negroes.

After consultation, the committee adjourned to meet at the call of the Chairman.

WEDNESDAY, *May* 30, 1860.

The committee met pursuant to adjournment.

Present—Messrs. Mason, Collamer, and Doolittle.

On motion, the committee adjourned to meet on Friday next, at 10 o'clock.

FRIDAY, *June* 1, 1860,

The committee met pursuant to adjournment.

Present—Messrs. Mason, Collamer, Fitch, and Doolittle.

The committee, after considering the letter heretofore submitted from J. H. Lane and W. A. Phillips, of Kansas, and examining, in connection with it, the testimony given by Charles Robinson,

Ordered, That certain portions of that testimony, either on hearsay or irrelevant, be omitted from the printed testimony.

On motion, the committee adjourned to meet at the call of the Chairman.

THURSDAY, *June* 7, 1860.

The committee met pursuant to the call of the Chairman.

Present—Messrs. Mason, Davis, Collamer, Fitch, and Doolittle.

The CHAIRMAN submitted a petition, referred by the Senate to this committee, from negroes in Massachusetts, for the action of the committee, as to the course proper to be pursued thereon.

After consultation,

Ordered, That the consideration of the subject be postponed until Monday next.

On motion, the committee adjourned to meet on Monday morning next, at 10 o'clock.

MONDAY, *June* 11, 1860.

The committee met pursuant to adjournment.

Present—Messrs. Mason, Davis, Collamer, Fitch, and Doolittle.

The committee, after consultation, agreed to direct the Chairman to submit to the Senate the petition from negroes in Massachusetts, with a recommendation that it be returned to the Senator who presented it.

After consultation as to making the final report, the committee adjourned until Thursday morning, 14th instant, at 9½ o'clock, a. m.

THURSDAY, *June* 14, 1860.

The committee met pursuant to adjournment.

Present—Messrs. Mason, Davis, Collamer, Fitch, and Doolittle.

The CHAIRMAN submitted a report, which he had prepared, on the subject of the investigation referred to this committee.

Mr. COLLAMER presented the views of himself and Mr. DOOLITTLE.

On motion, the Chairman was authorized to present the report to the Senate; and Mr. Collamer was empowered to present the minority report.

APPENDIX.

1. Colonel Lee's report.
2. Journal of "Provisional Constitutional Convention."
3. Provisional Constitution.
4. Orders for distribution of forces of "provisional army."
5. Specimen of commission.
6. Correspondence of John Brown, jr., and others.
7. Correspondence and other papers in evidence.

No. 1.

Colonel Lee to the Adjutant General.

HEADQUARTERS HARPER'S FERRY,
October 19, 1859.

COLONEL: I have the honor to report, for the information of the Secretary of War, that on arriving here on the night of the 17th instant, in obedience to Special Orders No. 194 of that date from your office, I learn that a party of insurgents, about 11 p. m. on the 16th, had seized the watchmen stationed at the armory, arsenal, rifle factory, and bridge across the Potomac, and taken possession of those points. They then dispatched six men, under one of their party, called Captain Aaron C. Stevens, to arrest the principal citizens in the neighborhood and incite the negroes to join in the insurrection. The party took Colonel L. W. Washington from his bed about 1½ a. m. on the 17th, and brought him, with four of his servants, to this place. Mr. J. H. Allstadt and six of his servants were in the same manner seized about 3 a. m., and arms placed in the hands of the negroes. Upon their return here, John E. Cook, one of the party sent to Mr. Washington's, was dispatched to Maryland, with Mr. Washington's wagon, two of his servants, and three of Mr. Allstadt's, for arms and ammunition, &c. As day advanced, and the citizens of Harper's Ferry commenced their usual avocations, they were separately captured, to the number of forty, as well as I could learn, and confined in one room of the fire-engine house of the armory, which seems early to have been selected as a point of defense. About 11 a. m. the volunteer companies from Virginia began to arrive, and the Jefferson Guards and volunteers from Charlestown, under Captain J. W. Rowen, I understood, were first on the ground. The Hamtramck Guards, Captain V. M. Butler; the Shepherdstown troop, Captain Jacob Rienahart; and Captain Alburtis's company from Martinsburg arrived in the afternoon. These companies, under the direction of Colonels R. W. Baylor and John T. Gibson, forced the insurgents to abandon their positions at the bridge and in the village, and to withdraw within the armory inclosure, where they fortified themselves in the fire-engine house, and carried ten of their prisoners for the purpose of insuring their safety and facilitating their escape, whom they termed hostages, and whose names are Colonel L. W. Washington, of Jefferson county, Virginia; Mr. J. H. Allstadt, of Jefferson county, Virginia; Mr. Israel Russell, justice of the peace, Harper's Ferry; Mr. John Donahue, clerk of Baltimore and Ohio railroad; Mr. Terence Byrne, of Maryland; Mr. George D. Shope, of Frederick, Maryland; Mr. Benjamin Mills, master armorer, Harper's Ferry arsenal; Mr. A. M. Ball, master machinist, Harper's Ferry arsenal; Mr. J. E. P. Dangerfield, paymaster's clerk, Harper's Ferry arsenal; Mr. J. Burd, armorer, Harper's Ferry arsenal. After sunset more troops arrived. Captain B. B. Washington's company from Winchester, and three companies from Fredericktown, Maryland, under Colonel Shriver. Later in the evening the companies from Baltimore, under General Charles C. Edgerton, second light brigade, and a detachment of marines, commanded by Lieutenant J. Green accompanied by Major Russell, of that corps, reached Sandy Hook, about one and a half mile east of Harper's Ferry. At this point I came up with these last-named troops, and leaving General Edgerton and his command on the Maryland side of the river for the night, caused the marines to proceed to Harper's Ferry, and placed them within the armory grounds to prevent the possibility of the escape of the insurgents. Having taken measures to halt, in Baltimore, the artillery companies ordered from Fort Monroe, I made preparations to attack the insurgents at daylight. But for the fear of sacrificing the lives of some of the gentlemen held by them as prisoners in a midnight assault, I should have ordered the attack at once.

Their safety was the subject of painful consideration, and to prevent, if possible, jeopardizing their lives, I determined to summon the insurgents to surrender. As soon after daylight as the arrangements were made Lieutenant J. E. B. Stewart, 1st cavalry, who had accompanied me from Washington as staff officer, was dispatched, under a flag, with a written summons, (a copy of which is hereto annexed, marked A.) Knowing the character of the leader of the insurgents, I did not expect it would be accepted. I had therefore directed that the volunteer troops, under their respective commanders, should be paraded on the lines assigned them outside the armory, and had prepared a storming party of twelve marines, under their commander, Lieutenant Green, and had placed them close to the engine-house, and secure from its fire. Three marines were furnished with sledge-hammers to break in the doors, and the men were instructed how to distinguish our citizens from the insurgents; to attack with the bayonet, and not to injure the blacks detained in custody unless they resisted. Lieutenant Stewart was also directed not to receive from the insurgents any counter propositions. If they accepted the terms offered, they must immediately deliver up their arms and release their prisoners. If they did not, he must, on leaving the engine-house, give me the signal. My object was, with a view of saving our citizens, to have as short an interval as possible between the summons and attack. The summons, as I had anticipated, was rejected. At the concerted signal the storming party moved quickly to the door and commenced the attack. The fire-engines within the house had been placed by the besieged close to the doors. The doors were fastened by ropes, the spring of which prevented their being broken by the blows of the hammers. The men were therefore ordered to drop the hammers, and, with a portion of the reserve, to use as a battering-ram a heavy ladder, with which they dashed in a part of the door and gave admittance to the storming party. The fire of the insurgents up to this time had been harmless. At the threshold one marine fell mortally wounded. The rest, led by Lieutenant Green and Major Russell, quickly ended the contest. The insurgents that resisted were bayoneted. Their leader, John Brown, was cut down by the sword of Lieutenant Green, and our citizens were protected by both officers and men. The whole was over in a few minutes.

After our citizens were liberated and the wounded cared for, Lieutenant Colonel S. S. Mills, of the 53d Maryland regiment, with the Baltimore Independent Greys, Lieutenant B. F. Simpson commanding, was sent on the Maryland side of the river to search for John E. Cook, and to bring in the arms, &c., belonging to the insurgent party, which were said to be deposited in a school-house two and a half miles distant. Subsequently, Lieutenant J. E. B. Stewart, with a party of marines, was dispatched to the Kennedy farm, situated in Maryland, about four and a half miles from Harper's Ferry, which had been rented by John Brown, and used as the depot for his men and munitions. Colonel Mills saw nothing of Cook, but found the boxes of arms, (Sharp's carbines and belt revolvers,) and recovered Mr. Washington's wagon and horses. Lieutenant Stewart found also at the Kennedy farm a number of sword pikes, blankets, shoes, tents, and all the necessaries for a campaign. These articles have been deposited in the government storehouse at the armory.

From the information derived from the papers found upon the persons and among the baggage of the insurgents, and the statement of those now in custody, it appears that the party consisted of nineteen men—fourteen white and five black. That they were headed by John Brown, of some notoriety in Kansas, who in June last located himself in Maryland, at the Kennedy farm, where he has been engaged in preparing to capture the United States works at Harper's Ferry. He avows that his object was the liberation of the slaves of Virginia, and of the whole South; and acknowledges that he has been disappointed in his expectations of aid from the black as well as white population, both in the Southern and Northern States. The blacks whom he forced from their homes in this neighborhood, as far as I could learn, gave him no voluntary assistance. The servants of Messrs. Washington and Allstadt, retained at the armory, took no part in the conflict, and those carried to Maryland returned to their homes as soon as released. The result proves that the plan was the attempt of a fanatic or madman, which could only end in failure; and its temporary success was owing to the panic and confusion he succeeded in creating by magnifying his numbers. I append a list of the insurgents, (marked B.) Cook is the only man known to have escaped. The other survivors of the expedition, viz: John Brown, A. C. Stevens, Edwin Coppic, and Green Shields, (*alias* S. Emperor,) I have delivered into the hands of the marshal of the western district of Virginia and the sheriff of Jefferson county. They were escorted to Charlestown by a detachment of marines, under Lieutenant Green. About nine o'clock this evening I received a report from Mr. Moore, from Pleasant Valley, Maryland, that a body of men had, about sunset, descended from the mountains, attacked the house of Mr. Gennett, and from the cries of murder and the screams of the women and children, he believed the residents of the valley were being massacred. The alarm and excitement in the village of Harper's Ferry was increased by the arrival of families from Sandy Hook, fleeing for safety. The report was, however, so improbable that I could give no credence to it, yet I thought it possible that some atrocity might have been committed, and I started with twenty-five marines, under Lieutenant Green, accompanied by Lieutenant Stewart, for the scene of the alleged outrage, about four and a half miles distant. I was happy to find it a false alarm. The inhabitants of Pleasant Valley were quiet and unharmed, and Mr. Gennett and his family safe and asleep.

I will now, in obedience to your dispatch of this date, direct the detachment of marines to return to the navy-yard at Washington in the train that passes here at $1\frac{1}{4}$ a. m. to-night, and will myself take advantage of the same train to report to you in person at the War Department. I must also ask to express my thanks to Lieutenant Stewart, Major Russell, and Lieutenant Green, for the aid they afforded me, and my entire commendation of the conduct of the detachment of marines, who were at all times ready and prompt in the execution of any duty.

The promptness with which the volunteer troops repaired to the scene of disturbance, and the alacrity they displayed to suppress the gross outrage against law and order, I know will elicit your hearty approbation. Equal zeal was shown by the president and officers of the Baltimore and Ohio Railroad Company in their transportation of the troops, and in their readiness to furnish the facilities of their well-ordered road.

A list of the killed and wounded, as far as came to my knowledge, is herewith annexed, (marked C;) and I inclose a copy of the "Provisional Constitution and ordinances for the people of the United States," of which there were a large number prepared for issue by the insurgents.

I am, very respectfully, your obedient servant,

R. E. LEE, *Colonel Commanding.*

Colonel S. COOPER,
Adjutant General U. S. Army, Washington City, D. C.

A.

HEADQUARTERS HARPER'S FERRY,
October 18, 1859.

Colonel Lee, United States army, commanding the troops sent by the President of the United States to suppress the insurrection at this place, demands the surrender of the persons in the armory buildings.

If they will peaceably surrender themselves and restore the pillaged property, they shall be kept in safety to await the orders of the President. Colonel Lee represents to them, in all frankness, that it is impossible for them to escape; that the armory is surrounded on all sides by troops; and that if he is compelled to take them by force he cannot answer for their safety.

R. E. LEE,
Colonel Commanding United States Troops.

B.

List of Insurgents.—14.

John Brown, of New York, commander-in-chief, badly wounded; prisoner.
Aaron C. Stevens, Connecticut, captain, badly wounded; prisoner.
Edwin Coppic, Iowa, Lieutenant, unhurt; prisoner,
Oliver Brown, New York, captain; killed.
Watson Brown, New York, captain; killed.
Albert Hazlett, Pennsylvania, lieutenant; killed.
William Leeman, Maine, lieutenant; killed.
Stuart Taylor, Canada, private; killed.
Charles P. Tidd, Maine, private; killed.
William Thompson, New York, private; killed.
Adolph Thompson, New York, private; killed.
John Kagi, Ohio, private; killed.
Jeremiah Anderson, Indiana, private; killed.
John E. Cook, Connecticut, captain; escaped.

Negroes.—5.

Dangerfield, Newly, Ohio; killed.
Louis Leary, Oberlin, Ohio; killed.
Green Shields, (alias Emperor,) New York, unhurt; prisoner.
Copeland, Oberlin, Ohio; prisoner.
O. P. Anderson, Pennsylvania, unaccounted for.

C.

List of the killed and wounded by the insurgents.—14.

Fontaine Beckham, railroad agent and mayor of Harper's Ferry; killed.
G. W. Turner, Jefferson county, Virginia; killed.
Thomas Boerly, Harper's Ferry; killed.
Heywood Shepherd, negro, railroad porter; killed.
Private Quinn, marine corps; killed.
Mr. Murphy; wounded.

Mr. Young; wounded.
Mr. Richardson; wounded.
Mr. Hammond; wounded.
Mr. McCabe; wounded.
Mr. Dorsey; wounded.
Mr. Hooper; wounded.
Mr. Woollet; wounded.
Private Rupert, marine corps; wounded.

Colonel Lee to the Secretary of War.

HARPER'S FERRY ARSENAL,
October 18, 1859.

SIR: Upon a more deliberate examination of the wounds of O. Brown, they are believed not to be mortal. He has three wounds, but they are not considered by the surgeon as bad as first reported. Please direct me what to do with him and the other white prisoners.

I am, very respectfully, your obedient servant,

R. E. LEE,
Colonel Commanding.

HON. SECRETARY OF WAR,
Washington, D. C.

No. 2.

Journal of the Provisional Constitutional Convention, held on Saturday, May 8, 1858.

CHATHAM, CANADA WEST,
Saturday, May 8, 1858.

10 a. m.—Convention met in pursuance to call of John Brown and others, and was called to order by Mr. Jackson, on whose motion Mr. William C. Monroe was chosen president; when,

On motion of Mr. Brown, Mr. J. H. Kagi was elected secretary.

On motion of Mr. Delany, Mr. Brown then proceeded to state the object of the convention, at length, and then to explain the general features of the plan of action in the execution of the project in view by the convention.

Mr. Delany and others spoke in favor of the project and the plan, and both were agreed to by general consent.

Mr. Brown then presented a plan of organization, entitled "Provisional Constitution and Ordinances for the People of the United States," and moved the reading of the same.

Mr. Kinnard objected to the reading until an oath of secresy be taken by each member of the convention; whereupon,

Mr. Delany moved that the following parole of honor be taken by all members of the convention: "I solemnly affirm that I will not in any way divulge any of the secrets of this convention, except to persons entitled to know the same, on the pain of forfeiting the respect and protection of this organization;" which motion was carried.

The president then proceeded to administer the obligation. After which, the question was taken on the reading of the plan proposed by Mr. Brown, and the same carried.

The plan was then read by the secretary. After which,

On motion of Mr. Whipple, it was ordered that it be now read by articles, for consideration.

The articles from one to forty-five, inclusive, were then read and adopted. On the reading of the forty-sixth, Mr. Reynolds moved to strike out the same. Reynolds spoke in favor, and Brown, Monroe, Owen Brown, Delany, Realf, Kinnard, and Kagi, against.

The question was then taken and lost, there being but one vote in the affirmative.

The article was then adopted.

The forty-seventh and forty-eighth articles, with the schedule, were then adopted in the same manner.

It was then moved by Mr. Delany that the title and preamble stand as read. Carried.

On motion of Mr. Kagi, the constitution as a whole was then unanimously adopted.

The convention then, at 1½, p. m., adjourned, on motion of Mr. Jackson, till 3 o'clock.

3 p. m.—Journal read and approved.

On motion of Mr. Delany, it was then ordered that those approving of the constitution, as adopted, sign the same. Whereupon the names of all the members were appended. (See No. [91].)

After congratulatory remarks by Messrs. Kinnard and Delany, the convention, on motion of Mr. Whipple, adjourned, at a quarter to four.

J. H. KAGI,
Secretary of the Convention.

CHATHAM, CANADA WEST,
Saturday, May 8, 1858.

6 p. m.—In accordance with and obedience to the provisions of the schedule to the constitution for the "proscribed and oppressed people" of the United States of America, to-day adopted at this place, a convention was called by the president of the convention framing that instrument, and met at the above-named hour, for the purpose of electing officers to fill the offices specially established and named by said constitution.

The convention was called to order by Mr. M. R. Delany, upon whose nomination Mr. William C. Munroe was chosen president, and Mr. J. H. Kagi, secretary.

A committee, consisting of Messrs. Whipple, Kagi, Bell, Cook, and Munroe, was then chosen, to select candidates for the various offices to be filled, for the consideration of the convention.

On reporting progress and asking leave to sit again, the request was refused, and the committee discharged.

On motion of Mr. Bell, the convention then went into the election of officers, in the following manner and order:

Mr. Whipple nominated John Brown for commander-in-chief, who was, on the seconding of Mr. Delany, elected by acclamation.

Mr. Realf nominated J. H. Kagi for secretary of war, who was elected in the same manner.

On motion of Mr. Brown, the convention then adjourned to 9, a. m., on Monday, the 10th.

MONDAY, *May* 10, 1858.

9 a. m.—The proceedings of convention on Saturday were read and approved.

The president announced that the business before the convention was the further election of officers.

Mr. Whipple nominated Thomas M. Kinnard for president. In a speech of some length, Mr. Kinnard declined.

Mr. Anderson nominated J. W. Loguen for the same office. The nomination was afterwards withdrawn, Mr. Loguen not being present, and it being announced that he would not serve, if elected.

Mr. Brown then moved to postpone the election of president for the present. Carried.

The convention then went into the election of members of congress. Messrs. Alfred M. Ellsworth and Osborn Anderson were elected.

After which the convention went into the election of secretary of state, to which office Richard Realf was chosen.

Whereupon the convention adjourned to 2½ p. m.

2½ p. m.—Convention again assembled, and went into a balloting for the election of treasurer and secretary of the treasury. Owen Brown was elected as the former, and George B. Gill as the latter.

The following resolution was then introduced by Mr. Brown, and unanimously passed:

Resolved, That John Brown, J. H. Kagi, Richard Realf, L. F. Parsons, C. P. Tidd, E. Whipple, C. W. Moffett, John E. Cook, Owen Brown, Steward Taylor, Osborn Anderson, A. M. Ellsworth, Richard Richardson, W. H. Leeman, and John Lawrence, be, and are hereby, appointed a committee, to whom is delegated the power of the convention to fill by election all the offices specially named in the provisional constitution which may be vacant after the adjournment of this convention.

The convention then adjourned *sine die*.

J. H. KAGI,
Secretary of the Convention.

No. 3.

Provisional Constitution and Ordinances for the people of the United States.

Preamble.

Whereas slavery, throughout its entire existence in the United States, is none other than a most barbarous, unprovoked, and unjustifiable war of one portion of its citizens upon another portion—the only conditions of which are perpetual imprisonment and hopeless servitude or absolute extermination—in utter disregard and violation of those eternal and self-evident truths set forth in our Declaration of Independence:

Therefore, we, citizens of the United States, and the oppressed people who, by a recent decision of the Supreme Court, are declared to have no rights which the white man is bound to respect, together with all other people degraded by the laws thereof, do, for the time being, ordain and establish for ourselves the following Provisional Constitution and Ordinances, the better to protect our persons, property, lives, and liberties, and to govern our actions:

Article I.

Qualifications for membership.

All persons of mature age, whether proscribed, oppressed, and enslaved citizens, or of the proscribed and oppressed races of the United States, who shall agree to sustain and enforce the Provisional Constitution and Ordinances of this organization, together with all minor children of such persons, shall be held to be fully entitled to protection under the same.

Article II.

Branches of government.

The provisional government of this organization shall consist of three branches, viz: legislative, executive, and judicial.

Article III.

Legislative.

The legislative branch shall be a Congress or House of Representatives, composed of not less than five nor more than ten members, who shall be elected by all citizens of mature age and of sound mind connected with this organization, and who shall remain in office for three years, unless sooner removed for misconduct, inability, or by death. A majority of such members shall constitute a quorum.

ARTICLE IV.

Executive.

The executive branch of this organization shall consist of a President and Vice-President, who shall be chosen by the citizens or members of this organization, and each of whom shall hold his office for three years, unless sooner removed by death or for inability or misconduct.

ARTICLE V.

Judicial.

The judicial branch of this organization shall consist of one Chief Justice of the Supreme Court and of four associate judges of said court, each constituting a circuit court. They shall each be chosen in the same manner as the President, and shall continue in office until their places have been filled in the same manner by election of the citizens. Said court shall have jurisdiction in all civil or criminal causes arising under this constitution, except breaches of the rules of war.

ARTICLE VI.

Validity of enactments.

All enactments of the legislative branch shall, to become valid during the first three years, have the approbation of the President and of the Commander-in-chief of the army.

ARTICLE VII.

Commander-in-chief.

A Commander-in-chief of the army shall be chosen by the President, Vice-President, a majority of the Provisional Congress, and of the Supreme Court, and he shall receive his commission from the President, signed by the Vice-President, the Chief Justice of the Supreme Court, and the Secretary of War, and he shall hold his office for three years, unless removed by death or on proof of incapacity or misbehavior. He shall, unless under arrest, (and until his place is actually filled as provided for by this constitution,) direct all movements of the army and advise with any allies. He shall, however, be tried, removed, or punished, on complaint of the President, by at least three general officers, or a majority of the House of Representatives, or of the Supreme Court; which House of Representatives, (the President presiding,) the Vice-President, and the members of the Supreme Court, shall constitute a court-martial for his trial; with power to remove or punish, as the case may require, and to fill his place, as above provided.

Article VIII.

Officers.

A Treasurer, Secretary of State, Secretary of War, and Secretary of the Treasury, shall each be chosen, for the first three years, in the same way and manner as the Commander-in-chief, subject to trial or removal on complaint of the President, Vice-President, or Commander-in-chief, to the Chief Justice of the Supreme Court, or on complaint of the majority of the members of said court or the Provisional Congress. The Supreme Court shall have power to try or punish either of those officers, and their places shall be filled as before.

Article IX.

Secretary of War.

The Secretary of War shall be under the immediate direction of the Commander-in-chief, who may temporarily fill his place in case of arrest or of any inability to serve.

Article X.

Congress, or House of Representatives.

The House of Representatives shall make ordinances providing for the appointment (by the President or otherwise) of all civil officers, excepting those already named; and shall have power to make all laws and ordinances for the general good, not inconsistent with this Constitution and these ordinances.

Article XI.

Appropriation of money, &c.

The Provisional Congress shall have power to appropriate money or other property actually in the hands of the treasurer, to any object calculated to promote the general good, so far as may be consistent with the provisions of this constitution; and may, in certain cases, appropriate for a moderate compensation of agents, or persons not members of this organization, for any important service they are known to have rendered.

Article XII.

Special duties.

It shall be the duty of Congress to provide for the instant removal of any civil officer or policeman, who becomes habitually intoxicated, or who is addicted to other immoral conduct, or to any neglect or unfaithfulness in the discharge of his official duties. Congress shall also be a Standing Committee of Safety, for the purpose of obtaining important information; and shall be in constant communication with the Commander-in-chief; the members of which shall each, as also the President, Vice-President, members of the Supreme Court, and Secretary of State, have full power to issue warrants, returnable as Congress shall ordain (naming witnesses, &c.,) upon their own information, without the formality of a complaint. Complaint shall be immediately made after arrest, and before trial; the party arrested to be served with a copy at once.

Article XIII.

Trial of President and other officers.

The President and Vice-President may either of them be tried, removed, or punished, on complaint made to the Chief Justice of the Supreme Court, by a majority of the House of Representatives; which house together with the Associate Judges of the Supreme Court, the whole to be presided over by the Chief Justice in case of the trial of the Vice-President, shall have full power to try such officers, to remove or punish as the case may require, and to fill any vacancy so occurring, the same as in the case of the Commander-in-chief.

Article XIV.

Trial of members of Congress.

The members of the House of Representatives may, any and all of them, be tried, and, on conviction, removed or punished, on complaint before the Chief Justice of the Supreme Court, made by any number of the members of said house exceeding one-third; which house, with the Vice-President and Associate Judges of the Supreme Court, shall constitute the proper tribunal with power to fill such vacancies.

Article XV.

Impeachment of Judges.

Any member of the Supreme Court may also be impeached, tried, convicted, or punished by removal or otherwise, on complaint to the President, who shall in such case, preside; the Vice-President, House of Representatives, and other members of the Supreme Court, constituting the proper tribunal, (with power to fill vacancies,) on complaint of a majority of said House of Representatives, or of the Supreme Court; a majority of the whole having power to decide.

Article XVI.

Duties of President and Secretary of State.

The President, with the Secretary of State, shall, immediately upon entering on the duties of their office, give special attention to secure from amongst their own people, men of integrity, intelligence, and good business habits and capacity, and, above all, of first-rate moral and religious character and influence, to act as civil officers of every description and grade, as well as teachers, chaplains, physicians, surgeons, mechanics, agents of every description, clerks, and messengers. They shall make special efforts to induce, at the earliest possible period, persons and families of that description to locate themselves within the limits secured by this organization; and shall, moreover, from time to time, supply the names and residence of such persons to the Congress, for their special notice and information, as among the most important of their duties; and the President is hereby authorized and empowered to afford special aid to such individuals, from such moderate appropriations as the Congress shall be able and may deem advisable to make for that object. The President and Secretary of State, and in all cases of disagreement the Vice-President, shall appoint all civil officers, but shall not have power to remove any officer. All removals shall be the result of a fair trial, whether civil or military.

Article XVII.

Further duties.

It shall be the duty of the President and Secretary of State to find out (as soon as possible) the real friends as well as enemies of this organization in every part of the country; to secure among them innkeepers, private postmasters, private mail contractors, messengers, and agents, through whom may be obtained correct and regular information constantly; recruits for the service, places of deposit and sale, together with all needed supplies; and it shall be matter of special regard to secure such facilities through the northern States.

Article XVII.

Duty of the President.

It shall be the duty of the President, as well as the House of Representatives, at all times, to inform the Commander-in-chief of any matter that may require his attention, or that may affect the public safety.

Article XIX.

Duty of President, continued.

It shall be the duty of the President to see that the provisional ordinances of this organization, and those made by the Congress, are promptly and faithfully executed; and he may, in cases of great urgency, call on the Commander-in-chief of the army or other officers for aid; it being, however, intended that a sufficient civil police shall always be in readiness to secure implicit obedience to law.

Article XX.

The Vice-President.

The Vice-President shall be the presiding officer of the Provisional Congress, and in cases of tie shall give the casting vote.

Article XXI.

Vacancies.

In case of the death, removal, or inability of the President, the Vice-President, and, next to him, the Chief Justice of the Supreme Court shall be the President during the remainder of the term; and the place of the Chief Justice, thus made vacant, shall be filled by Congress from some of the members of said court; and the places of the Vice-President and Associate Justice, thus made vacant, filled by an election by the united action of the Provisional Congress and members of the Supreme Court. All other vacancies, not heretofore specially provided for, shall, during the first three years, be filled by the united action of the President, Vice-President, Supreme Court, and Commander-in-chief of the army.

Article XXII.

Punishment of crimes.

The punishment of crimes not capital, except in case of insubordinate convicts or other prisoners, shall be (so far as may be) by hard labor on the public works, roads, &c.

Article XXIII.

Army appointments.

It shall be the duty of all commissioned officers of the army to name candidates of merit, for office or elevation, to the Commander-in-chief, who, with the Secretary of War, and, in cases of disagreement, the President, shall be the appointing power of the army; and all commissions of military officers shall bear the signatures of the Commander-in-chief and the Secretary of War. And it shall be the special duty of the Secretary of War to keep for constant reference of the Commander-in-chief a full list of names of persons nominated for office or elevation by the officers of the army, with the name and rank of the officer nominating, stating distinctly, but briefly, the grounds for such notice or nomination. The Commander-in-chief shall not have power to remove or punish any officer or soldier, but he may order their arrest and trial at any time by court-martial.

Article XXIV.

Courts-martial.

Courts-martial for companies, regiments, brigades, &c., shall be called by the chief officer of each command, on complaint to him by any officer, or any five privates in such command, and shall consist of not less than five nor more than nine officers, non-commissioned officers and privates, one half of whom shall not be lower in rank than the person on trial, to be chosen by the three highest officers in the command, which officers shall not be a part of such court. The chief officer of any command shall, of course, be tried by a court-martial of the command above his own. All decisions affecting the lives of persons, or office of persons holding commission, must, before taking full effect, have the signature of the Commander-in-chief, who may also, on the recommendation of at least one third of the members of the court-martial finding any sentence, grant a reprieve or commutation of the same.

Article XXV.

Salaries.

No person connected with this organization shall be entitled to any salary, pay, or emolument, other than a competent support of himself and family, unless it be from an equal dividend made of public property, on the establishment of peace, or of special provision by treaty; which provision shall be made for all persons who may have been in any active civil or military service at any time previous to any hostile action for liberty and equality.

Article XXVI.

Treaties of peace.

Before any treaty of peace shall take full effect it shall be signed by the President and Vice-President, the Commander-in-chief, a majority of the House of Representatives, a majority of the Supreme Court, and a majority of all the general officers of the army.

Article XXVII.

Duty of the military.

It shall be the duty of the Commander-in-chief and all officers and soldiers of the army to afford special protection, when needed, to Congress or any member thereof, to the Supreme Court or any member thereof, to the President, Vice-President, Treasurer, Secretary of State, Secretary of the Treasury, and Secretary of War; and to afford general protection to all civil officers or other persons having right to the same.

ARTICLE XXVIII.

Property.

All captured or confiscated property, and all property the product of the labor of those belonging to this organization and of their families, shall be held as the property of the whole, equally, without distinction, and may be used for the common benefit, or disposed of for the same object; and any person, officer, or otherwise, who shall improperly retain, secrete, use, or needlessly destroy such property, or property found, captured, or confiscated, belonging to the enemy, or shall willfully neglect to render a full and fair statement of such property by him so taken or held, shall be deemed guilty of a misdemeanor, and, on conviction, shall be punished accordingly.

ARTICLE XXIX.

Safety or intelligence fund.

All money, plate, watches, or jewelry captured by honorable warfare, found, taken, or confiscated, belonging to the enemy, shall be held sacred to constitute a liberal safety or intelligence fund; and any person who shall improperly retain, dispose of, hide, use, or destroy such money or other article above named, contrary to the provisions and spirit of this article, shall be deemed guilty of theft, and, on conviction thereof, shall be punished accordingly. The treasurer shall furnish the Commander-in-chief at all times with a full statement of the condition of such fund, and its nature.

ARTICLE XXX.

The Commander-in-chief and the treasury.

The Commander-in-chief shall have power to draw from the treasury the money and other property of the fund provided for in article twenty-ninth; but his orders shall be signed also by the Secretary of War, who shall keep strict account of the same subject to examination by any member of Congress or general officer.

ARTICLE XXXI.

Surplus of the safety or intelligence fund.

It shall be the duty of the Commander-in-chief to advice the President of any surplus of the safety and intelligence fund, who shall have power to draw such surplus (his order being also signed by the Secretary of State) to enable him to carry out the provisions of article seventeenth.

Article XXXII.

Prisoners.

No person, after having surrendered himself or herself a prisoner, and who shall properly demean himself or herself as such, to any officer or private connected with this organization, shall afterward be put to death, or be subject to any corporeal punishment, without first having had the benefit of a fair and impartial trial; nor shall any prisoner be treated with any kind of cruelty, disrespect, insult, or needless severity; but it shall be the duty of all persons, male and female, connected herewith, at all times and under all circumstances, to treat all such prisoners with every degree of respect and kindness that the nature of the circumstances will admit of, and to insist on a like course of conduct from all others, as in the fear of Almighty God, to whose care and keeping we commit our cause.

Article XXXIII.

Voluntaries.

All persons who may come forward, and shall voluntarily deliver up their slaves, and have their names registered on the books of the organization, shall, so long as they continue at peace, be entitled to the fullest protection of person and property, though not connected with this organization, and shall be treated as friends, and not merely as persons neutral.

Article XXXIV.

Neutrals.

The persons and property of all non-slaveholders, who shall remain absolutely neutral, shall be respected so far as the circumstances can allow of it, but they shall not be entitled to any active protection.

Article XXXV.

No needless waste.

The needless waste or destruction of any useful property or article by fire, throwing open of fences, fields, buildings, or needless killing of animals, or injury of either, shall not be tolerated at any time or place, but shall be promptly and properly punished.

Article XXXVI.

Property confiscated.

The entire personal and real property of all persons known to be acting either directly or indirectly with or for the enemy, or found in arms with them, or found willfully holding slaves, shall be confiscated and taken whenever and wherever it may be found in either free or slave States.

ARTICLE XXXVII.

Desertion.

Persons convicted on impartial trial of desertion to the enemy, after becoming members, acting as spies, or of treacherous surrender of property, ammunition, provisions, or supplies of any kind, roads, bridges, persons, or fortifications, shall be put to death, and their entire property confiscated.

ARTICLE XXXVIII.

Violation of parole of honor.

Persons proven to be guilty of taking up arms after having been set at liberty on parole of honor, or, after the same, to have taken any active part with or for the enemy, direct or indirect, shall be put to death, and their entire property confiscated.

ARTICLE XXXIX.

All must labor.

All persons connected in any way with this organization, and who may be entitled to full protection under it, shall be held as under obligation to labor in some way for the general good; and persons refusing or neglecting so to do, shall, on conviction, receive a suitable and appropriate punishment.

ARTICLE XL.

Irregularities.

Profane swearing, filthy conversation, indecent behavior, or indecent exposure of the person, or intoxication or quarreling, shall not be allowed or tolerated, neither unlawful intercourse of the sexes.

ARTICLE XLI.

Crimes.

Persons convicted of the forcible violation of any female prisoner shall be put to death.

ARTICLE XLII.

The marriage relation, schools, the Sabbath.

The marriage relation shall be at all times respected, and families kept together, as far as possible; and broken families encouraged to reunite, and intelligence offices established for that purpose. Schools and churches established, as soon as may be, for the purpose of religious and other instructions; for the first day of the week, regarded as a day of rest, and appropriated to moral and religious instruction and improvement, relief of the suffering, instruction of the young and ignorant, and the encouragement of personal cleanliness; nor shall any persons be required on that day to perform ordinary manual labor, unless in extremely urgent cases.

Article XLIII.

Carry arms openly.

All persons known to be of good character and of sound mind and suitable age, who are connected with this organization, whether male or female, shall be encouraged to carry arms openly.

Article XLIV.

No person to carry concealed weapons.

No person within the limits of the conquered territory, except regularly appointed policemen, express officers of the army, mail carriers, or other fully accredited messengers of the Congress, President, Vice-President, members of the Supreme Court, or commissioned officers of the army—and those only under peculiar circumstances—shall be allowed at any time to carry concealed weapons; and any person not specially authorized so to do, who shall be found so doing, shall be deemed a suspicious person, and may at once be arrested by any officer, soldier, or citizen, without the formality of a complaint or warrant, and may at once be subjected to thorough search, and shall have his or her case thoroughly investigated, and be dealt with as circumstances on proof shall require.

Article XLV.

Persons to be seized.

Persons within the limits of the territory holden by this organization, not connected with this organization, having arms at all, concealed or otherwise, shall be seized at once, or be taken in charge of some vigilant officer, and their case thoroughly investigated; and it shall be the duty of all citizens and soldiers, as well as officers, to arrest such parties as are named in this and the preceding section or article, without the formality of complaint or warrant; and they shall be placed in charge of some proper officer for examination or for safe-keeping.

Article XLVI.

These articles not for the overthrow of government.

The foregoing articles shall not be construed so as in any way to encourage the overthrow of any State government, or of the general government of the United States, and look to no dissolution of the Union, but simply to amendment and repeal. And our flag shall be the same that our fathers fought under in the Revolution.

Article XLVII.

No plurality of offices.

No two of the offices specially provided for by this instrument shall be filled by the same person at the same time.

Article XLVIII.

Oath.

Every officer, civil or military, connected with this organization shall, before entering upon the duties of his office, make solemn oath or affirmation to abide by and support this provisional constitution and these ordinances; also every citizen and soldier, before being fully recognized as such, shall do the same.

Schedule.

The president of this convention shall convene, immediately on the adoption of this instrument, a convention of all such persons as shall have given their adherence by signature to the constitution, who shall proceed to fill, by election, all offices specially named in said constitution, the president of this convention presiding, and issuing commissions to such officers elect; all such officers being thereafter elected in the manner provided in the body of this instrument.

No. 4.

Headquarters War Department, Provisional Army,
Harper's Ferry, October 10, 1859.

General Orders, No. 1.]

ORGANIZATION.

The divisions of the provisional army and the coalition are hereby established, as follows:

1—*Company.*

A company will consist of fifty-six privates, twelve non-commissioned officers, (eight corporals, 4 sergeants,) three commissioned officers, (two lieutenants, a captain,) and a surgeon.

The privates shall be divided into bands or messes of seven each, numbering from one to eight, with a corporal to each, numbered like his band.

Two bands will comprise a section. Sections will be numbered from one to four. A sergeant will be attached to each section, and numbered like it.

Two sections will comprise a platoon. Platoons will be numbered one and two, and each commanded by a lieutenant designated by like number.

2.—*Battalion.*

The battalion will consist of four companies complete.

The commissioned officers of the battalion will be a chief of battalion, and a first and second major, one of whom shall be attached to each wing.

3.—*The Regiment.*

The regiment will consist of four battalions complete.

The commissioned officers of the regiment will be a colonel and two lieutenant colonels, attached to the wings.

4.—*The Brigade.*

The brigade will consist of four regiments complete.

The commissioned officer of the brigade will be a general of brigade.

5.—*Each General Staff.*

Each of the above divisions will be entitled to a general staff, consisting of an adjutant, a commissary, a musician, and a surgeon.

6—*Appointment.*

Non-commissioned officers will be chosen by those whom they are to command.

Commissioned officers will be appointed and commissioned by this department.

The staff officers of each division will be appointed by the respective commanders of the same.

[This document is in the handwriting of J. H. Kagi.]

No. 5.—*Commission.*

GREETING:

HEADQUARTERS, WAR DEPARTMENT,
Near Harper's Ferry, Maryland.

Whereas *Oliver Brown* has been nominated a *captain* in the army established under the provisional constitution,

Now, therefore, in pursuance of the authority vested in us by said constitution, we do hereby appoint and commission the said *Oliver Brown a captain.*

Given at the office of the Secretary of War, this day, *October* 15, 1859.

JOHN BROWN,
Commander in Chief.

J. H. KAGI, *Secretary of War.*

[This document is printed in the original, with the exception of the words in italics and the figures, which are in the writing of Kagi, with the exception of the signature of John Brown, which is in his own hand.]

No. 6.

Correspondence of John Brown, jr., and others, referred to in report.

CHAMBERSBURG, PA., *Monday, July* 18, 1859.

DEAR SIR: I have just received the following:

"COLLINSVILLE, *July* 12, 1859.

"We are in receipt of a letter from ———, in which he wishes the price list of Collins & Co.'s tools forwarded to you.

"I have made inquiry of their agent concerning the matter. He says that their business is all done through their commission house in New York, and to them he wished me to refer you. Their address is, Collins & Co., 212 Water street, New York.

"Yours, respectfully,
"CHARLES H. BLAIR.

"Messrs. J. SMITH & SONS."

I wrote to Tidd one week ago to-day; several days before receiving your letter directing me to do so, and inclosing letter to H. Lindsley, which I forwarded by first mail.

None of your things have yet arrived. The railroad from Harrisburg here does no freight business itself, that all being done by a number of forwarding houses, which run private freight cars. I have requested each of these (there are six or eight of them) to give me notice of the arrival of anything for you.

I am, your obedient servant,

J. HENRIE.

J. SMITH & SONS,
Harper's Ferry, Va.

[This is indorsed "J. Henrie's letter." It is identified by R. Realf as Kagi's writing and Brown's indorsement. Kagi was addressed and wrote as J. Henrie.]

[All the letters signed John Smith are from John Brown, jr.]

ASHTABULA, ASHTABULA COUNTY, OHIO,
Monday, July 18, 1859.

DEAR FATHER: Yours, dated at Chambersburg, Pennsylvania, July 5, and mailed at Troy, New York, July 7, and also yours of the 8th, with inclosed drafts for $100, I received in due season; am here to-day to get drafts cashed; have now got all my business so arranged that I can devote my time, for the present, entirely to any business you may see fit to intrust me; shall immediately ship your freight, as you directed, most probably by canal, from Hartstown (formerly Hart's Cross Roads, Crawford county) to the river at Rochester, Pennsylvania, (formerly Beaver,) thence by railroad *via* Pittsburg, &c., as you directed; shall hold myself in readiness to go north on any business you choose to direct or confide in my hands; all well; have two or three letters from "N. E.," which I will forward to "J. H."

In haste, your affectionate son,

JOHN SMITH.

CHAMBERSBURG, *Friday, July* 22.

DEAR SIR: I received the within, and another for Oliver, to-day; I thought best not to send the other; it is from his wife. There are other reasons, which I need not name now; have here no other letters from any one.

J. HENRIE.

John Smith's letter to J. Henrie.

WEST ANDOVER, OHIO,
Saturday, July 23, 1859.

DEAR SIR: Your favor of July 16, inclosing a brief note from J. Smith & Co. is received; will preserve the *list*, but, as yet, I have received no letter with *instructions*, or as to *when, how, &c.*

Please say to Esquire Smith that I yesterday forwarded to canal at Hartstown, Pennsylvania, eleven boxes "hardware and castings" from King & Brothers. They are numbered and marked thus: X1 to 11; "By R. Rd. *via* Pittsburg and Harrisburg; J. Smith & Sons, Chambersburg, Pa.;" shall send balance hardware, &c., on Monday next. X8 and X9 are those which were on store with E. A. F. at Lindenville; Mr. Smith will remember. His household goods I shall send along as fast as possible. The letter asking me to retain the drafts came too late; I had got them cashed.

Write often, directing to John Smith, under cover to Horace Lindsley, as before. Let me know if those goods come through safely.

Please say to Mr. S—— I am still ready to serve.

Very respectfully, &c.,

JOHN SMITH.

J. HENRIE, Esq.,
Chambersburg, Pennsylvania.

West Andover, Ashtabula County, Ohio,
Wednesday, July 27, 1859.

Friend Henrie: I yesterday went to Hartstown with the balance of the hardware and castings. They consist, all told, of fifteen boxes, numbered 1 to 15, thus: X1, X2, &c., and marked J. Smith & Sons, Chambersburg, Pa., by R. Rd., *via* Pittsburg and Harrisburg.

The household stuff will soon follow. These latter boxes will be numbered A, B, &c.

It is almost impossible to get teams to do hauling, for, owing to the drouth, grass is drying up, and every horse and man is busy. You may be assured it has cost no small amount of labor, both of head and hands, to get this lot of freight so far on its way "all right." I inclose to you some cards of King & Brothers; you may find them of some use to you. If they succeed in disposing of that territory, you will of *course* need all the cast-iron patterns for their post that I have sent you.

Let me know of the safe arrival of this freight.

All well, in haste, your friend,

JOHN SMITH.

West Andover, Ashtabula County, Ohio,
Sunday evening, August 7, 1859.

Friend J. H.: I leave to-morrow (Monday) for my northern tour; have succeeded admirably in getting the freight started in good shape, in short, "all right;" saw Mr. W. yesterday; Wm. H. L. was here a day or two since; they will start in a couple of weeks, unless they hear from you in the meantime to the contrary; have written you three letters before this; have received the drafts for two hundred; the last shall probably get cashed in Rochester, perhaps at Ashtabula. If you wish to communicate with me before I return, write to my wife under cover to Mr. L., as heretofore, and she will forward to me at Chatham.

I yesterday gave W. $6, which, in addition to the $20 which our friend S. gave him, will enable the three to meet their traveling expenses. Shall write you quite often while away.

The first lot of freight of fifteen boxes I presume has reached you ere this; the last (six boxes and one chest) will not be many days behind them.

All well, very truly, &c.

JOHN SMITH.

Chambersburg, Pennsylvania,
A. M., Thursday, August 11, 1859.

Messrs. J. Smith & Sons: Oaks & Caufman have notified me that they have received fifteen boxes of freight, marked to your address, with about $85 (eighty-five dollars) charges, all told.

I await your directions in the matter.

Respectfully,

J. HENRIE.

AKRON, OHIO, *August* 12, 1859.

GENTLEMEN: I sent the box of clothing, yesterday, as directed. By mistake at the post office, your letter was not put into our box, and I did not get it till it was advertised; this is the reason why the box was not sent sooner. Our box at the post office is No. 412. All well.

Yours respectfully, J.

J. SMITH & SONS.

[This is a letter from Jason Brown.]

AKRON, OHIO, *August* 25, 1859.

Your letter of the 17th I got yesterday. I had sent the box some time ago, and wrote you at the time, directing the box as you told me, and the line as above. Your first letter I did not get till it was advertised; by mistake at the office, it was not put into our box (No. 412.) We are glad to hear that you are all well, and your prospects so good. Ellen is sick. She was confined about two weeks ago, over a month before the time. The child was born dead. Ellen is quite weak and feeble yet, but I think she will get about before long.

Your friend, J.

J. SMITH & SONS.

CHAMBERSBURG, PA., *Saturday, August* 27, 1859.

I to-day received the inclosed letter and check ($50.) One box of freight from Akron has arrived; weight about 275 pounds; charges $3 50.

The goods remaining at O. & K.'s, and those at E. & Co.'s, have been started; were taken from here yesterday morning. They should have arrived at your place last night.

The box, I neglected to say, is at O. & K.'s.

I also send letter from John Smith.

J. HENRIE.

ISAAC SMITH.

No. 7.

Correspondence and other papers in evidence before the committee.

ANDOVER, ASHTABULA COUNTY, OHIO,
Friday, September 2, 1859.

FRIEND HENRIE: I reached home day before yesterday, and have since been *busy* writing to "*our folks*," both in C. and nearer home. Have sent off letters to DeB. at D., to C——m, and to Buxton, and to Hamilton; to P——r, in N. Y.; and this morning to F. B. S., at

Concord, Mass. In all of these letters, I have forwarded the latest word from your region.

Friend L——y, at Ob——, will be on hand soon. Mr. C. H. L——n will do all he can here, but his health is bad. "J. D. H." I did not see, but L——n thought would be right on. Mrs. Sturtevant is a *working* woman. Anything she can do, she will take hold of in earnest. Write her, if you get time. Jas. Smith is marrying a wife, "and therefore cannot come." John L——n, at Ob——, brother of C. H. L., sympathizes strongly, and will work hard; Ralph, also, I think.

I shall start out soon, to try to get some means in the way father suggested when here, to help on the cause. In the meantime, I wish he would remit me some more means, say $25 or $30, as I had only enough left to get back with, and I have to *purchase* the material to winter my little stock on, since I was absent and on this business during the haying season.

Am greatly rejoiced that the fifteen boxes freight are *all through safe*, as that was the most important part. Surely, as father says, "a good Providence seems to lead us." How was our "R——r" friend pleased? You say he returned; I wish to know in what "frame of mind."

Inclosed is a letter to W——e, which came under cover to me. Don't fail to keep me fully advised, as through me you can reach the *faithful wherever I have been*. I will write very often. The last letter I sent you from Sandusky, O.

My warmest regards to each and all.

Yours, JOHN.

WEST ANDOVER, ASHTABULA COUNTY, OHIO,
Thursday morning, *September* 8, 1859.

FRIEND HENRIE: I yesterday evening received yours of Friday, September 2, and I not only hasten to reply, but hasten to lay its contents before those who are interested. Through those associations which I formed in C., I am, through the corresponding secretaries of each, able to reach *each individual member at the shortest notice by letter*.

I am devoting my whole time to our company business; shall immediately go out organizing, and raising funds. From what *I*, even, had understood, I had supposed you would not think it best to commence opening the coal banks before spring, unless circumstances should make it imperative. However, I suppose the reasons are satisfactory to you, and, if so, those who own smaller shares ought not to object. I hope we shall be able to get on in season some of those old miners of whom I wrote you; shall strain every nerve to accomplish this.

You may be assured that what you say to me will reach those who may be benefited thereby, and those who would take stock, in the shortest possible time. So, don't fail to keep me posted. My *initials*, simply, under cover to Horace, will answer just as well, and perhaps better. Please remember this.

Did the last shipment of six boxes and one chest of household goods safely arrive? How did the mining *prospect* seem to strike our R——r friend; in short, was his faith increased in the practicability and profit of the work, and how much stock did he take?

I some think of exhibiting a specimen of the fence at Cleveland fair in October—about the first of the month, I believe—and I may direct you to write me there, in care of the friends with whom you used to board. When in Cleveland, I made their acquaintance; am pleased with them. Mrs. S. thought *she* could do something, even though her husband was too much absorbed in other business. She might, I think, invest profitably, and would be a good stockholder. You might drop her a line through me, if you think better than to her direct.

I feel that it is *all important* you should have that wire from the East, and hope you will not have to make any fence without it. The specimens put up here are beautiful. Our castings cost us here not less than three cents per pound. If our plan succeeds, I think the cost might be materially lessened.

Last night we had a smart frost; cannot say how much the corn is injured. No piece that I have seen is out the way of frost yet.

There is a general dearth of news in this region. By the way, I notice, through the "Cleveland Leader," that "Old Brown" is again figuring in Kansas. Well, every dog must have his day, and he will no doubt find the end of his tether. Did you ever know of such a high-handed piece of business? However, it is just like him. The Black Republicans, some of them, may wink at such things, but I tell you, friend Henrie, he is too salt a dose for many of them to swallow, and I can already see symptoms of division in their ranks. We are bound to roll up a good, stiff majority for our side this fall. I will send you herewith the item referred to, which I clipped from the "Leader."

Give best regards to all, and believe me faithfully yours,

JOHN.

P. S. Direct to "J. B., jr.," under cover to Horace, until further notice.

J.

BOSTON, *December* 23, 1858.

DEAR SIR: I have heard vaguely of your general purpose, and have been seeking definite information for some time past, and now Mr. Redpath and Mr. Hinton have told me of your contemplated action, in which I earnestly wish to join you to act in any capacity you wish to place me as far as my small capacities go.

I am now about starting for Hayti with Mr. Redpath to pass the winter there, and I shall return in *time* for all movements. In case you should accept my services, I would return at any time you might wish me to, and in the spring at any rate.

Is there anything it would be well for me to study meanwhile? Of course I shall pay all my expenses, and shall acquire the use of the proper tools for the work which I have bought.

Any letters addressed to the care of my grandfather, Francis Jackson, 31 Hollis street, Boston, will be safe, and will be forwarded to me.

I already consider this the whole present business of my life. I am entirely free from any family ties which would impede my action. I was much disappointed in not meeting you in Kansas last winter, with a letter of recommendation from Wendell Phillips.

Immediately upon my return in the spring, I should wish to be employed in any manner to be of service to you; and, if convenient, to go through your system of training which I propose studying.

Yours,

FRANCIS J. MERIAM.

[Identified by James Jackson as the handwriting of Meriam.]

DEAR FRIEND: Our friend from Concord called with your note. I begin the investment with fifty dollars, inclosed, and will try to do more through friends.

DOCTOR.

[The corner with the date is torn off. Indorsed by Brown, "Dr. S. G. H's letter."]

JEFFERSON, OHIO, *May* 26, 1859.

MY DEAR SIR: I shall be absent during the next week, and hope to be at home during the summer. Shall be happy to see you at my house.

Very truly,

J. R. GIDDINGS.

JOHN BROWN, Esq.

SPRINGFIELD, *August* 27, 1859.

DEAR FRIEND: Yours of the 18th has been received and communicated. S. G. H. has sent you fifty dollars in a draft on New York, and I am expecting to get more from other sources, perhaps some here, and will make up to you the $300, if I can, as soon as I can, but I can give nothing myself just now, being already in debt. I hear, with great pleasure, what you say of the success of the business, and hope nothing will occur to thwart it. Your son John was in Boston a week or two since, and I tried to find him, but did not; and, being away from Concord, he did not come to see me. He saw S. G. H., G. L., S. W. P., F. J., &c., and everybody liked him. I am very sorry I could not see him. All your Boston friends are well. Theodore Parker is in Switzerland, much better, it is thought, than when he left home. Henry Stevens of this place is dead—July 28. I reached here yesterday and have seen few people as yet; here I expect letters from those to whom I have written.

I conclude that your operations will not be delayed if the money reaches you in course of the next fortnight, if you are sure of having it there. I cannot certainly promise that you will; but I think so.

Harriet Tubman is probably in New Bedford, sick. She has staid here in N. E. a long time, and been a kind of missionary.

Your friends in C. are all well; I go back there in a week.

God prosper you in all your works.

I shall write again soon.

Yours, ever, F.

[Indorsed in Brown's hand. F. B. Sanborn's letter.]

SPRINGFIELD, *August* 30, 1859.

DEAR FRIEND: I inclose you a draft for fifty dollars on New York, bought with money sent by Mrs. Russell. Dr. Howe has already sent you fifty dollars, and G. S., of P., writes me he has sent, or will send, one hundred dollars. The remainder will, perhaps, come more slowly; but I think it will come. I have sent your letter to G. S. Please acknowledge the receipt of these sums.

Yours, ever, F.

[Indorsed in Brown's hand. "F. B. S——s letter"—meaning F. B. Sanborn.]

SYRACUSE, NEW YORK,
Thursday, *August* 11, 1859.

FRIEND J. HENRIE: Day before yesterday I reached Rochester. Found our "Rochester friend" absent at Niagara Falls. Yesterday he returned, and I spent remainder of day and evening with him and Mr. E. Morton, with whom friend Isaac is acquainted.

The friend at Rochester will set out to make you a *visit* in a few days. He will be accompanied by that "other young man," and, if it can be brought around, also, by the woman that the Syracuse friend could tell me of. The *son* will probably remain back for awhile. I gave "Fred'k" $22 to defray expenses. If alive and well, you will see him ere long. I found him in rather low spirits—left him in high. Accidentally met at R—— Mr. E. Morton. He was much pleased to hear from you; was anxious for a copy of that letter of instructions to show our friend at "Pr.," who, Mr. M. says, *has his whole soul absorbed in this matter*. I have just made him a copy and mailed him at R., where he expects to be for two or three weeks. He wished me to say to you that he had reliable information that a certain noted colonel, whose name you are all acquainted with, is now in Italy. By the way, the impression prevails generally, that a certain acquaintance of ours headed the party that visited St. J., in Mo., lately. Of course I don't try to deny that which bears *such* earmarks.

Came on here this morning. Found L. gone to Boston, Massachusetts, and also said woman. As T. does not know personally those persons in C., to whom it is necessary to have letters of introduction, and he thinks I had better get him to go with me there. I have made up my mind, notwithstanding the extra expense, to go on to Boston. L. is expecting to visit C. soon, any way, and his wife thinks would contrive to go immediately. I think, for other reasons, also, I had better go on to Boston. Morton says our particular friend, Mr. S—n, in that city, is especially anxious to hear from you; has his heart and hand both engaged in the cause. Shall try and find him. Our Rochester friend thinks the woman, whom I shall see in Boston, "whose services might prove invaluable," had better be helped on.

I leave this eve. in the 11.35 train from here; shall return as soon as possible to make my visit at C. Will write you often. So far, all is well.

Keep me advised as far as consistent.

Fraternally yours,

JOHN SMITH.

SYRACUSE, NEW YORK,
Thursday, August 17, 1859.

FRIEND HENRIE: I am here to-day, so far on my way back from Boston, whither I went on Friday last. Found our Syracuse friend there; but his engagements were such that he could not possibly leave until yesterday morning. We reached here about twelve o'clock last night. While in Boston I improved the time in making the acquaintance of those staunch friends of our friend Isaac. First called on Dr. H., who, though I had no letter of introduction, received me most cordially. He gave me a letter to the friend who does business on Milk street. Went with him to his home in Medford and took dinner. The last word he said to me was, "Tell friend" (Isaac) "that we have the *fullest confidence in his endeavor*, whatever may be the result." I have met no man on whom I think more implicit reliance may be placed. He views matters from the standpoints of reason and principle, and I think his firmness is unshakable.

The friend at Concord I did not see; he was absent from home. The others here will, however, communicate with him. They were all, in short, *very much* gratified, and have had their faith and hopes much strengthened. Found a number of *earnest* and *warm* friends, whose sympathies and *theories* do not exactly harmonize, but, in spite of themselves, their *hearts will lead* their *heads*. Our Boston friends thought it better that our old friend from Syracuse should accompany me in my journey northward. I shall leave in an hour or two for Rochester, where I will finish this letter.

I am very glad I went to Boston, as *all* the friends were of the opinion that our friend "I." was in another part of the world, if not in another sphere. Our cause is *their* cause, in the fullest sense of the word.

ROCHESTER, THURSDAY EVE.,
August 17, 1859.

On my way up to our friend's house I met his son Lewis, who informs me that his father left here on Tuesday, *via* New York and Philadelphia, to make you a visit. Mr. L. will come on to-night in the 1.30 train, when we shall go right on north. That other young friend went on from here to visit you yesterday. He will take a more direct route.

Do not know as I have any thing further to say now. My warmest regards to all our friends.

Faithfully yours,

JOHN.

$10 00.] BROOKLYN, *August* 18, 1859.

ESTEEMED FRIEND: I gladly avail myself of the opportunity offered by our friend Mr. F. Douglass, who has just called upon us previous to his visit to you, to inclose to you for the good cause in which you are such a zealous laborer a small amount, which please accept with my most ardent wishes for its and your benefit. The visit of our mutual friend Douglass has somewhat revived my rather drooping spirits in the cause; but seeing such ambition and enterprise in him, I am again encouraged with best wishes for your welfare and prosperity and the good of your cause.

I subscribe myself your sincere friend,

MRS. E. A. GLOUCESTER.

Please write to me, with best respects to your son.

[Indorsed, in Brown's hand, "E. A. Gloucester's letter."]

WESTPORT, NEW YORK, *April* 16, 1859.

DEAR SIR: I am here waiting a conveyance to take me home; have been quite prostrated almost the whole time since you left me at John's with the difficulty in my head and ear, and with the ague, in *consequence* am now some better; had a good visit at Rochester, but did not effect *much;* had a first rate time at Peterboro; got of Mr. S. and others $160 nearly, and a note (which I think a good one) for $285. Mr. S. wrote eastern friends to make up at least $2,000, saying he was in for one fifth the amount. I feel encouraged to believe it will soon be done, and wish you to let our folks all round understand how the prospects are. Still it will be some days (and it may be weeks) before I can get ready to return. I shall *not* be idle. If you have found my writing case and papers, please forward them *without delay by express* to Henry Thompson, North Elba, Essex county, New York, *care of James A. Allen, Westport, New York.*

Your friend in truth,

B.

J. H. KAGI, Esq.

BOSTON, MASSACHUSETTS, *May* 16, 1859.

DEAR SIR: I should have acknowledged the receipt of yours of April 21, to Henry Thompson, together with writing case and papers, (all safe so far as I now see,) and also yours of 27th April to me, but for being badly down with the ague, so much so as to disqualify me for everything nearly. I have been here going on two weeks, and am getting better for two days past, but am very weak. I wish you to say to our folks, *all as soon as may be*, that there is *scarce a doubt* but that all will set right in a very few days more, so that I can be on my way back. They must none of them think I have been slack to *try* and urge forward a delicate and very difficult matter. I *cannot now* write you a long letter, being obliged to neglect replying to others, and also to put off some very important correspondence. My reception has been everywhere most cordial and cheering.

Your friend in truth,

JOHN BROWN.

J. H. KAGI, Esq.

KEENE, N. Y., *June* 9, 1859.

DEAR SIR: After being delayed with sickness and other hindrances, I am so far on my way back, and hope to be in Ohio within the coming week. Will you please advise the *friends* all of the fact, and say to them that as soon as I do reach, I will let them know where I will be found. I have been middling successful in my business.

Yours, in truth, JOHN BROWN.

J. HENRIE, Esq.

CHAMBERSBURG, PA., *June* 30, 1859.

DEAR SIR: We leave here to-day for Harper's Ferry, *via* Hagerstown. When you get there you had best look on the hotel register for I. Smith & Sons, without making much inquiry. We shall be looking for cheap lands near the railroad in all probability. You can write I. Smith & Sons, at Harper's Ferry, should you *need* to do so.

Yours, in truth, I. SMITH.

JOHN HENRIE, Esq.

TESTIMONY

Taken before the Select Committee, appointed by the United States Senate, to inquire into the facts and circumstances connected with the invasion and seizure of the Harper's Ferry Armory in October, 1859.

JANUARY 5, 1860.

JOHN C. UNSELD sworn and examined.

By the CHAIRMAN:

Question. Will you please to state your age, and where you reside?

Answer. I was fifty-four years old last fall. I reside about a mile from Harper's Ferry, Virginia, in Washington county, Maryland.

Question. Were you acquinted with the late John Brown, who was executed, by sentence of the law, in Jefferson county, Virginia, in December last?

Answer. I had been acquainted with him since the 4th day of July, on which day I became acquainted with him, but by the name of Smith. He informed me that his name was Smith.

Question. State the circumstances under which you made his acquaintance; and when, and where?

Answer. It was about two thirds of a mile from Harper's Ferry, Virginia, on the edge of the mountain, in Maryland, on the 4th day of July last, between eight and nine o'clock. I was going to Harper's Ferry, and met him there and saluted him, saying, "Good morning, gentlemen; how do you do?" There were four of them together; his two sons, Watson and Oliver—he told me their names—and a Mr. Anderson.

Question. State whether he told you his name, and what name he gave?

Answer. I said, "Well, gentlemen," after saluting them in that form, "I suppose you are out hunting mineral, gold, and silver?" His answer was, "No, we are not, we are out looking for land; we want to buy land; we have a little money, but we want to make it go as far as we can." He asked me the price of land. I told him that it ranged from fifteen dollars to thirty dollars in the neighborhood. He remarked, "That is high; I thought I could buy land here for about a dollar or two dollars per acre." I remarked to him, "No, sir; if you expect to get land for that price, you will have to go further west, to Kansas, or some of those Territories where there is government land"—"Congress land" perhaps I said.

Question. Did he state his business?

Answer. He did afterwards.

Question. Give the whole conversation.

Answer. I then asked him where they came from. His answer was, "from the northern part of the State of New York." I asked him what he followed there. He said farming, and the frost had been so heavy lately, that it cut off their crops there; that he could not make anything, and had sold out, and thought he would come further South and try it awhile. Then, I think, I left him.

Question Did he tell you what business he was engaged in?

Answer. He told me he was farming, and the frost had cut off his crops.

Question. But what business he was going to follow?

Answer. I then left and went to Harper's Ferry, and on my return afterwards I met the same party in the same vicinity. He then said to me, "I have been looking round your country up here, and it is a very fine country, a very pleasant place, a fine view; the land is much better than I expected to find it; your crops are pretty good." He was around where they were cutting grain. He then asked me, "Do you know of any farm that is in the neighborhood for sale?" I answered him, "I did; that there was a farm about four miles from there, owned by the heirs of Dr. Kennedy, that was for sale." He then remarked to his company, and to me also, "I think we had better rent awhile until we get better acquainted, and they could not take the advantage of us by the purchase of land;" and said to me, did I know of any property to rent. I told him perhaps he might rent that; I did not know; but it was for sale I knew. He then asked me the direction to it. I told him the direction, and the distance.

Question. Did he inform you what occupation he expected to pursue after he bought or rented land in that neighborhood?

Answer. I will tell the story. He then remarked to Watson Brown and Anderson, "Boys, as you are not very well, you had better go back and tell the landlord at Sandy Hook that we shall not be there to dinner; that we will go on up and look at the place; but you can do as you please." Finally Watson looked around at Anderson, and I did not hear him say anything; but then he turned round and answered, "Well, we will go along." "Well," said I, "if you go on with me up to my house, I can then point you the road exactly." They went up, and I asked them to come in and take dinner. They thanked me, and would not, and did not drink. "Well," said I, "if you follow up this road along the foot of the mountain, it is shady and pleasant, and you will come out at a church up here about three miles, and then you can see the house by looking from that church right up the road that runs to Boonesborough, or you can go right across and get into the county road and follow that up." He sat and talked with me awhile, and I finally asked him what he expected to follow there. I perhaps remarked to him, he could not more than make a living on the farm. "Well," said he, "my business has been buying up fat cattle, and driving them on to the State of New York, and selling them, and we expect to engage in that again." They left me then, and went on. So in the course of about three days I think, I met him again on the road between my house and Harper's Ferry, and he said, "Well, I think that place will suit me; now just give me a description where I can find the widow and the administrator." I told him that the widow lived in Sharpsburgh, a small town about ten miles from there, and the administrator (Fiery) lived between five and six miles north of Sharpsburgh; and he told me he would go and see them. I met him again in a few days after that, and he told me he had rented the two houses on the Kennedy farm. He said, "I intend going up in a few days, or sending one of the boys up to pay the rent." The following week I met him again, and he had a receipt—I presume it was. He had a paper, and said, "Well, we have got the houses, and paid the rent; we pay thirty-five dollars for the two houses, pasture for a cow and horse, and firewood, from now until the first day of March next, and here is the receipt." I remarked to him, "I do not want to see the receipt, it is nothing to me." That is about all I know of him about that time.

Question. Did he tell you then what his name was, and the name of his two sons?

Answer. He told me his name was Smith. He did not give me any given name of himself but "these are my two sons, Watson and Oliver."

Question. What were the ages of the two sons apparently?

Answer. I would judge that Oliver was about 30, and Watson perhaps 25 or 27.

Question. What was the age of Anderson?

Answer. I think he was rather younger, from his appearance; perhaps 22 or 23 years old.

Question. Did Smith, *alias* Brown, afterwards live at the houses that he had rented on the Kennedy farm?

Answer. Yes, sir.

Question. Will you state whether you ever, and how often were you, at Brown's house while he lived there?

Answer. I cannot state how often, but I was frequently there. I was there nearly every week up to the first of October. I was there every ten days at least.

Question. What took you there?

Answer. I just went up to talk to the old man; but sometimes, at the request of others, on business, about selling him some horses or cows.

Question. Were you ever in the house?

Answer. No, sir. He often invited me in. Indeed, nearly every time I went there he asked me to go in, and remarked to me frequently, "We have no chairs for you to sit on, but we have trunks and boxes." I declined going in, but sat there on my horse and chatted with him.

Question. Can you tell me whether he purchased any stock or farming implements of any kind?

Answer. He purchased one cow, one horse, a small wagon, and three hogs.

Question. Did he cultivate the land?

Answer. No, sir. He cut some hay there that he had permission to cut after he removed there.

Question. Can you state whether his family increased or diminished, as to the number of persons that were around him, during the time you knew him there?

Answer. Not to my own personal knowledge, with the exception of two females and another son. They came after he did.

Question. Did you know the name of that other son?

Answer. Watson told me his name was Owen.

Question. Who were the two females?

Answer. One was his daughter, and the other his daughter-in-law, the wife of Watson, as his son Watson told me. I never heard the name of Brown until after he was taken, and Dr. Murphy, the paymaster at the Ferry, told me that some United States officer had told him that he was old Ossawattomie Brown, of Kansas.

Question. Was his daughter unmarried, as far as you heard?

Answer. Watson told me she was a single girl.

Question. How long did those women remain there?

Answer. I think they left, or I missed them, about the 1st of October. They came, I suppose, about the 15th or 20th of July.

Question. Did Brown mingle much in the neighborhood in society?

Answer. He did not. I do not know that he was ever in any person's house but one, and he was a man by the name of Nicholls. He boarded there a day or two, and those females boarded there from Saturday night until Tuesday morning, when they came on. He was in my yard frequently, perhaps four or five times. I would always ask him in, but he would never go in, and, of course, I would not go in his house.

Question. Was any thing said or done by this person, or any of his party, which led you to suppose what was his real object in coming to that part of the country?

Answer. Nothing, only what he told me, that he followed buying up fat cattle and driving them to New York and selling them. He told others in the neighborhood the same thing. There was nothing which induced me to suppose that his purpose was anything different from what he stated to me. I frequently missed him from there, and sometimes I would find him at home and the boys away. I would remark to him, "where are the boys?" "Well," said he, "they are away somewhere." Twice I went there and found none of the men there, but the two ladies, and I sat there on my horse—there was a high porch on the house, and I could sit there and chat with them—and then I rode off and left them. They told me there were none of the men at home, but did not tell me where they were.

Question. How soon after you first saw him there, did he take possession of the house?

Answer. It was a very short time; it was the following week, at farthest. He told me that he was an old surveyor, in one instance; that he had surveyed land in Ohio, and New York, and Kansas Territory; that he followed that, and he said, "I have a little instrument that I carry in my hand, about the size of a small bucket, that has a magnet that will tell where there is any iron ore; sometimes I carry that; it has a needle to it; if the ore is in front of me the needle will point to it, and as I come there it will turn."

Question. Had you any knowledge when he was joined by the men who were afterwards found with him at Harper's Ferry?

Answer. No, sir; I had not.

Question. What was the distance of your residence from where he lived?

Answer. About four miles.

Question. What was the distance from Smith's (*alias* Brown's) house to Harper's Ferry?

Answer. About five miles.

Question. Was his house on or near a public road?

Answer. It was within about 300 yards of the public road.

Question. In sight?

Answer. Very plain; it makes a very pretty show for a small house; I mean the house he resided in.

Question. You stated that he rented two houses; do you know to what purpose he put the one he did not live in, or did he live in both?

Answer. He told me, and I thought that he had rented it for his son to live in. One time I went there, after the females had come on, and inquired for them, and one of the females answered me, "they are across there at the cabin, you had better ride over and see them." I replied it did not make any difference, and I would not bother them, and I rode back home.

Question. What were the distances of the two houses apart?

Answer. About 600 yards; one on one side of the road, and the other on the other; the house they called the cabin is hid by shrubbery in the summer season pretty much; it is a swampy piece of ground, and going from Harper's Ferry to Boonesborough you cannot see it until you get by; indeed, you could not see it from the other house when they went there.

Question. How large was that cabin-house, as you call it—how many rooms had it?

Answer. Only one room and a garret.

Question. When you first saw Smith or Brown did he tell you how long he had been there—when he came?

Answer. He told me that he had come in the cars to Harper's Ferry the evening before, which was the 3d of July; that when they got out of the cars he inquired—I do not know whether he said "he" or "we"—where they could get board the cheapest, and were informed they could get it cheaper at Sandy Hook, about a mile below the Ferry, and they consequently went there and took board, and this morning were walking out to take a view of the country. Sandy Hook is a small village about a mile below Harper's Ferry, on the Maryland side of the Potomac river.

By Mr. COLLAMER:

Question. Did you ever know from July until October about Brown's receiving any boxes or anything of that kind?

Answer. I heard of him receiving one load of boxes with very heavy things in them.

Question. But you did not see them yourself, or know anything about them?

Answer. No, sir.
Question. Did he tell you anything about them?
Answer. No, sir.

By the CHAIRMAN:

Question. When did you first hear of the invasion and seizure of the armory at Harper's Ferry by this man and his party?
Answer. On Monday, the 17th of October, about 9 o'clock in the morning.
Question. Did you go to the Ferry after you heard it?
Answer. I did not go the Ferry until Tuesday morning.

By Mr. COLLAMER:

Question. Did Brown, in passing to and from the Ferry and his house, go by your house?
Answer. No, sir; if he had, I should have been one of the first ones taken. The neighbors wanted me to get on my horse and go away, but I would not. There was no slaveholder in the neighborhood but myself, except Byrne, whom they had.

By the CHAIRMAN:

Question. Are you a slaveholder?
Answer. Yes, sir.
Question. What is your occupation in life?
Answer. I am living off what I have saved. I suppose you might consider me a mechanic, but I am not engaged in any business. I have made some money, and I am living off the interest of my money. I am living on a farm, but I am not farming it. I own the farm, and rent it out.
Question. When did you first learn, and how did you learn, that this man Smith was not really named Smith?
Answer. I have already stated that Dr. Murphy was the first one who informed me, on Tuesday about noon, that some United States officer—I do not remember his name—had said to him that that must be old Ossawattomie Brown, from Kansas. That was the first news I had of it. That was after the attack on Harper's Ferry.
Question. Did you see Brown after he was captured?
Answer. Yes, sir; I saw him the Monday before he was hung.
Question. Was he the same man to whom you have referred as passing by the name of Smith?
Answer. Yes, sir; he was.
Question. Did you have any conversation with him?
Answer. Yes, sir.

By Mr. COLLAMER:

Question. This was a short time before his execution?
Answer. Yes, sir; the Monday before. He was executed on the 2d of December, and it was on the 28th or 29th of November that I saw him in jail.

By the CHAIRMAN:

Question. In that conversation did he recognize you as his acquaintance in Maryland?

Answer. He did, and also a little son I had that was with me.

Question. Give the conversation.

Answer. I asked him why he made his attack on Virginia, and at the place he did, Harper's Ferry. His answer was: "I knew there were a good many guns there that would be of service to me, and, if I could conquer Virginia, the balance of the Southern States would nearly conquer themselves, there being such a large number of slaves in them."

Question. Have you any further information that you consider important in the inquiry before the committee, as to Brown's object in going there, or what he did after he got there, as derived from Brown himself or any of his party?

Answer. Nothing that I know of, except as to the taking of those arms, Sharp's rifles and pistols.

Question. What do you know about that?

Answer. I was present when they were taken. You asked me, some time ago, when I went to the Ferry. I told you Tuesday morning. When I got there, I saw Captain Butler, of the Hamtramck Guards, a volunteer company, of Jefferson county, Virginia. I asked him to take his company over to the school-house, on the other side of the river; that the men were in the school-house. He remarked, "My company is dismissed," and I left him, and passed on. I supposed Cook, Tidd, and the negroes they had taken were in the school-house, and had arms with them, guns, pistols, &c. I heard so from others.

Question. State what you did as to sending any party or going with any party to the school-house?

Answer. I had learned from others the day before that Cook and others were at the school-house where my child, a school-boy, attended. On Tuesday, after passing from Captain Butler, I came across Captain Rowan, another captain of a volunteer company of Jefferson county, and I said to him, "Mr. Rowan, take your company and go over to the school-house; they are over there; the danger is there." He remarked, "I will, if John will go," pointing to Mr. Avis, who stood about a rod from him. I asked him "What do you say, John?" "Well," said he, "I will see about it directly," and he walked across the street as though he were going to attend to some business. That did not suit me, and I went on around the hotel, and I came across Captain Rhinehart, who was captain of the cavalry of Jefferson county, and I asked him to take his company and go to the school-house and capture those that were there. He said, "My company is dismissed." I turned around—I was back of the railroad now on the back porch at the tavern—coming back, I met Mr. Faulkner at the front door. I asked him if he knew Colonel Lee. He replied that he did. "Well," said I, "I wish you would get him to send a company to the school-house; they are over there now; here is a man, Pitcher, who says he just came over, and they opened the door and pointed a gun at him." "Well," said he, "come along and I will get them." He was then in the superintendent's office, or under that roof in one of the offices.

He went in and saw Colonel Lee. While he was in the house Colonel Baylor, of the Virginia militia, came to the door, and I said to him, "Colonel, send a company over to Maryland to the school-house." Said he, "I have no right to send a company to Maryland." Said I, "The devil you have not; I would send them anywhere at such a time as this." "I will not do it," said he, and he turned away and left me. I was outside and he was inside. Mr. Faulkner came out and said, "Colonel Lee says they have gone an hour ago." I asked "What company has gone." "The Baltimore Greys," he replied. "Well now," said I, "they have not gone at all, for I have just come over, and there was one man standing in the street who had a uniform on, and I hailed him and asked what company he belonged to, and he told me the Baltimore Greys. Where is your captain, I asked. He has gone up on Camp Hill to get his breakfast. Well, said I, come with me and show him to me. We started up on foot and overtook him before he got to the top of the hill. I told him my business, and he said that after he got something to eat he would go with me. I went back to the ferry, and passing the armory gate I met Captain Simmes, of the company belonging to Frederick, Maryland, who was just going in the gate. I asked him "Are you captain of that company?" "Yes," said he. "Well," said I, "come take them, and go to the school-house and capture those fellows." He said "I cannot do it now." I then began to get a little out of humor. I passed on to the square and I met Mr. Boteler, member of the House of Representatives. I said to him, "Aleck, do you know Colonel Lee?" He said he did. "Now," said I, "I wish you would go and see him and get him to send a company over to this school-house. Nobody will go there, and those fellows are over in the school-house." He went and saw him and came back and said, "Colonel Lee says they have been gone an hour and a half, or two hours ago." Said I, "they have not gone at all; I know they have not; where is Colonel Lee?" He pointed him out to me, and I went up to him and saw him. He said to me, "My dear friend, they have been gone two hours ago." "No," said I, "colonel, they have not gone; I have come from there." He asked me, "Are you certain of it, and will you pilot men there." Yes," said I. "Then," said he, "come and I will get you a company." He came down towards the gate and met the lieutenant of this Baltimore company and asked him why did he not go to the school-house in Maryland, as he had ordered him. "Why," said he, "they told me the order was countermanded." While he was talking to him, the captain came up and the colonel said to him, "Why did you not go to the school-house when I ordered you?" "Why," said he, "my men were hungry, and I thought a short time would not make any difference, and we went on Camp Hill to get breakfast, and when we came back they told me the order was countermanded." The colonel said, "I did not countermand it, and nobody else had authority to do do so. Now," said he, "I want you to get your company and go with this man who says he will pilot you there." We started and went across the river to the school-house, which is in Maryland, about a mile from Harper's Ferry. When we got there two of the men of the company and myself opened the front door, went in, and found a number of boxes there. The people had gone away; there was no person there. The door was fastened with a chain and the chain was run through a staple and a stick in it. That fastening was outside. We pushed it open. I think we pushed it four times before we got it open. We thought they were behind the door, and when we would give it a push it would fly shut; but after three or four pushes it stayed back, when we pushed the things behind it away a little.

Question. State now what you found in the school-house.

Answer. We found a number of boxes. I think there must have been about fifteen small boxes, about four feet long and a foot square. We opened about five of them, as well as I recollect. They contained Sharp's rifles, and then we opened a large box that contained a number of pistols and some powder flasks.

By Mr. DOOLITTLE:

Question. Were these rifles and pistols new, fresh, as if they had never been used?

Answer. They were new, I think.

By Mr. COLLAMER:

Question. Were they unsoiled?

Answer. They never had been used, judging from the appearance of them.

By the CHAIRMAN:

Question. Did you count them to ascertain their number?

Answer. No, sir; we did not. We opened the boxes and distributed them to the Baltimore company, and there were some young men who came there. Every man who was there got a gun and a good many a pistol. After that the captain said "Let us take these to the Ferry, and when we get there we will open the balance and distribute them." I made some remark to him and he came to the door and asked me if we could not get a wagon. There was a man there who lived close by, and I said to him, "Mr. Beck, hitch up your horse and bring these things to the Ferry." He started off, and after he started I looked down a ravine rather south from the school-house, and I saw a wagon down there among the bushes. I remarked to some one, "Down there is a wagon; now, come along with me and we will go down and see what is there." We went down and found a very large wagon and three horses—one horse tied to the wagon and the other two loose. We caught them, hooked two of them up to the wagon, drove it to the school-house, and put a large number of these things in that wagon, and some of the boxes in Mr. Beck's wagon. After we got them in, nobody appeared to be willing to drive the horses, and I said to some one, "Here, get on my horse, and I will drive them." One of the soldiers got on my horse and I got on the wagon and drove those horses down to the Ferry, with the guns in the wagon.

Question. Did you bring off all the guns that you found there?

Answer. All but what we distributed. We thought we had a right to them after going there.

Question. The people who were there helped themselves out of the boxes?

Answer. Everybody that was there, I believe, got a gun, and I have frequently remarked that anybody who says he was there and did not get a gun does not tell the truth, for I carried a number of them out of the house to give to some fellows myself.

Question. Can you form an idea how many guns and pistols were distributed?

Answer; I do not know accurately, but I think there were between forty and fifty persons, each of whom got a gun and most of them a pistol.

Question. Were there no other arms than rifles and pistols?

Answer. I saw nothing else there at all, except one sword.

Question. Were there no other arms of any other kind?

Answer. No, sir.

Question. Were there any pikes there?

Answer. No, sir.

Question. Were all the arms that were found at the school-house which were not taken by the people who were there, carried down to the Ferry?

Answer. Yes, sir.

By Mr. COLLAMER:

Question. To whom were those arms delivered at the Ferry?

Answer. To Mr. Kitzmiller, who was then acting as superintendent of the armory. I think there were also at the school-house some few grubbing hoes and a few picks and shovels—not many.

Question. Were they new?

Answer. Yes, sir; they had never been used.

By the CHAIRMAN:

Question. Were you present when the pikes were taken?

Answer. Yes, sir.

Question. State what knowledge you have of any pikes being found, and where they were found, and the circumstances attending it, without going into detail?

Answer. I raised a parcel of men at the Ferry to go with me. The superintendent said we should have guns. When we got the company formed, Dr. Murphy said we should not have any guns; they were in his charge, and he could not give any more out; he had given out a large number the day before. I then went to Colonel Lee again, and said I, "Colonel, the company that was with me at the school-house have left me, and I want another company to go to Brown's house." "Well," said he, "if you will pilot them there, I will give you another company." So he hunted up Lieutenant Green, of the marines, and told him to take these men with me up to Brown's residence. "How far is it?" he asked. Said I, "it is about five miles." We started; I went up with him to the Kennedy farm. He took his company, I do not know how many; probably all they could spare there. We took along the wagon and horses we found at the school-house, which I learned was Lewis Washington's wagon. Mr. Washington told me so himself at the arsenal, and we took it by his permission to the Kennedy farm. When we got to the farm-house, we ascertained that there had been some citizens from the neighborhood of Sharpsburg at the house before we got there. We did not find anybody in the house; it was deserted. We found there a number of trunks, carpet-bags, and a large quantity of paper of different kinds—"Patriotic Volunteer," I believe it is termed on the outside. It is a drill-book for soldiers, gotten up by Forbes, I believe. There was a number of them in a large box, but no furniture there at all save one table and a cook stove. We found this pamphlet in a map of Kansas Territory. The map my little boy tore up.

[The witness produced the printed paper, which is entitled "The Laws of Kansas; Speech of Schuyler Colfax, of Indiana, in the House of Representatives, June 21, 1856," which is left with the committee, and identified by the name of the chairman written on it.]

Question. State whether you found any arms there, what they were, and what you did with them?

Answer. We found no arms at all at the house where he lived.

Question. Did you bring off the trunks, the papers, &c.?"

Answer. We put the trunks and papers in the wagons. Some of them were destroyed and carried off by citizens around, but there were a good many taken to the Ferry—some trunks; I do not think any boxes were.

Question. State what you found at the cabin, if you went there?

Answer. Lieutenant Green and myself went in the cabin. He placed one of the soldiers at the door. In the lower part of the house we found a quantity of bed-clothing, such as comforters and canvass for tents, and some axes. There were two cast-iron hominy mills, as I was informed they were, and a good deal of clothing boxed up—new clothing; but the boxes had been opened when we got there. This was clothing for men, and some boots.

Question. Can you give an idea of the amount of clothing—the quantity?

Answer. No, sir.

By Mr. COLLAMER:

Question. Can you give us the size of the boxes or the amount of them?

Answer. I cannot. The clothing was all given away up there and carried off by the citizens of the neighborhood. The boxes had been opened before we got there. There was a pile of counterpanes that looked to be new and very good, that was piled up, I suppose, between two and three feet high, doubled up and piled nicely, laid outside the boxes. There were some knives and forks and spoons, also new, which had never been used. I had a number of them in my hands. I picked them up and threw them down.

By the CHAIRMAN:

Question. What else?

Answer. In the upper part of the house, in the loft, we found, as I supposed, about a thousand pikes.

Question. What was done with all those things you found there?

Answer. They were put in the wagon that we took up there. A

number of the pikes were distributed there. Green gave the men a great many. He told me to break the window open and throw them out. I helped him to throw out a good many until I got tired, and I told him I would not throw any more out. He said "send up a couple of soldiers and I will tell them to throw them out." He told the neighbors who were present that they could have as many as they wanted. He said to them, at first, "you can have five a piece;" afterwards he told them ten a piece, and finally, he said, "you may have fifty a piece." They took as many as they wanted and the rest were put in the wagon.

Question. Who took them?

Answer. The citizens of the neighborhood.

By Mr. Doolittle:

Question. Had those pikes handles?

Answer. They had handles on them. There were two straw ticks on the floor, and on turning them up I found two pikes under them, one under each, without handles.

Mr. Davis. You spoke of picks being found in the school-house; you did not mean pikes?

Answer. No, sir; the picks were for grubbing. They were what we call grubbing hoes in our country. He had both.

By the Chairman:

Question. Did you find any picks or shovels in the cabin?

Answer. There were a very few. Apparently they were taken there to be used. There were perhaps half a dozen shovels, short and long handles together, at his house, but they were carried off by the citizens?

Question. Were they all taken to the Ferry?

Answer. All that were not distributed, we carried to the Ferry.

By Mr. Doolittle:

Question. Were these pikes in boxes, or loose?

Answer. They were lying loose, piled up in a corner, as though you would put something up here to hold them from rolling down. They were piled up in one corner right close to where the window had been, but it was nailed up. Handles had been put on the pikes by Brown's men, as I was told by Cook afterwards.

By the Chairman:

Question. To whom were they delivered at the Ferry?

Answer. To Mr. Kitzmiller, the acting superintendent. They were taken to the store-room in the armory, just as the guns were.

Mr. Davis. I should like to know from Mr. Unseld whether he heard of these people being in the school-house from his little boy, or whether some other person told him that his boy was in the school-house, and that these people were there.

Answer. I heard it from the teacher.

JOHN C. UNSELD.

JANUARY 6, 1860.

TERENCE BYRNE sworn and examined:

By the CHAIRMAN:

Question. Will you please to state your age, and where you reside, and what your occupation is?

Answer. I am forty-two years of age. I reside in Washington county, Maryland, about three miles northwest of Harper's Ferry. I am engaged in farming.

Question. Are you a landholder and slaveholder?

Answer. I am a landholder jointly with my brother Joseph. I am a slaveholder.

Question. Will you state whether you formed the acquaintance of John Brown, recently executed in Jefferson county, Virginia, and by what name he passed when you formed his acquaintance, and when you formed it?

Answer. I was not personally acquainted with him.

Question. Do you know the place he lived at—the Kennedy farm?

Answer. Very well, sir.

Question. How far is your residence from that?

Answer. My place is not quite a mile and a half south of the Kennedy farm.

Question. Did you know there was such a man in the neighborhood, although you had not seen him personally?

Answer. I knew there was a man who had rented the Kennedy farm.

Question. By what name did he pass?

Answer. By the name of Smith. I had seen him frequently and passed him on the road. I knew him by sight, but not personally.

Question. Will you state whether you were taken into custody by any party of men, and who they were, and at what time, and where?

Answer. I would rather state it in my own way. On the morning of the 17th of October I left home on horseback early, between 5 and 6 o'clock, and I had progressed about a mile and a quarter when I passed a wagon on the road, driven by a colored man. Almost at the same time that I passed the tail-end of the wagon—I going in the direction of Harper's Ferry and the wagon towards the Kennedy farm—I heard a voice call out, "Mr. Byrnes, stop." I reined up my horse and looked back, and recognized John E. Cook on the ground. I had known Cook before in that neighborhood. He approached me, on the right side of my horse, and remarked to me, "I am very sorry to inform you that you are my prisoner," or something like that. I do not remember the exact words. I had left home with a view of riding a distance of about six miles. I looked at him and smiled, and said, "you are certainly joking." He said, "I am not." I looked down, and under his coat I saw a barrel of a rifle protruding, and he kept moving it and jerking it. I thought he wanted to attract my attention, from his actions, to his being armed; and almost a moment afterwards a second man approached me, whom I have learned since was C. P. Tidd, but at that time he was unknown to me. He presented his gun to me and said, "no parley here, or I will put a ball in you," or "through you;"

"you must go with us to your place; we want your negroes," or something like that. I told him if that was the case I would go back rather than that he should put a ball through me.

Question. Well, you went back where?

Answer. To my house.

Question. What did they do after you got there?

Answer. I passed my brother on the porch just before entering the door, and I whispered to him, "civil war," or something like that; perhaps I said "servile war." I walked in. I do not know whether Cook preceded me or not, but I know we all got into the room about the same time. Cook, Leeman, and Tidd seated themselves uninvited. I walked up and down the floor, and Cook commenced making a kind of speech, sitting down, what we term a higher-law speech. My mind was busy with the future, and I paid very little attention to what he said.

Question. What was the subject?

Answer. The subject of slavery. He said that all men were created equal. That was a quotation. I remember that distinctly. Just about the time he commenced, I asked my sister, who I saw was very much alarmed, where a cousin of mine was, who was then on a visit to my house. She answered that she was up stairs, and I told her to call her down and be witness to every thing that was said and done, as she was a lady of considerable nerve. I was too much excited to pay much attention to the speech. The first word my cousin said when she came down was, "Cowhide those scoundrels out of the house; why do you suffer them to talk to you?" I did not heed her, either. I do not recollect all her remarks.

Question. There were three men then, Cook, Tidd, and Leeman; were they all armed?

Answer. All armed.

Question. With what?

Answer. Sharp's rifles and revolvers.

Question. What requirements or demands did they make of you?

Answer. I am a little too fast. Just after my arrest on the road, on turning back, they made a proposition to me to this effect, that I had better be quiet and give up my slaves; or, if I would give up my slaves voluntarily, they would enter into an article of agreement with me. They said they would first take me before their captain, and they were certain that if I would give up my slaves voluntarily their captain would enter into an article of agreement with me to protect my person and property. I told them that was something I would not do, that I looked to the State government, or, if that failed, to the federal government to protect me in my person and property. They remarked they would have them any how.

Question. What demands did they make of you in the house, and how were they made?

Answer. They addressed my brother, in the house, and said: "Mr. Byrne we want your slaves." My brother's reply was, "Captain Cook you must do as I do when I want them—hunt for them." They were too early in the morning. My brother's servant and my own, two men, had left home the Saturday evening preceding, and had not returned yet, Monday morning. They did not get them. They did not want the negro women or children at that time.

Question. Did they become satisfied that the negro men were not at home?

Answer. Yes, sir.

Question. What did they do then?

Answer. They kept my brother and myself prisoners there. Two of them remained with us, and Tidd started with some five or six or seven negroes to the Kennedy farm, in Colonel Washington's wagon. I did not know whose wagon it was at the time, but Cook told me afterwards that it was Colonel Washington's wagon.

Question. They left two of the men at your house, holding you and your brother in their custody?

Answer. Yes, sir; Cook and Leeman remained.

Question. How long did they remain?

Answer. I do not know; I had no idea of the time. I did not notice the clock, though it was on the mantle.

Question. What time in the morning was it when they first arrested you?

Answer. I do not know exactly, but it was between five and six o'clock, a little after daylight.

Question. Can you give an idea of the time they remained, as near as you can come; whether they went away before noon?

Answer. Yes, sir. They left shortly after; say a late breakfast.

Question. Then they were not there more than three or four hours, if so much?

Answer. Not so much.

Question. Did they give you any reasons for their going away, when they did go away?

Answer. They said I would have to go to Harper's Ferry, that their orders were to take me to Harper's Ferry before their captain.

Question. Did they do so?

Answer. They took me to Harper's Ferry and placed me in the watch-house. It was between 9 and 10 o'clock on Monday morning when I got into the watch-house.

Question. Who took you?

Answer. I was detained at my place until the wagon went to the Kennedy farm and returned back. Tidd, who had charge of the wagon and the negroes, came to Cook and remarked that they were ready to proceed. I was escorted by them. We went first to the school-house, where the arms were deposited. We had to pass by it on the road.

Question. What was in the wagon?

Answer. There were boxes. They seemed to be well filled with something. I did not know at the time what was in them.

Question. Was the wagon heavily loaded?

Answer. It seemed to pull pretty heavily, but it was a damp morning.

Question. How many horses?

Answer. Four horses. It was a heavy farm wagon.

Question. Did you know any of the negroes who were with them?

Answer. I did not. I was told by Cook, in the morning, that they had possession of the armory, railroad bridge, and telegraph, and before night would have the canal; that Colonel Washington was a prisoner at Harper's Ferry, and that his fowling-piece was carried by one of the negroes. He did not name the colored man who had it.

Question. Did Cook tell you whose negroes they were?

Answer. He did not, that I recollect.

Question. What was done at the school-house? Was there anybody in the school-house when you got there?

Answer. Yes, sir; Mr. Currie, the teacher, and his pupils. The school was in session.

Question. What passed when they arrived at the school-house?

Answer. I do not know that I heard all that passed. I do not think I did.

Question. I do not mean so much in conversation as what was done.

Answer. Some of the party went in; I shall not be positive who, but one of the three, or perhaps two, went in and asked him to suspend school for a while, and then he could go on; they wanted to occupy one corner of the house, saying they wanted to deposit some boxes there, but I shall not be positive about that; but I know Mr. Currie came out, and I whispered across the fence to him—I did not go in—that I was a prisoner, and remarked to him, "You have nothing to fear, you are not a slaveholder." I did not know at the time whether he owned slaves or not. My object was to put him on his guard, and he whispered in my ear "I am."

Question. What was done there? What did these men do?

Answer. The wagon was unloaded there, and the boxes placed in the school-house.

Question. Did the children leave the school-house?

Answer. Some of them manifested a disposition to leave, and I think, perhaps, were told to stay, but I am not certain about that. Most of the children were there when I left. As soon as the wagon was unloaded, Cook or Tidd told Leeman to accompany me down to Harper's Ferry.

Question. Did you leave the rest with the wagon at the school-house, when you went away?

Answer. Tidd, Cook, and the negroes were left behind with the wagon at the school-house. I proceeded with Lehman about one hundred or one hundred and fifty yards south of the school-house, and I was met by one of the Brown party, whom I had known by the name of Thompson. He came up smiling. He was armed, and, I think, had a blanket over his shoulders. He extended his hand and said, "How are you, Byrne?" I said, "Good morning, Mr. Thompson; I am well; how are you?" I was then disposed to put on a cheerful face, and I asked him what was the news at Harper's Ferry. He said the people were more frightened than hurt, and he passed on. It commenced raining about the same time, and Leeman suggested that we get under a tree until the shower passed. We sat down on the side of the road. I had an umbrella, and proposed to him to sit up close to me, and my umbrella would be some protection to him. He did so. He remarked to me, "Our captain is no longer John Smith," or I. Smith, or J. Smith, or something like that, but was "John

Brown, of Kansas notoriety,'' I think he said, but I shall not be positive about that, for I was disposed to assume a character that I did not have at the time, that of cheerfulness. My mind was busy with the future. I was fearful of a bloody civil war. I was under the impression that, unless they were there in great numbers, they would not be foolish enough to make an attack on the borders of two slaveholding States.

Question. Did you have any further conversation with Leeman at that time, as to Brown's objects or purposes in coming there?

Answer. I did not feel disposed to question him at all; but he appeared to be very serious, had very little to say while at the house, and I am inclined to think he was meditating his escape from them, judging from his manner. He took a seat by the side of the fireplace in my house and put his head against the mantel and drew his cap down. He wore a cloth cap, I think. Cook asked him if he was hungry. He said yes, he was a little hungry. Cook then asked him if he was sleepy, and I think he answered in the affirmative.

Question. Did you proceed to the Ferry afterwards?

Answer. We sat under this tree, Leeman and myself, until the shower had almost ceased, and we started. Whether Thompson overhauled us at that point, or at a point further down near Harper's Ferry, I do not recollect; but I know that Thompson and Leeman were with me almost all the distance from the school-house to the Virginia side of the bridge. There Thompson stopped, and Leeman passed through the town as far as the watch-house with me.

Question. Was Leeman armed all that time?

Answer. Yes, sir.

Question. With what?

Answer. With a Sharp's rifle.

Question. Did he carry it exposed or concealed?

Answer. It was a damp, rainy morning, and he carried it under his coat or blanket. A portion of the barrel protruded.

Question. Did you meet any armed men after you got on the Virginia side of the bridge, and before you got to the watch-house?

Answer. I passed two armed men on the bridge—a white man and a colored man. I think the white man was a son of Brown, or Smith, as he was called.

Question. Did you pass any after you left the bridge?

Answer. Not that I recollect.

Question. Did they speak to you in any way?

Answer. This white man had a mit on, and as soon as he saw me he took it off and shook hands with me.

Question. Had you seen him before?

Answer. I think I had.

Question. Did he call you by name?

Answer. I do not recollect that he did.

Question. What time did you get to the Ferry?

Answer. I do not recollect exactly, but I think it was between nine and ten o'clock in the morning.

Question. The watch-house is in the inside of the armory yard, adjoining the engine-house?

Answer. Yes, sir; under the same roof. They are adjoining rooms. The watch-house is on the west end of the building.

Question. Did anybody accost you or speak to you when you arrived at the watch-house?

Answer. I was marched up to the door. I do not recollect whom I first spoke to, but I recollect my remark, "Good morning, gentlemen; I hope I am in good company," or something like that.

Question. When did you see Brown first at the watch-house?

Answer. Almost immediately after my arrival. I saw him moving about in front of the engine-house.

Question. Did he come and speak to you?

Answer. He did not, and I did not ask to be taken to him.

Question. He did not speak to you on your coming?

Answer. No, sir.

Question. Did he at any time while you remained there?

Answer. Not until after I first addressed him. He put his hand on me and said, "I want you, sir." He went around to different ones, and I think he selected five hostages in the first place out of the watch-house. I was one of the second batch that was taken out. He just walked around and put his hand on or pointed to us; I think he put his hand on me and said, "I want you, sir."

Question. What did he want with you?

Answer. We were taken in the engine-house and pointed to the back part of the room, and told to stand there.

Question. Then Brown came into the watch-house some time after you got there, and selected five men, you among them?

Answer. Ten altogether, five the first time, and five the second time. I was one of the second five.

Question. He took those ten into the engine-house and told you to take your places in the back part of the engine-house?

Answer. I do not know whether he said "back part." The Brown party occupied the front part, and if we had taken any other position than we did, we should have been in their way.

Question. What time were you put in the engine-house?

Answer. I think it was after the middle of the day on Monday, but I could not say positively.

Question. How long did you remain in the engine-house?

Answer. Until sometime Tuesday morning, until we were rescued by the marines.

Question. Did you have any conversation with Brown while you were in the engine-house, or did you hear him conversing with any of the rest of the party on the subject of what brought him there, or what he expected or intended to do?

Answer. Yes, sir; at different times there was a great deal said. I cannot recollect one fifth part.

Question. Can you recollect anything that would disclose what his object was in coming to the Ferry; what his purpose was; what he was after; what his object was in taking the prisoners and keeping them there; what his general object was?

Answer. At one time I heard him remark: "Gentlemen, if you knew my past history, you would not blame me for being here," or something to that effect. He then went on to state that he had gone to Kansas a peaceable man, and was hunted down like a wolf by the pro-slavery men from Virginia and Kentucky, and he lost some members of his family; I think he said a son; "and now," said he, "I am here." At that time he did not say for what purpose. One son of his was laying on my right who had been wounded on Monday about the middle of the day on the street. He seemed to suffer intensely, and complained very much. He asked to be dispatched, or killed, or put out of his misery, or something of that kind, I think, and Brown remarked to him, "No, my son, have patience; I think you will get well; if you die, you die in a glorious cause, fighting for liberty," or "freedom," or something like that.

Question. Can you recollect anything that passed tending to show what his object was in coming to Harper's Ferry with a body of armed men?

Answer. I do not recollect that I heard him say, but I know his men said they were there for the purpose of giving freedom to the slaves.

Question. Did I understand you to say that Brown's son, who was wounded, had been shot in the street and came into the engine-house wounded, or was he shot while in the engine-house?

Answer. He was wounded in the street and came into the watch-house, and afterwards went into the engine-house before I was in there. I first saw him in the watch-house after he was wounded. There was some firing before I was taken into the engine-house, and he asked for his rifle, and moved in himself from the watch-house to the engine-house; but when I went in the engine-house he was laying down on the floor, and I heard his father remark that he had exerted himself too much.

Question. Did he die before your rescue?

Answer. No, sir; I think he was brought out alive. He was speechless, though.

Question. Was there much firing by the party in the engine-house; much shooting at persons outside?

Answer. There was a good deal of firing on Monday evening.

Question. Can you tell in what way the party inside fired out—through the doors, or through windows, or through loop-holes, or how?

Answer. Some through the doors, and some through port holes, or loop-holes.

Question. Were those loop-holes made after you got there, or before?

Answer. I know that some of them were made after I was taken in.

Question. Was anybody of Brown's party killed in the engine-house while you were there?

Answer. They were killed about the door. When the firing was going on, I kept as close to the floor as I could. I got down, and did not see much until there was a cessation of the firing; but there were two of Brown's party killed on Monday evening. I do not think they died until some time during the night. They were shot about the door of the engine-house in which they were.

Question. Can you tell us how many of Brown's party were in there?

Answer. I do not think I can; not when I was first taken in; but I can tell you how many there were on Tuesday morning.

Question. How many of Brown's party, dead and alive together, were in there when you were rescued by the marines?

Answer. There were two dead, one in a dying condition, (Brown's son,) and five or six active men, including a negro, at the time the attack was made by the marines.

Question. How many negroes were there in Brown's party in the engine-house, that Brown brought with him, not negroes of the neighborhood?

Answer. But one.

Question. What was his name?

Answer. Shields Green, I understood.

Question. At what time were you rescued, and when, and how?

Answer. I do not know what time it was. I had no time-piece, and did not inquire after I got out.

Question. I do not speak accurately as to hours of course, but was it in the morning, or the middle of the day, or at night?

Answer. It was Tuesday morning.

Question. How was the rescue made?

Answer. By the marines; but I cannot tell how it was done, because we were inside, and they were outside. We first heard a hammering at the door, and then the Brown party commenced firing at the door. The door was closed, and an engine run against it at tha time. They barricaded it as well as they could. There was a cessation for a moment or two, and during this time one of Brown's men turned round to him and said, "Captain, I believe I will surrender." His answer was, "Sir, you can do as you please." This man was then down on his knees, and he got upon his feet, and turned round to me and said, "Hallo 'surrender' for me." I hallooed at the top of my voice, and Mr. Daingerfield hallooed at the same time, "One man surrenders;" but we could not make ourselves heard on the outside. Coppic was further over to the left, and partly sheltered by an engine. A portion of his body was sheltered. He said to this man, "Get down on your knees, sir, or your head will be shot off." But he did not heed them until they commenced hitting on the door, and then he got down.

Question. Were any propositions made to you, or to any of the other prisoners in the engine-house, by Brown, as to your being redeemed by putting a slave in your place, or anything of that sort?

Answer. No sir. I heard of that afterwards. No such proposition was made to me. I did not hear it there.

By Mr. Collamer:

Question. How many prisoners did you find in the watch-house when you were first put in there?

Answer. I do not know. I made no count. I think there were

twenty, or perhaps twenty-five. It might have exceeded even that number.

By the CHAIRMAN:

Question. Then, as I understand you, Brown came in and selected ten, five each at two different times, and took them out of there into the engine-house?

Answer. Yes, sir.

TERENCE BYRNE.

DANIEL WHELAN sworn and examined.

By the CHAIRMAN:

Question. State your age and where you live?

Answer. I live at Harper's Ferry; I am about thirty-nine years of age.

Question. What was your business at the Ferry at the time of the invasion by John Brown?

Answer. I was a watchman at the armory gate on Sunday night.

Question. In whose service were you?

Answer. In the United States service.

Question. State when you first saw or heard or knew anything of Brown's party; what occurred when they came there?

Answer. The first time I ever saw them I heard the noise of their wagon coming down the street from the depot, and then I advanced about three yards out from the watch-house door, and observed the wagon standing facing the armory gate.

Question. Was the gate locked?

Answer. Yes, sir; I went and I advanced a little closer; I thought it was Mr. Mason, the head watchman; there were two men at the padlock striving to open it; I told them to "hold on;" I went to the gate, and when I observed it was not Mr. Mason, I drew aside at the gate and looked until I observed them, and saw they were strangers; when they all came into the yard I think there was about twenty-five men; they asked me to open the gate; I told them I could not open the gate by any means; "Open the gate," said they; I said "I could not if I was stuck," and one of them jumped up on the pier of the gate over my head, and another fellow ran and put his hand on me and caught me by the coat and held me; I was inside and they were outside, and the fellow standing over my head upon the pier, and then when I would not open the gate for them, five or six ran in from the wagon, clapped their guns against my breast, and told me I should deliver up the key; I told them I could not; and another fellow made answer and said they had not time now to be waiting for a key, but to go to the wagon and bring out the crowbar and large hammer, and they would soon get in; they went to the little wagon and brought a large crowbar out of it; there is a large chain around the two sides of the wagon-gate going in; they twisted the crowbar in the chain and they opened it, and in they ran and got in the wagon; one fellow took me; they all gathered about me and looked in my face; I was nearly scared to death with so many guns about me; I did not know the minute or the hour I should drop; they told me to be very quiet and still and make no noise or else they would put me to eternity; one of them ordered the wagon to be marched in, and all were in the wagon except four who had me; they took the wagon down the yard and passed the horses' heads to the gate where Colonel Barbour's office is; after that, the head man of them, Brown, ordered all the men to dispatch out of the yard, but he left a man at each side of the big gate along with himself; he himself still had me and Bill Williams, the watchman whom he brought down off the Potomac bridge; those other two men were at the gate, and then he said "I came here from Kansas, and this is a slave State; I want to free all the negroes in this State; I have possession now of the United States armory, and if the citizens interfere with me, I must only burn the town and have blood."

Question. Were you the only watchman in the armory yard?

Answer. There was another above in the upper end, but they did not go near him until about 1 o'clock.

Question. How far was the upper end from the gate?

Answer. About 300 yards, I guess.

Question. You saw nothing of him until about 1 o'clock in the morning?

Answer. Not until the train came down, and he was coming down to see where I was, and Brown met him and marched him into the watch-house.

Question. What time in the night was it when Brown's party appeared there at the gate?

Answer. To the best of my knowledge it was a quarter before eleven o'clock on Sunday night, the 16th of October.

Question. What did they do with you after they took you?

Answer. They kept me in the yard and began to question me about all the officers. I told them as well as I could, and the leader said he would have all those gentlemen in the morning; and with that, before he took me into the watch-house, they had old Mr. Williams down from the rifle-works. He was the other watchman up at the rifle-works. They also brought in two or three young fellows off the street. The men scattered out of the armory yard and brought them in. I had a sword in my hand, and when they all came to view me Cook took that out of my hand. I knew Cook well. There were two old muskets in the watch-house, and they took them and put them into the wagon, and I could get no person to tell me anything about them since.

Question. There were no watchmen in the armory yard except you at the gate, and one man at the far end, about 300 yards off?

Answer. That was all.

Question. Was the gate kept locked always at night?

Answer. Always. I had the key on Monday when Mr. Daingerfield was marched out, and he asked who was the watchman last night, I said "I was the watchman." He said, "why don't you open this gate?" "I could not open it," said I. "Have you the key?" "Yes," said I, "I have the key." "Well," said Daingerfield, on Monday, about 8 or 9 o'clock, when he was taken prisoner, "you had better open the gate." I was going to open the little gate by the word of Mr. Daingerfield, and Mr. Brown struck up, took the two keys, and said he was the man who could open it, and kept the keys. They were picking them up, and brought in Mr. Allstadt and Mr. Washington there, and their negroes, their wagons and horses.

Question. Did they keep you confined in the watch-house, or leave you go about the yard?

Answer. They kept me until I was taken out of it by the force of Martinsburg or the Charlestown company, I do not know which.

Question. What time of day was that?

Answer. About three o'clock on Monday

his
DANIEL × WHELAN.
mark.

Attest:
D. F. MURPHY.

JOHN D. STARRY, sworn and examined.

By the CHAIRMAN:

Question. Will you state what is your age, where you reside, and what your profession is?

Answer. I am thirty-five years of age. I reside at Harper's Ferry. I am a practising physician.

Question. Will you state at what time you first heard of the presence of an armed party at Harper's Ferry; where you heard it; and what occurred when you first became aware of it?

Answer. On Sunday night, the 16th of October, about half past one o'clock, I heard a shot fired in the direction of the Baltimore and Ohio railroad bridge, the iron span of the bridge, and immediately afterwards a cry of distress, as if somebody had been hurt. At the same time I heard considerable confusion about the Baltimore and Ohio railroad train—the starting point just opposite the hotel. I jumped out of my bed. My room is nearly opposite the railroad bridge. I went to the window and saw two armed men passing from the bridge towards the armory gate. These men were low fellows. While I was standing there, a tall man came from the direction of the armory gate, and met them near the Winchester railroad. Some noise about the hotel attracted his attention, and he turned and went towards the armory gate again. About that time some of the passengers came out from between the hotel and the railroad station, and the tall man said to them, "The first man that fires at me I will shoot," or, "the first man who interrupts me," or some such expression as that. In a very short time I was in the street, and there was some firing going on between the railroad party or citizens and that man. I did not know who fired first. There were several shots passed between them. I was then going across the street towards the railroad office. When I got there I found the negro porter, Hayward, shot, the ball entering from behind, through the body, nearly on a line with the base of the heart, a little below it. He told me that he had been out on the railroad bridge looking for a watchman who was missing, and he had been ordered to halt by some men who were there, and, instead of doing that, he turned to go back to the office, and as he turned they shot him in the back. I understood from him that he walked from there to the office, and when I found him he was lying on a plank upon two chairs in the office.

Question. Will you state in whose employment that negro was?

Answer. He was in the employment of the Baltimore and Ohio Railroad Company, and it was his duty to be up whenever the train arrived to attend to baggage, and receive whatever baggage was put off for the station, and attend to everything about the office during the absence of Mr. Beckham, the agent. He was a free negro, and had permission of the county court to remain in Jefferson county. I believe he did not belong to the county.

Question. Did you examine his wound?

Answer. Yes, sir. I found he was shot in the back, nearly on a line with the base of the heart, a little below it, and the ball came out in front?

Question. How long did he live?

Answer. I saw him about daylight; he was still living. I understood he died between twelve and one o'clock on Monday, the next day. Soon after that, which was probably about two o'clock in the morning, I stood at the corner of the railroad station and saw three men, who, I supposed, were the three I had first seen, coming from the armory gate, and I stood at the corner of the depot until they got within five or six feet of me. I then passed back the angle of the station until I got to the office-door and went in, and said to the passengers, and others who were there, "here go these three men now whom I saw go into the armory yard, and I will go down to the armory and see what is going on."

Question. Could you see whether those men were armed?

Answer. Yes, sir; I knew they were armed. I stood until they were very close to me. I went then to the armory gate, and before I got to the gate I called for the watchman. I was ordered to halt. I did so, and inquired of the men who halted me, what had become of the watchmen. I wanted to inquire why they allowed persons to go in and out of that gate, when they knew they were shooting down those whom they met in the street. I did not understand it, and I asked for Medler and Murphy, the watchmen. The fellow told me that there were no watchmen there; that he did not know Medler or Murphy, but, said he, "there are a few of us here." I did not say anything more to him, but turned and went up the street, and came off on the Winchester railroad, and down to the railroad office again. Soon after that, I was on the platform, and some of that party from the bridge hailed me to know if that train was coming over—the train which they had stopped. I told them I thought it was very doubtful; I did not think it would come over until after daylight; we did not understand their movements, and should like to know what they were doing. He said to me, "Never mind, you will find out in a day or two." I asked him if he expected to stay there a day or two. He made no reply to that. I passed on around the railroad office or post office, I do not remember which. That was about three o'clock, I suppose. I watched them from that time until daylight, sometimes very close to them, and sometimes further off. About four o'clock I heard a wagon coming down the street. I did not know what that meant, and I watched them as closely as I could. About five minutes after five o'clock, I saw a four-horse team driving over the Baltimore and Ohio railroad bridge. I did not know whose it was. In that wagon there were three men standing up in the front part, with spears in their hands, white men, and two were walking alongside, armed with rifles. I did not see any negroes. I saw but these men. I understood afterwards there were negroes with them, but I did not see them. About daylight, as these strangers seemed to have possession of the public works there, I determined to get on my horse and go and notify Mr. Kitzmiller, acting superintendent of the armory, of the condition of things there, but before I did that I went to the island of Virginius, and roused up Mr. Welch and others there. I knew there were a good many men about the mill and cooper-shop there. I told them the condition of things as well as I could. I met no one on the way. I then got my horse and came out into Shenandoah street, and had to go perhaps fifty yards before I made the turn of the street leading to the hill. About the time I was making that turn, I saw three of these men coming across from the armory gate towards the arsenal. They had just made a few steps from the gate into the street. I did not know whether their intention was to stop me or not. They made a sort of half turn, and I was out of their sight in a moment. I went to Mr. Kitzmiller and informed him that the armory was in possession of an armed band. I then passed up to Bolivar, and roused up some of the people, and went from there to Hall's Works, and found three of these men there armed. I rode up to the fence, which was probably twenty-five or thirty steps from where they were. They stepped out in front of one of the buildings, and marched down inside of the fence fifty or sixty yards, and out into the public street, and down towards the armory. I went back to the hillside then, and tried to get the citizens together, to see what we could do to get rid of these fellows. They seemed to be very troublesome. When I got on the hill I learned that they had shot Boerley. That was probably about 7 o'clock. Boerley was an Irishman, living there, a citizen of the town. He died very soon afterwards.

Question. Tell us about that incident; did you see Boerley?

Answer. I did not see him.

Question. Did you see him after he was dead?

Answer. No, sir. Dr. Claggett, who is here, saw him after he was dead, and was with him when he died.

Question. Do you know anything of the killing of Mr. Turner?

Answer. No, sir; I will go on with what I was stating; I had ordered the Lutheran church bell to be rung to get the citizens together to see what sort of arms they had; I found one or two squirrel rifles and a few shot guns; I had sent a messenger to Charlestown in the meantime for Captain Rowan, commander of a volunteer company there: I also sent messengers to the Baltimore and Ohio railroad to stop the trains coming east, and not let them approach the Ferry, and also a messenger to Shepherdstown. When I could find no guns fit for use, and learned from the operatives and foremen at the armory that all the guns that they knew of were in the arsenal and in possession of these men, I thought I had better go to Charlestown myself, perhaps; I did so, and hurried Captain Rowan off. When I returned to the Ferry, I found that the citizens had gotten some guns out of one of the workshops—guns which had been placed there to keep them out of the high water—and were pretty well armed. I assisted, from that time until some time in the night, in various ways, organizing the citizens and getting them to the best place of attack, and sometimes acting professionally.

Question. State the position of the armory and armory yard in reference to the rivers?

Answer. It is just at the confluence of the two rivers. After passing across the bridge, these men had about 60 yards to go to get to the armory gate, down the street, in front of the hotel. They would go up the Potomac river. The arsenal is rather up the Shenandoah river from there. It is probably about 60 yards from the armory gate to the arsenal gate on the Shenandoah side.

Question. Where are Hall's rifle works?

Answer. About half a mile up the Shenandoah river.

Question. These armed parties were in possession of those three points?

Answer. Yes, sir; and the Baltimore and Ohio railroad bridge also.

Question. Were you aware of the killing of any other person than this free negro you have mentioned?

Answer, No, sir; I did not see the others; I saw Mr. Turner after he was dead, and also Mr. Beckham; I did not know that Mr. Turner was shot until after he was dead.

Question. Did you examine Turner to see how he was killed?

Answer. No; I did not make an examination; I saw him after he was dead.

Question. Did you examine Beckham to see in what way he was killed?

Answer. Yes, sir; Mr Beckham was killed by a rifle ball. He was shot in the right breast.

Question. Where was the body when you saw it?

Answer. In his room. He had been removed from the place where he was killed and carried to his sleeping room near his office.

Question. Did you see this man Brown during that night, so as to identify him, that you know of?

Answer. I do not think I did; I asked him afterwards if he was at the armory gate when I was there, but he said he was not, and did not know why I had not been taken prisoner.

Question. Had you any arms?

Answer. None at all.

Question. Will you state where your chamber was, in what part of the town?

Answer. Nearly opposite the mouth of the Baltimore and Ohio railroad bridge, within 50 steps of the mouth of the Baltimore and Ohio railroad bridge, in a building across from the hotel; I was awake at the time the shot was fired and the cry of distress heard. My first idea was that some one had been shot at the train.

Question. When you first went out was the train there?

Answer. Yes, sir; it had attempted to cross the bridge before Hayward was shot, and was ordered back again by the conductor.

Question. Did you see any of Brown's party killed?

Answer. I saw a man shot in the Potomac river on Monday, I suppose about one o'clock. He was shot from near the small bridge, at the upper end of the trussel work, or from the hill side. He was attempting to cross the Potomac river from the Virginia to the Maryland side.

Question. Have you any means of knowing how many of them were killed except those in the engine-house?

Answer. I saw part of the fight at Hall's works; I went to put on some dry clothes on at half past three o'clock, and that fight was then over. A yellow fellow was brought down on the bank of the river and citizens were tying their handkerchiefs together to hang him; I put my horse between the armory wall and the fence and held him there until I allowed the officer to get off some 25 or 30 steps with the prisoner; I said to them that two or three of Brown's men were in Hall's works, and if they wanted to show their bravery they could go there. They did so. They were the citizens and neighbors of the Ferry. I organized a party about half past two or three o'clock, and sent them over there, with directions to commence the fight as soon as they got near enough; that party was under the command of a young man named Irwin. He went over, and at the first fire Kagi, and the others who were with him in Hall's works, went out the back way towards the Winchester railroad, climbed out on the railroad and into the Shenandoah river. They were met on the opposite side by a party who were there and driven back again, and two of them were shot; Kagi was killed, and a yellow fellow, Leary, was wounded and died that night; and the yellow fellow Copeland was taken unhurt.

Question. How many of the Brown party did you see dead, including those who were in the engine-house?

Answer. Four dead and Stevens wounded, and the yellow fellow Leary wounded. I saw ten of Brown's party dead altogether, including those in the engine-house.

Question. How many of those ten were negroes?

Answer. I only give you the names of the negroes as given to me by Stevens—Leary and Anderson and Daingerfield Newby were the negroes killed. Anderson was of very light color, but was given to me by Stevens, one of the party, as a colored man.

Question. Do you know the number of citizens who were killed?

Answer. Four; three white men and the negro Hayward. Hayward first, Boerley, Mr. Turner, and Mr. Beckham. Beckham was the last shot, about four o'clock in the evening.

Question. Were there any of the citizens wounded?

Answer. Edward McCábe was wounded. There were some of the Berkeley men wounded, who were acting as military. I do not know any other citizen of Harper's Ferry who was wounded but McCabe.

JOHN D. STARRY.

GEORGE W. CHAMBERS sworn and examined.

By the CHAIRMAN:

Question. Will you be good enough to state your age, where you reside, and your occupation?

Answer. My age is thirty-one years; I reside in Harper's Ferry, Virginia; I am a liquor merchant, and have a restaurant in connection with my establishment.

Question. Will you state whether you were at Harper's Ferry at the time of Brown's invasion?

Answer. I was.

Question. Please to state when you were first aware that Brown and a party of men were there, and what made you aware of it, and what you first saw in relation to it?

Answer. The first I knew of it was between one and two o'clock on Sunday night or Monday morning. The shooting of the negro on the bridge first aroused me, and, inquiring about it, I heard that a party of armed men had possession of the bridge, and had killed this negro man. I had no idea who the party were. They stopped the train due for Baltimore, and a great many passengers were concentrated in the depot. They were giving their views very generally upon the matter, and the general impression seemed to be that it was a railroad strike. I then went home, and did not know anything of it until next morning. I saw armed negroes walking about the street, and I saw Cook just about daylight. They drove Colonel Washington's four-horse wagon out of the armory yard, and I saw Cook and another white man in advance of it, and two negroes in the wagon.

Question. Did you see any of the citizens of Harper's Ferry killed?

Answer. I did not. I left Mr. Beckham a few moments before he was killed, not over five minutes I am sure.

Question. Did you see Mr. Turner when he was killed?

Answer. I did not; I did not see him at all.

Question. Nor Boerley?

Answer. No, sir; I did not see Boerley; he was killed in the morning, early, between seven and eight o'clock.

Question. Did you see the dead bodies of either of those men?

Answer. I saw Mr. Beckham after he was dead; I went and moved his head; I came over the trussel-work a few minutes after he was shot, and he seemed to be lying on his head, his neck twisted. I thought perhaps he was not dead, and ran up and laid him straight on the railroad track, and then came off. I believe his son-in-law, Mr. Hough, afterwards took him down to his house. He lived just about fifteen steps from where he was killed.

Question. Did you see any of the armed party that night?

Answer. No, sir; the first I knew of it was between one and two o'clock.

Question. And, thinking it was a railroad strike, you went back home?

Answer. Yes, sir; I went back to my house. I live on the point at the junction of the two railroads, just opposite the depot. I saw Mr. Beckham just before he was killed and just after he was killed; he attempted to go up the trusseling and I tried to hold him back, saying he had no arms. I could see none. He might have had a pistol in his pocket, but I could not tell; he had no visible arms. When I came back, a few minutes afterwards, he was lying there as stated.

By Mr. COLLAMER:

Question. What was the object of his going there?

Answer. I do not know; I believe he was very much excited; I told him it was very foolish for him to go that way.

By the CHAIRMAN:

Question. Will you state whereabouts Beckham was killed?

Answer. He was killed on the trusseling just above the hotel and near the water station; he was shot from the engine-house in the armory-yard. There is trusseling-work that runs along the river, and the engine-house is situated in the yard, perhaps thirty yards from the trusseling-work. The bridge runs in front of the armory-yard.

By Mr. COLLAMER:

Question. Were Brown's folks in possession of the engine-house at that time, and firing from it?

Answer. Yes, sir.

By the CHAIRMAN:

Question. Were many shots fired in the direction of the Baltimore and Ohio railroad trussel-work from the engine-house?

Answer. Yes, sir; a great many.

Question. Did any pass on the houses fronting on the railroad?

Answer. Yes, sir. The water station I speak of has the marks of ten or a dozen bullets in it. Mr. Beckham was killed just above it.

Question. What was Mr. Beckham's probable age, and what was his occupation?

Answer. I judge he must have been at his death sixty years old. He was agent of the Baltimore and Ohio Railroad Company at Harper's Ferry, and had the general superintendence of their business there. He was the mayor of the town also.

G. W. CHAMBERS.

LEWIS W. WASHINGTON sworn and examined.

By the CHAIRMAN:

Question. Will you please to state your age, and where you reside, and what your occupation is?

Answer. I am about forty-six years of age. I reside in Jefferson county, Virginia. I am a farmer.

Question. Are you a landholder and slaveowner?

Answer. Yes, sir.

Question. How far is your residence from Harper's Ferry?

Answer. It is about five miles.

Question. Will you state whether you saw an armed party at your house, who they were, what their business was, and what brought them there, on the night of Sunday, the 16th of October last?

Answer. There was a body at my house, five of whom I saw, and the other I did not see. They appeared at my chamber door about half past one o'clock in the morning. My name was called in an under tone, and supposing it to be by some friend who had possibly arrived late, and being familiar with the house, had been admitted in the rear by the servants, I opened the door in my night-shirt and slippers. I was in bed and asleep. As I opened the door there were four armed men with their guns drawn upon me just around me. Three had rifles, and one a large revolver. The man having a revolver held in his left hand a large flambeau, which was burning. The person in command turned out to be Stevens. He asked me my name, and then referred to a man of the name of Cook, who had been at my house before, to know whether I was Colonel Washington. On being told that I was, he said, "You are our prisoner." I looked around, and the only thing that astonished me particularly was the presence of this man Cook, who had been at my house some three or four weeks before that. I met him in the street at Harper's Ferry as I was passing along. He came out and addressed me by name, and said, "I believe you have a great many interesting relics at your house; could I have permission to see them if I should walk out some day?" I said, "Yes." At that time I supposed he was an armorer, engaged in the public works at Harper's Ferry, almost all of whom know me, though I do not know them; but I am familiar with the faces of most of them. I had not seen this man before, or I should have recognized him. He came out to my house about four weeks before this attack. While there he was looking at a pistol that General Lafayette had presented to General Washington about the period of the revolution. He asked me if I had ever shot it. I told him I had. He asked, "Does it shoot well?" I told him I had not shot it for six or eight or ten years, that I had merely tried it, and cleaned it, and put it in the cabinet, and, I remarked, it would never be shot again. He was very curious about arms. He finally told me that he belonged to a Kansas hunting party, and found it very profitable to hunt buffaloes for their hides. He unbuttoned his coat and showed me two revolvers, and said, he was in the habit of carrying them in his occupation, that he had been attacked with chills and fevers some time ago, and was wearing them to accustom his hips to their weight. He asked if I was fond of shooting. I said I formerly was; and then he said, "You would possibly like to try these?" We went in front of my house, and under a tree we stuck up a target, and fired some twenty-four shots. He then told me that he had a rifle, a twenty-two shooter, that he would like me to look at, as he saw I had some fondness for fire-arms. He said to me, "When you come down to the Ferry, if you will call, I should like you to see it and try it." I was at the Ferry, it so happened, ten or fifteen days from that period, and inquired for him. I happened to know his name in this way: he did not introduce himself when he came, but in taking up his large revolver, (the size used in the army,) I found "John E. Cook" engraved on the breech of it on a brass plate, and he said, "I engraved that myself; I borrowed the tools from a silversmith, a bungler, and thinking I could do it better myself, I did it." Then, said I, "I presume that is your name?" and he said, "Yes." When I asked for him at the Ferry, they told me he had left, and I supposed, in all probability, he had gone to Kansas, as he told me he intended to go in a few days. Believing that he had gone to Kansas, I was surprised to find him among the number at my house.

Question. You say that he had before asked permission to go to your house and see certain relics, and that he did go there; did you show him those arms?

Answer. Yes; he saw and handled them.

Question. What did they consist of?

Answer. The sword presented by Frederick the Great to General Washington, which he used as his dress sword, and one of the pistols presented to him by Lafayette.

Question. How did they come into your possession?

Answer. They descended to my father, and from him to me. My grandfather had the first choice of five swords left by the general.

Question. Shortly after midnight of the 16th of October, you were in bed and heard your name called at your chamber door, and opened it, and found an armed party with their arms presented towards you?

Answer. Yes, sir. I looked around at every gun to see if it was cocked, and found that they were all cocked.

Question. Who composed that party?

Answer. I only knew Cook's name at the time. I afterwards learned the others. The party consisted of Stevens, Cook, Tidd, Taylor, and the negro man Shields Green. There was a sixth man whom I did not see; but Cook afterwards told me his name was Meriam. He was engaged in hitching up the horses, as I understood.

Question. How did they get in your house?

Answer. They broke in the rear door of the house, and in that way reached the back entry that enters my dining room. They attacked it with the end of a fence rail used as a battering ram.

Question. You did not hear them?

Answer. No, sir; that is about fifty feet from my chamber, with about five feet of walls interposing.

Question. Where is your chamber?

Answer. On the front of the house on the first floor.

Question. Was there any other white person in the house besides yourself?

Answer. No, sir; they asked me directly for my overseer. I told them he was not there; that his family did not reside on my place, and he went to his own house every night.

Question. What did your family consist of?

Answer. My daughter had left the morning before for Baltimore; she had been spending the summer with me. Mr. William Turner and his two daughters were with me the night preceding. I was then alone.

Question. Was your daughter the only member of your family?

Answer. I have two daughters, one of whom has never resided with me, and the other was with me temporarily only, spending a few months in summer. She resides with her grandmother. She is a young lady grown. She had gone off the morning before, Saturday, with Mr. Turner and his daughters to Baltimore. This attack was on Sunday night or Monday morning, at the change of hours. After looking around I observed that each man had two revolvers sticking in his belt in front besides the rifle. I remarked to them, "you are a very bold looking set of fellows, but I should doubt your courage; you have too many arms to take one man." I said to one of them, "I believe with a pop-gun I could take either of you in your shirt tail." At that time the fire began falling from the flambeau, and I asked them to come in my room and light my candles, so as to prevent my house from being burnt. After going in, and while dressing myself, I said, "Possibly you will have the courtesy to tell me what this means; it is really a myth to me." Stevens spoke up and said, "We have come here for the purpose of liberating all the slaves of the South, and we are able (or prepared) to do it," or words to that effect. I went on deliberately and dressed myself, and went into the dining room, thinking that possibly there was a better fire there; the fire in my chamber had gone out. I went into the dining room, and when I first got in, Stevens said to me, "You have some fire-arms, have you not?" I replied, "Yes, but all unloaded." He said, "I want them," and Cook made a signal to him that he had seen a very handsome gun in my closet. It was a gun which I had imported from England, and thinking he was a workman in the armory, I showed it to him, to get his opinion. I opened my closet in the dining room, and they took out the guns.

Question. What guns were they?

Answer. A shot gun and a rifle, and an old pistol of Harper's Ferry make of 1806, which was merely kept as a curiosity. They took them. Then Stevens said to me, "Have you a watch, sir?" I replied, "I have." Said he, "Where is it?" I said, "It is on my person." Said he, "I want it, sir." Said I, "You shall not have it." Said he, "Take care, sir." He then asked, "Have you money?" I remarked, "It is very comfortable to have a good deal of it these times; money is rather scarce." Then he made the same remark to me that he did before, "Take care, sir." I then said to him, "I am going to speak very plainly; you told me your purpose was philanthropic, but you did not mention at the same time that it was robbery and rascality. I do not choose to surrender my watch." He yielded the point; did not insist on it. I told him there were four there with arms, and they could take it, but I would not surrender it. Then he said to me, "I presume you have heard of Ossawatomie Brown?" I said, "No, I have not." "Then," said he, "you have paid very little attention to Kansas matters." I remarked to him that I had become so much disgusted with Kansas, and everything connected with it, that whenever I saw a paper with "Kansas" at the head of it I turned it over and did not read it. "Well," said he, "you will see him this morning," speaking apparently with great glorification. After some little time they announced to me that my carriage was ready at the door.

Question. Did they inquire about plate?

Answer. Yes; they saw in my cabinet a camp-service that belonged to General Arista in the Mexican war; I had taken it out of the case where it belonged and placed it in the cabinet; it is of very rare and beautiful workmanship; Stevens said "I do not know but we shall want that," but afterwards he said he did not know but that it was plated-ware, instead of silver. After some little time, one came and announced that the carriage was at the door. I went out, and found the fellow, Shields Green; they called him "Emperor;" it was the first time I had seen him; he drove the carriage to the door, and as soon as I went out I found my large farm wagon with four horses hitched behind the carriage. I said to the men "These horses" (referring to the carriage horses) "will not drive in that way; they are high-spirited horses; they are on the wrong side;" Tidd, I think, went up and said "This horse is reined too short." One horse is slightly shorter than the other, and they had got the small harness on the large horse; we got on some little distance when the horses refused to work; by the by, this Emperor, as they termed him, Shields Green, was ordered off the seat when the carriage was about leaving the house, and my house servant, one of my slaves, was put in his place; Cook was on the back seat with me, and Tidd by the side of the driver; the other men were in the wagon behind; I only saw the wagon indistinctly, and did not know who was being placed in it.

Question. Did they tell you anything about taking your negroes?

Answer. They said "We ordered your wagon to take your servants;" and I supposed they were going to take women and all, but it seems they did not want women. I did not know until I got in my field who was in the wagon. When the carriage horses refused to pull, I said "These horses must be shifted;" I got down and put my foot on the wheel, and one of my servants came to help shift the horses, the servant whom they afterwards had in Maryland and who returned; the carriage horses were shifted in the field, and they went very well until they reached some point on the road; in the hurry of putting the harness on, the hames came loose near the top of the hill near Mr. Allstadt's house.

Question. What direction did they take on leaving your house?

Answer. The direction of Harper's Ferry by the usual road that led to the Ferry.

Question. Where was your first stopping place?

Answer. At the house of Mrs. Henderson, widow of Richard Henderson; they stopped the carriage just in front of the house; there were four or five daughters in the house who had recently slot their father, and I remarked to the party in front of me "There is no one here but ladies, and it would be an infamous shame to wake them up at this hour of the night." Tidd jumped out, went to the wagon, and made some remark, and they went on; they went on to Allstadt's; I heard them take a fence rail from opposite the house; we stopped on the main road in front of the house; I did not hear any directions given there; a portion of the party was left with me in my carriage; Allstadt's inclosure bordering on the pike has a post and rail fence around it; the road on the opposite side of the pike has one of our Virginia worm fences, and from this fence I heard rails moving; being familiar with the sound, I knew what they were taking; they then went towards Allstadt's house, and I heard the jar of the rail against the door, and in a few moments there was a shout of murder and general commotion in the house; I thought first it was his servants hallowing murder, but he told me afterwards it was his daughters; finding this commotion going on, they put their heads out of the window and hallooed murder; one of these fellows drew his rifle on them and ordered them to go in and shut the window; I supposed of course what their purpose was; they took a number of negroes from him, I do not know exactly how many, and Allstadt was placed in the wagon with the negroes and taken to Harper's Ferry; they mentioned to him, as he afterwards informed me, that I was in my carriage; we then proceeded on to Harper's Ferry. Up to that time I supposed it was merely a robbing party who possibly had some room at the Ferry; I did not look on the thing as very serious at all until we drove to the armory gate, and the party on the front seat of the carriage said "All's well," and the reply came from the sentinel at the gate "All's well;" then the gates were opened and I was driven in and was received by old Brown; the carriage drove into the armory yard nearly opposite the engine-house.

Question. What did Brown say? How did he know who you were?

Answer. I presume he knew who had been sent for, and he at once assumed who I was.

Question. Did he address you by name?

Answer. He did not at that moment, but as "sir." He said, "You will find a fire in here, sir; it is rather cool this morning." Afterwards he came and said, "I presume you are Mr. Washington." He then remarked to me, "It is too dark to see to write at this time, but when it shall have cleared off a little and become lighter, if you have not pen and ink, I will furnish them to you, and I shall require you to write to some of your friends to send a stout, able-bodied negro; I think after a while, possibly, I shall be enabled to release you, but only on the condition of getting your friends to send in a negro man as a ransom." Then he said, "I shall be very attentive to you, sir, for I may get the worst of it in my first encounter, and if so, your life is worth as much as mine. I shall be very particular to pay attention to you. My particular reason for taking you first was that, as the aid to the governor of Virginia, I knew you would endeavor to perform your duty, and perhaps you would have been a troublesome customer to me; and, apart from that, I wanted you particularly for the moral effect it would give our cause, having one of your name as a prisoner."

Question. Did he tell you what his purpose was; what "cause" he was in?

Answer. He spoke generally of it. He said, perhaps, "this thing must be put a stop to," or something of that sort. He used general terms.

Question. "This thing," alluding to what?

Answer. Alluding to slavery.

Question. Did you see your negroes after they were brought there?

Answer. Yes, sir.

Question. What was done with them?

Answer. They were brought in to the fire. The engine-house and the watch-house are divided by a wall. I should suppose the engine-house to be, perhaps, twenty-two or twenty-four feet square. The engine-house being partitioned off, is of course about twenty-two or twenty-four feet, as the other may be, the one way, by about ten the other. The stove was in the small watch-house. The engine-house and watch-house are divided. They are under the same roof—a wall between them. There is no communication between them through that wall. The servants were all taken into the engine-house, and we into the watch-house, but they came in repeatedly to warm themselves, each negro having a pike in his hand.

Question. How many of your negroes did they take, including your house servants?

Answer. My servants were almost all away, that being Sunday night. They took two of mine, and one, the husband of one of my servants.

Question. Did they take but three negro men of yours, altogether?

Answer. Only three there. One other heard something was wrong, and got in the wagon at Allstadt's. I understood that was the point where he overtook them. That man who joined them at Allstadt's did not belong to me, but to Dr. Fuller. He was hired at my house.

Question. Do you know what use was made of your negroes afterwards, by the party at the Ferry?

Answer. In a short time after they first appeared with these pikes in their hands, I saw my house-servant walking about without one. My other servant was taken, with my team, over to Maryland, as I afterwards understood, to remove the arms from the Kennedy farm to the school-house.

Question. Did any of the servants remain with you in the engine-house or watch-house?

Answer. Yes, sir; my house-servant was in the engine-house with me all the time.

Question. Did they put him to any use at all?

Answer. Not at all. They made a servant of Allstadt's drill some port-holes.

Question. How many servants did they bring from Allstadt's?

Answer. I do not know; five or six perhaps.

Question. How many of yours and Allstadt's together were with you in the engine-house?

Answer. There was one of mine and one of Allstadt's that I know, and a servant I have known for some time, one of Mr. Daniel Moore's, who resides near Allstadt. He was arrested on the bridge or in the Ferry. He had a wife there, possibly. I do not recollect exactly the number of Mr. Allstadt's servants there.

Question. Did they put any of the slaves they had captured to any work in the engine-house?

Answer. None, except one servant of Mr. Allstadt, named Phil. Old Brown said to him, "you are a pretty stout looking fellow; can't you knock a hole through there for me?" There were some mason's tools with which he effected it. The holes were loop-holes to shoot through.

Question. Did they make more than one loop-hole?

Answer. Yes, sir; four I think.

Question. How long were you detained in the engine-house?

Answer. I went in there about twelve o'clock on Monday, noon, and I was in there until Tuesday at seven. I was taken into the watch-house first, but he took us out as hostages about eleven or twelve o'clock on Monday.

Question. What time did you arrive at the watch-house?

Answer. I suppose about half past three; some time before daylight on Monday morning.

Question. After being in there until about midday on Monday, they took you out and carried you into the engine-house. Did they take any others with you?

Answer. Nine others.

Question. Did he say for what reason you were taken out and carried to the engine-house?

Answer. He did not specify it at that time, but I understood it very well from the remarks he had made early in the morning. He just came and said, "I want you to walk with me;" and we went from one room to the other.

Question. What was the largest number of persons that he had as prisoners at any time in the watch-house?

Answer. I should say, at a rough estimate, perhaps thirty-odd; between thirty and forty.

Question. Who were they?

Answer. They were principally the armorers, the workmen of the armory, and officers of the armory; for instance, Mr. Kitzmiller, who was acting as superintendent at the time in the absence of Mr. Barbour, Mr. Daingerfield, who was the paymaster's clerk, and Mr. Mills, the master armorer, and several others, operatives, and some who were not. One was the watchman on the bridge, I believe, and one was an old man who rang the bell.

Question. They were all citizens of the Ferry and workmen there?

Answer. Yes, sir.

Question. Were any of those men armed?

Answer. None.

Question. Did you find any of them there when you first went there?

Answer. Yes, sir; perhaps four or five.

Question. Were they brought in in a body or brought in singly?

Answer. Generally one or two at a time. As they made their appearance they were arrested, as I understood.

Question. Will you state whether you heard any conversation of Brown's during the night in the engine-house, in which he disclosed his purpose in coming there?

Answer. I think two or three different times, possibly, he made remarks to the effect that he came for the purpose of freeing the slaves, and that he meant to carry it out. I heard a remark made by Stevens pretty early. He was talking to a young man, and asked him what his view in reference to slavery was, and this young man said, "of course, being born south, my views are with the south on that subject." Stevens asked him if he was a slaveholder. He said he was not. "Well," said Stevens, "you would be the first fellow I would hang, for you defend a cause not to protect your own interest in doing so," and he used an oath at the time.

Question. Did you hear anything from Brown from which you could learn whether he expected assistance, and where it was to come from?

Answer. I do not know that I heard any such expressions. I supposed at that time he was very strong. I supposed from his actions the force was a large one. Some one asked him the number of his force, and he made an evasive answer. Said he, "I cannot exactly say. I have four companies—one stationed" at such a place, and so on. He used the term "companies."

Question. What points did he designate?

Answer. The arsenal was one, Hall's works was another, and some other point in the yard.

By Mr. DOOLITTLE:

Question. They were companies at or about Harper's Ferry?

Answer. Yes, sir.

By the CHAIRMAN:

Question. Can you tell how many of Brown's party you found in the engine-house when you went there?

Answer. Up to a certain period they were in and out until the firing became very severe in the street. There were eight, I think, of his party in the engine-house.

Question. I mean from the time they were beleagured so that they could not get out?

Answer. Then I think there were eight.

Question. How many of them were negroes?

Answer. One, I think.

Question. Was there not more than one negro?

Answer. Yes, but not with us. There was only one negro of his party in the engine-house. There were several slaves, but only one of his party.

Question. Do you know what his name was?

Answer. Shields Green.

Question. What was his color?

Answer. Black.

Question. Will you state whether that negro, Shields Green, was armed?

Answer. Yes, sir, like the rest, with a rifle and revolver, and a butcher knife in his sheath.

Question. Did he use his arms; did he fire?

Answer. Yes, sir, very rapidly and dilligently. I do not know with what effect.

Question. What was his deportment?

Answer. It was rather impudent in the morning. I saw him order some gentlemen to shut a window, with a rifle raised at them. He said, "Shut that window, damn you; shut it instantly." He did it in a very impudent manner. But when the attack came on, he had thrown off his hat and all his equipments, and was endeavoring to represent himself as one of the slaves.

Question. Will you state at what time you were delivered from their custody?

Answer. I suppose it was about half-past seven o'clock on Tuesday morning.

Question. Did Brown give any reasons for keeping you gentlemen confined there?

Answer. Yes. He alluded to the fact that through us he expected to gain his terms. He was very anxious towards the last. He was very solicitous to have some capitulation by which he could gain his terms, and was very obstinate in reference to his terms.

Question. Did you hear what his terms were?

Answer. Yes, sir, there were several. One was that he was to be permitted to leave the Ferry, and take all his prisoners to a point about half a mile or three-quarters of a mile above the Ferry, on the Maryland side, unmolested; and at this point he promised to release the prisoners.

Question. Was that refused?

Answer. Yes, sir.

Question. Now will you state in what manner you were ultimately rescued?

Answer. By the marines.

Question. How did they do it?

Answer. They broke in the door, and entered with a charge. In the excitement of the moment there was a gun or two fired, I believe, in the act of breaking in the door.

Question. A gun or two fired, by whom?

Answer. By both parties.

Question. While you were confined there during Monday, was there much firing from the engine-house?

Answer. A good deal.

Question. Did you know of anybody being killed?

Answer. I did not know at the time. I knew the parties who were killed, but I did not know the fact at the time.

By Mr. Collamer:

Question. Was there firing upon the engine-house also?

Answer. There was firing upon it and from it.

By the Chairman:

Question. Did you see any of the citizens who were killed at the Ferry?

Answer. I think not.

Question. Were you acquainted with George W. Turner?

Answer. Intimately.

Question. Was he killed there?

Answer. He was killed there, I believe. He was killed in the street; not near us.

Question. Were you at his funeral?

Answer. He was merely entombed for a short time, and was buried recently at Charlestown. I was at that funeral?

Question. Will you state where he lived?

Answer. He lived at a place called Wheatland, about five miles from Charlestown, and about eight miles or eight miles and a half from my house.

Question. Were you on terms of intimate relations?

Answer. Yes, sir.

Question. What was his character as a citizen and a gentleman?

Answer. Very fine. None better. He was a graduate of West Point, and a distinguished officer of the army.

Question. Was he a man of fortune?

Answer. Yes, sir.

Question. A landholder and slaveholder?

Answer. Yes, sir.

Question. Did you know Mr. Beckham, who was killed?

Answer. Yes, sir; for many years.

Question. What was his character as a citizen?

Answer. Very good indeed. He was an estimable man. He was mayor of the town, and had been for many years employed by the Baltimore and Ohio Railroad Company as their agent.

Question. Did you know Boerley, who was killed?

Answer. I knew him slightly. I had known him some years merely to speak to him.

Question. Do you know what his business was?

Answer. I think he kept a small grocery store.

Question. Did you know the negro, Hayward, who was killed?

Answer. Yes, sir.

Question. Did you know whether he was free or slave?

Answer. I understood he was free.

Question. What was his position in life?

Answer. He was the porter of the railroad station, and attended to the baggage. He was always remarkably civil.

Question. Was he esteemed and considered a man of respectability in his position?

Answer. Very much so. He was very trustworthy.

Question. Did you get back all your slaves?

Answer. Yes, sir; except the servant that was drowned at Hall's works. The others made their escape from those men who armed them in Maryland, and came down to the river, and were put across by a white woman in a boat, and were at home when I got there. They must have gone back on Tuesday night, I imagine. I did not go back until Wednesday evening. I remained at the Ferry with the governor two days.

Question. Did you find your negroes at home when you went back?

Answer. Yes, sir.

Question. Did you get back your wagon and horses?

Answer. After a while. The wagon was used afterwards in bringing over arms to the Ferry. On Thursday one of my horses was running up in the mountain, and I went over and got him, and took the negro boy who showed me where he had hidden my gun that they had given him to arm himself when he escaped. This was a double-barreled shot gun.

Question. You lost none of your negroes?

Answer. No, sir.

Question. But a man whom you had hired from Dr. Fuller was drowned in the canal?

Answer. Yes, sir.

Question. Did it excite any spirit of insubordination amongst your negroes?

Answer. Not the slightest. If anything, they were much more tractable than before.

Question. Had you any reason to believe that there was any alarm amongst them when they were carried off; had you any knowledge of that?

Answer. No; I could not see what transpired when they were taken; it was out of my sight.

By Mr. COLLAMER:

Question. What became of your carriage and carriage horses?

Answer. They were left in the yard, and I went to Brown and told him that if those horses remained there, some time they would get off and break the carriage all to pieces. The clerk of the hotel happened to be there, and I asked him to have those horses taken to the stable. The carriage was a good deal shot to pieces. The carriage remained in the armory yard; the horses were put in the tavern stable, and, I believe, they were something like myself, they did not get anything to eat or to drink for a good while. I got nothing to eat for forty hours. I ate nothing from Sunday at dinner until Tuesday at 10 o'clock. Brown, on Monday morning, came and invited me to breakfast; he had some breakfast ordered in the yard from the tavern. I went to several of the prisoners and suggested the impropriety of touching it, "for," said I, "you do not know what may be in it; the coffee may be drugged for the purpose of saving a guard over you." I advised them not to take it.

By the CHAIRMAN:

Question. I understood you to say that they carried off a pistol and sword belonging to your family relics; did you recover them?

Answer. I recovered the sword; Brown carried that in his hand all day Monday, and when the attacking party came on he laid it on a fire engine, and after the rescue I got it.

Question. By whom did you say that sword had been given to General Washington?

Answer. By Frederick the Great.

LEWIS W. WASHINGTON.

JOHN H. ALLSTADT sworn and examined.

By the CHAIRMAN:

Question. Will you state your age, and where you reside, and what your occupation is?

Answer. I am fifty-one; I reside two and a half miles above Harper's Ferry, in the county of Jefferson, State of Virginia; I am a farmer.

Question. Are you a landholder and an owner of slaves?

Answer. Yes, sir.

Question. Will you state whether a party of armed men came to your house at some time in October, what they did after they got there, and who they were, as far as you learned?

Answer. There was a party of men came there on the 17th of October, at three o'clock in the morning.

Question. What was the first information you had of their being present?

Answer. The first information I had of them was a rapping at our chamber door. I was in bed. I immediately got up, and inquired who was there; they told me to open the door directly or they would burn me up. I did not open the door, but at that moment they bursted the door open with a rail. The door was locked. When the door was bursted open I could see out; I had gotten up by that time, and my wife had gotten up also. I tried to shut the door. I saw five or six men with arms, rifles, standing right at the door, but three of them came into the room and told me to dress myself directly. I asked them their object. They told me they intended to free the country of slavery. I asked what they were going to do with me. They said they were going to take me to Harper's Ferry; that they had the armory in their possession, and they told me they had Colonel Washington. They asked me if there were any more men about the house. I told them none but my son. In the meantime my son had come down stairs, and they seized him by the collar, and held him until I dressed myself. My son is eighteen or nineteen years old. When I dressed myself they told me to march on, and when I went to the door they had all my black men and boys—they were all men except one—at the door waiting for me. I mean my slaves. There were seven of them. They were all grown but one. We were ordered out to the turnpike, which was just across the yard, and ordered to get into a four-horse wagon—my son, myself, and my negroes. I recognized the wagon to be Colonel Washington's. I inquired of them where Colonel Washington was. They said he was in his carriage, and that was right in front of us, driving down by the side of the fence, and they remarked they were ready.

Question. Were all those men armed?

Answer. Yes, sir.

Question. What with?

Answer. With Sharp's rifles. The carriage moved up on the pike and we followed. Four walked in front of the wagon armed, and two in the wagon. We went on in that way until we got over the hill near Bolivar. There is a skirt of woods to the left. They told the wagon to halt, and they went up into the woods and held a consultation, and came back and said, "Boys, mind; we may have a little fight," or something of that account, and then told them to drive on. They went on before and drove to Harper's Ferry, and drove into the armory yard. There my son and myself were ordered out. John Brown delivered us over to a man, who took us into the watch-house.

Question. Did he take all your negroes into the watch-house with you?

Answer. No, sir; not all of them; they were backwards and forwards; sometimes some of them would come into the watch-house

They had armed the negroes with spears, and they would occasonally walk in to the stove and they would go out again, though the most of them were placed in the engine-house. As I walked out, I could see them in the engine-house, standing there, some of their men with them. When I went into the watch-house I met with Colonel Washington; that was the first time I had seen him; I asked him what this meant; he said he did not know.

Question. How did these men get into your house, do you know?

Answer. They bursted the chamber-door open with a rail.

Question. How did they get into the house?

Answer. We lie in the front; our chamber was in the front.

Question. Did the chamber-door open out of doors?

Answer. Yes, sir; out on the porch. The chamber is on the first floor; the door of that is not the front door, either; the front door is next to the turnpike; that is not the room we occupy; we occupy the back room at the other end of the house; they went round to the other end of the house.

By Mr. Davis:

Question. Was the front door open or shut?

Answer. It was locked.

By the Chairman:

Question. Did you recognize or know any of those men who appeared at your chamber-door when it was broken open?

Answer. I did not; I did not know any of them.

Question. Had you known Cook before?

Answer. No, I never knew him before; I had seen Brown before.

Question. Where had you seen him?

Answer. I had seen him at Harper's Ferry, on the street; and I had seen him also at the cars when the cars would land there; I inquired who he was; he was walking up and down; he was a stranger to me, and I asked who that old gentleman was; they told me his name was Smith; I recognized him when we got to the armory yard as being that Smith, but they called him Brown then.

Question. When you first saw him at the cars, how long was that before this affair?

Answer. I had seen him at different times, perhaps a month before that, and perhaps I saw him not two weeks before that; I do not recollect exactly; I saw him at different times.

Question. You say that your negroes had pikes put in their hands and were walking about the engine-house and the watch-house from time to time. Do you know what other use was made of them by Brown or his party?

Answer. No, sir; I do not. The negroes laid the pikes down at last, and did not use them any more at all; they had not them in their hands except in the first instance.

Question. Did you get all your negroes back?

Answer. All but one.

Question. What became of him?

Answer. He was taken to Charlestown, to the jail.

Question. What ultimately became of him?

Answer. He died.

Question. How? From what cause?

Answer. I do not know. He was frightened very much, I suppose, and exposed very much that day; it was a very bad day; it rained very hard; I suppose he was exposed to the rain and cold; he was taken sick after he had been in jail a few days, and died.

Question. Were you kept in the engine-house until you were rescued by the marines?

Answer. Yes, sir; Brown came in and selected three men; I do not know what ones they were exactly, and he took them out; I think Mr. Daingerfield was one; I do not know the others; he took them out of the watch-house; I did not know what he was going to do with them; after awhile, he came back and he came up to me and tapped me on the shoulder, saying "I want you," and pointed out two or three others, "I want you," "I want you," and we followed him up and he took us right out of the watch-house into the engine-house; there we were kept all the time.

Question. You remained there until you were released by the marines on Tuesday morning?

Answer. Yes, sir.

Question. At what time did you get home?

Answer. I suppose it must have been the middle of the day on Tuesday.

Question. Were your negroes at home them?

Answer. No, sir; there were three of my negroes in the engine-house with us; they got home pretty soon after I did; there were three in the mountain in Maryland who were sent over with the wagon; two of them got home that evening pretty soon after; I do not know what time; the middle of the evening, I suppose; the other one got home in the evening, but he was at Harper's Ferry; they had brought him down there, I understood.

Question. Then your negroes all got home that night, except one who was taken to Charlestown?

Answer. Yes, sir.

Question. Did you see him in jail?

Answer. Yes, sir; I took cold in the engine-house; it was very cold there through the night, and when we wanted to sit down we had either to sit on the engines or sit on the brick floor; I was very hoarse when I came out of the engine-house on Tuesday; I thought I had better take care of myself, or else I might be taken sick; I did not go to Charlestown for some few days; I do not know how long; I saw the negro there when I went; he was very sick when I went there, so much so that I could not move him home.

Question. Do you know why he was taken to jail?

Answer. I inquired why he was taken to jail, and they said they did not know they had committed him to jail; the magistrate had committed him to jail, and I would see further about him when I went to town.

Question. Did you hear of any charges being made against him?

Answer. No, sir.

Question. The negro died from sickness in jail?
Answer. Yes, sir; he was too sick to carry home.
Question. What was his age?
Answer. He was about twenty years old; he was a very valuable fellow; the most valuable one I had.

By Mr. Davis:

Question. A negro of good character?
Answer. Yes, sir.
Question. What was the conduct of those negroes at the time they were walking about with pikes in their hands? Did they appear hostile to you?
Answer. Not at all; they did not appear hostile to any one.
Question. Did they appear to understand their condition?
Answer. No, sir; I do not know that they did.

By the Chairman:

Question. Did you hold any conversation with your negroes while you were in the watch-house or engine-house?
Answer. No, sir; I did not.
Question. What was their conduct after they got home. Were they submissive and tractable as usual, or were they insubordinate?
Answer. They were pretty much as usual, except that they seemed to be pretty much frightened. For instance, there was one of the troops from Charlestown called there one night, and I could see that they were frightened at that time; that was two or three weeks afterwards.
Question. Where did your negroes lodge; where were their cabins; how far from your house?
Answer. They were very close, almost adjoining the house. The porch ran from our room to the kitchen.
Question. After you dressed yourself and came out, did you find your negroes on the road?
Answer. They were guarded there right at the porch in front of the door.

By Mr. Davis:

Question. And a standing guard over them?
Answer. Yes, sir.

By the Chairman:

Question. Did the white men tell you what they intended to do with the negroes?
Answer. No, sir; they did not say what their object was.
Question. Did they disclose no reason for taking them along with you to the Ferry?
Answer. No, sir.
Question. While you were a prisoner did you hear Brown at any time say what he intended to do with the negroes; what his object was in catching them?
Answer. No, sir; I did not.

By Mr. DOOLITTLE:

Question. Or his object in coming there to Harper's Ferry and taking the armory?

Answer. No, sir; only that he intended to free the country of slavery—that was all.

By Mr. DAVIS:

Question. Were they shooting out of the engine-house when you were in there?

Answer. Yes, sir.

Question. Whom were they shooting at?

Answer. I could not tell that. We were kept back and they were in front and would shoot out at the door.

Question. Did they say whom they were shooting at?

Answer. I supposed they were shooting at the men. There was one of them said that he was shooting at a man; he had shot several times at a man peeping round the water station on the railroad, and he remarked to me that he thought he would take six inches of the wood. He said he had not hit him, and he thought he would take six inches of the wood, but Brown told him perhaps he had better not do that. He went back to his position and shot three or four times afterwards, and then he said, "That is the time I brought him?"

By Mr. DOOLITTLE:

Question. At this time they were shooting out; were there shots also fired in?

Answer. Yes, sir.

Question. Was there shooting both ways?

Answer. Yes, sir.

By the CHAIRMAN:

Question. Was anybody killed in the engine-house from shots fired outside?

Answer. Yes, sir.

Question. How many?

Answer. Two were killed.

Question. Was that during Monday?

Answer. Yes, sir.

By Mr. DOOLITTLE:

Question. I suppose they were killed by being shot through the loop-holes?

Answer. No, sir; nobody was shot through the loop-holes. They were shot at the door.

By the CHAIRMAN:

Question. The door was open to enable them to fire out, I suppose?

Answer. Yes, sir; the door was partly open to enable them to fire out, and they fired out there. They would go to the door and fire. They fired a great deal out of those loop-holes.

JOHN H. ALLSTADT.

JANUARY 10, 1860.

Colonel ROBERT E. LEE, United States army, sworn and examined.

By the CHAIRMAN:

Question. Will you state whether you made an official report to the Secretary of War of the execution of the orders which sent you to Harper's Ferry at the time of the difficulty there in October last?

Answer. I did.

Question. Did you make that report from the Ferry?

Answer. It was written at the Ferry, except the concluding part; I brought it down with me; I did not entirely conclude it at the Ferry; I had written as far as to say that I would take advantage of the morning train to go to Washington. When I got to Washington I finished the report, and then handed it to the adjutant general.

Question. Will you state as briefly as you can from whom you derived the list which is appended to your report as a list of insurgents?

Answer. Lieutenant Stuart brought from the Kennedy farm a roll, or what purported to be a roll of the conspirators; I endeavored to account for all the persons on that roll, to see whether they were among the killed or prisoners, or where they were; I was present when Governor Wise, that same evening, (Tuesday evening,) was questioning Brown and Stevens as to the number of persons that were engaged with them in this conspiracy. From their report I checked off several names that I found on this roster or roll; I directed Lieutenant Stuart to go and examine the dead bodies that were then uninterred to see if he could ascertain who were among the killed; I think, the next day, I got from Mr. Andrew Hunter the verification or the identification of some of the names upon that roll, of which he had made a record as Brown and Stevens reported the names to Governor Wise. These are, as far as I recollect, the means that I took to ascertain the truth of this roster or roll as it purported to be.

Question. Did you see the bodies of any of the citizens who were killed there?

Answer. I did not. They had been killed on Monday, before my arrival; I did not arrive there until Monday night, and they had all been taken care of by that time.

Question. Had you any opportunity of judging, or means of knowing, what the condition of the armory and the public property was there in reference to police to protect and defend it?

Answer. I had not. I know nothing more than that there were some watchmen employed at night, as I understood, merely as a guard to give an alarm in case of fire.

Question. Was there any military force kept there by the United States for the preservation of the works?

Answer. None that I am aware of.

Question. Were you able to get into communication with any of the persons in authority at the armory at the time of your arrival?

Answer. I was not able to do so at the time of my arrival. I did not reach the Ferry until about eleven or twelve o'clock on Monday night, and I found the village in possession of the State troops; but I did not have any intercourse with those in authority at the armory that I recollect. If I did, I did not know them. I understood that the superintendent, Mr. Barbour, was absent on business, and many of the master workmen were the prisoners of the conspirators. I did not know whom to apply to.

R. E. LEE,
Brevet Colonel, Lieut. Col. Second Cavalry.

THEODORE RYNDERS sworn and examined.

By the CHAIRMAN:

Question. Will you state whether you are a deputy of Isaiah Rynders, marshal of New York?

Answer. He has what he calls two regular deputies who do his principal work. There are eight of us who are deputized when we have a warrant. We simply serve warrants.

Question. Will you state whether you brought this trunk to Washington, and when? [Referring to a black trunk.]

Answer. Yes, sir. I left New York in the six o'clock train yesterday, and arrived here this morning with it.

Question. Did Mr. Rynders send the summons back; did he give you any papers except a letter to me?

Answer. No, sir; he kept the summons. I was with him when he went to serve the summons on the proprietor of the hotel.

Question. State whether you went with him when he went to serve that summons, and when, and where?

Answer. It was, I should think, about noon yesterday, at the corner of Broadway and Ninth street, at the European House.

Question. Who was the keeper of the house?

Answer. He has the name in the summons. I forget it now.

Question. You were with Mr. Rynders at the time?

Answer. Myself and Mr. Thompson.

Question. State what took place when you arrived at the house?

Answer. When I arrived there I was with Mr. Thompson. The marshal stayed behind. Mr. Thompson is his first deputy. He inquired for the name of the landlord that you mentioned; and it was a name something similar to that. It turned out to be the same party.

Question. What conversation took place between Mr. Thompson and him?

Answer. It was a colored man he spoke to at the door. The proprietor was out. He said he would be in at half past one o'clock. I went down to the office, and came back just about the time the marshal had met these gentlemen, and they were up in the room talking when I went in. They were in the gentleman's private parlor, I believe it was. The keeper was relating to the marshal how he came by the trunk. It seems there was a lady there who had kept the house previous to this gentleman coming there, and he took charge of it as a sort of agent for this lady. He told the marshal how he came by the trunk; how the major boarded there and did not pay his board.

Mr. COLLAMER. Is not that man to be here—the man who gave the trunk?

Mr. MASON. I issued a summons for him to come here and bring the trunk. I received a letter from the marshal this morning brought by this gentleman, just before I left home, informing me he had got the trunk, but that the man himself declined coming unless he is wanted. The landlord did not want to come, so Rynders reported, unless it was necessary for him to come.

Mr. COLLAMER. If there is any materiality in knowing where the trunk came from, this is no testimony of it—all talk—what he heard one man say to another.

Mr. MASON. It is only introductory. I only want to prove where the trunk was found, and in what condition it was when found, and that this was the trunk found.

WITNESS. I went with the marshal when he got the trunk, and it had just that rope on it. I went with the marshal to this man's library—the keeper of the hotel. He had it covered up with some other things.

By the CHAIRMAN:

Question. Did the keeper say the trunk was in his library?

Answer. His study he called it.

Question. You went with him and the marshal into the study at the time the trunk was first seen by you?

Answer. Yes, sir.

Question. Did he point it out as the trunk of Colonel Forbes?

Answer. Yes, sir. It was not opened at all. The marshal took it just as it was, tied up with a rope. The lock was broken at the time we saw it. The marshal sealed it.

Question. Was it opened while you were there?

Answer. No, sir; and it has not been opened since we got it in our possession.

Question. It is in the same condition now that it was when you first saw it, except that it is sealed up?

Answer. Yes, sir.

Question. You say that this keeper of the hotel said it had been left there—by whom?

Mr. DOOLITTLE. One moment. What the keeper of the hotel said to the marshal, or to any one, would not be any testimony really of the fact. You summoned him, and he will be here probably, and can state the fact directly. I do not know whether there is any materiality in it.

WITNESS. I was sitting in the room during the conversation.

Mr. MASON. I do not know that it would be really material, but still this is not strictly a judicial examination; and although I should be opposed to taking any proof that was irrelevant, or any proof that was of a loose character, yet I should not think that this sort of testimony direct, of a conversation with parties who knew the fact, would be rejected. However, I shall not press it.

WITNESS. I think the proprietor of the hotel would be the most proper person, because the marshal did not open it, and none of us have seen the contents. The marshal took it with the rope on it, and sealed it up as it was.

Mr. COLLAMER. If there is danger of any simulation, if the contents are material as deriving their importance from having been in the possession of Forbes, it would be proper to trace it directly to him. There may be, however, in the papers themselves, if there are any in it, intrinsic evidence of where they came from; but really, if it derives its importance from having been in Forbes's possession, there must be some proof of that. This does not prove anything.

Mr. DAVIS. I suppose all the witness can prove is, that this is the trunk he saw there, said to be Forbes's trunk.

Mr. COLLAMER. He does not know that it is Forbes's trunk; but he can say that he brought the trunk here in the condition which he received it.

Mr. DAVIS. I am rather inclined to take the view of the chairman; but I supposed the statement as to this being delivered as the trunk of Forbes was to be the justification of the committee in seeing what was inside of it, and that that was its whole value.

Mr. MASON. I did not design to prove anything by this witness, except the fact that this trunk was found at that house which he has described in New York; that he was present with the marshal when the trunk was shown to them by the keeper of the house, and was brought from there in the condition in which they found it, without its having been opened.

Mr. COLLAMER. It may not be necessary to go into it further.

Mr. DAVIS. There may be nothing in it.

Mr. COLLAMER. If there is something in it, there may be intrinsic evidence of where it came from, as in the case of a letter from Brown.

WITNESS. The proprietor told us it was Forbes's trunk.

THEODORE RYNDERS.

ARCHIBALD M. KITZMILLER sworn and examined.

By the CHAIRMAN:

Question. Will you state where you reside, and what your occupation is?

Answer. I reside in Harper's Ferry; I am the chief clerk to the superintendent of the armory.

Question. Will you state whether the superintendent, Colonel Barbour, was at the Ferry at the time of Brown's invasion; and if not, who was acting as superintendent at the time?

Answer. I was the acting superintendent from the 6th until the 21st of October, in the absence of the superintendent, who was on duty elsewhere.

Question. Will you state whether this list is not in your handwriting? [The list referred to is appended to this witness's testimony.]

Answer. It is not in my handwriting, but was made under my supervision; it is in the handwriting of our book-keeper; it is a correct list; I wrote the caption of it, and I saw that the articles in it were verified.

Question. Will you state whether you, by my direction as chairman of the committee, brought specimens of the arms referred to in that list, and what they are—now present in the committee room?

Answer. I did not bring a specimen of all the arms and military equipments which are named in that list, but I have brought a Sharp's rifled carbine; a pike taken from the rendezvous at the Kennedy farm; one of Ames's pistols, made at Chicopee, Massachusetts, by the Massachusetts Arms Manufacturing Company; a box of double water-proof percussion caps; a box of anti-corrosive percussion caps, (London;) a small japanned powder flask; half a dozen ball cartridges for Sharp's carbines; one box of Sharp's patent pellets or primers; also a faggot. There were one hundred of them; most of them were in the wagon brought to the armory that Brown came over with.

Question. Will you state whether the articles mentioned in that list, the rifles and pistols and other things, except the pikes, were taken from the boxes in which they were brought?

Answer. They were taken from the boxes under my sight within the arsenal building and against my wish.

Question. Why against your wish?

Answer. I protested against it, but could not prevent it; I did not want the boxes taken into the arsenal at all.

Question. All I want to get at is whether these were in the boxes?

Answer. They were in the boxes, and I saw them opened.

Question. Where were these faggots or torches found?

Answer. There were many of them in the wagon within the armory yard.

Question. Which wagon?

Answer. The wagon belonging to John Brown and his confreres.

Question. Were they anywhere else except in the wagon?

Answer. I do not know that they were; this particular faggot was taken by Mr. Allstadt, and he gave it to me; the boys took them about; there was also a piece of punk.

By Mr. DAVIS:

Question. Did you see any of those faggots lighted?

Answer. No, sir.

Question. From your knowledge of such things, do you think they would blaze?

Answer. Yes, sir.

Question. Would it throw off sparks?

Answer. If it was hickory it would, but I think it is a mingling of hickory and pine—hickory to retain the fire, and pine to ignite the hickory.

By the CHAIRMAN:

Question. Will you state whether that pike now present is a specimen of the weapon described in your list as "handled spears?"

Answer. It is; that is one of the same kind brought from the rendezvous at the Kennedy farm.

Question. In this list there are ten kegs of gunpowder; do you know the weight of those kegs?

Answer. They weighed about twenty-five pounds each.

By Mr. DAVIS:

Question. Does the witness suppose the handles to these spears to have been made in the neighborhood of Harper's Ferry?

Answer. No, sir; from the best information I can get, they were made to accommodate the spears.

By Mr. FITCH:

Question. What is the cost of a single one of those weapons?

Answer. I do not know the cost of them all; they sell a Sharp's carbine at about twenty-five dollars; Ames's pistol sells for fifteen dollars; they are thirty dollars a pair; I brought with me the lids of two boxes, one of which contained the Sharp's rifled carbines, and the other contained the percussion caps; the one containing the carbines is marked "T. B. Eldridge, Mt. Pleasant, Iowa;" part of the lid is taken off; the other lid is addressed—

20¼ M Caps.] [From S. & L., B.

F. J. MERRIAM,

204 *Barnum's Hotel, Baltimore.*

The remainder of the lid of the box containing the rifles was sawed off.

ARCHIBALD M. KITZMILLER.

Copy of the list referred to in Mr. Kitzmiller's testimony.

List of arms, military stores, mining tools, and stationery now in store at the Harper's Ferry armory, there deposited by a party of Maryland troops, and citizens of Virginia and Maryland, taken by them from the rendezvous of John Brown and other outlaws of Maryland:

No.	*Articles.*
102	Sharp's carbines.
102	Massachusetts Arms Company pistols.
58	Massachusetts Arms Company powder flasks.
4	large powder flasks.
10	kegs gunpowder.
23,000	percussion rifle caps.
1,500	percussion pistol caps.
1,300	ball cartridges for Sharp's rifle, some slightly damaged by water.
160	boxes Sharp's primers.
14	pounds lead balls.
1	old percussion pistol.
1	major general's sword.
55	old bayonets.
12	artillery swords.
483	handled spears.
175	broken handles for spears.
16	picks.
40	shovels.
1	tin powder can.
1	sack coat.
1	pair cloth pants.
1	pair linen pants.

Canvass for tent.
1 portmonnaie.
625 envelopes.
1 pocket map, Kentucky.
1 pocket map, Delaware and Maryland.
3 gross steel pens.
5 inkstands.
21 lead-pencils.
34 penholders.
2 boxes wafers.
47 small blank books.
2 papers pins.
5 pocket combs.
1 ball hemp twine.
1 ball cotton twine.
50 leather water caps.
1 pound emery.
2 yards cotton flannel.
1 roll sticking plaster for wounds.
½ ream post paper.
2 bottles medicine.
1 large trunk.
1 one-horse wagon.
3 blankets.

A. M. KITZMILLER.

ARMISTEAD M. BALL sworn and examined:

By the CHAIRMAN:

Question. Will you state where you reside, and what is your occupation?

Answer. I reside at Harper's Ferry, Virginia. I am master armorer of the United States armory at that place.

Question. Were you there at the time of the invasion by Brown and his armed party?

Answer. I was.

Question. Were you one of his prisoners?

Answer. I was.

Question. At what time in the day, and whereabouts were you taken?

Answer. I was taken near the arsenal, about five o'clock on the morning of Monday, the 17th of October, 1859.

Question. By how many men?

Answer. I was approached by three men bearing three Sharp's rifles. They presented them to my breast, and said I must march into the armory yard.

Question. Will you state whether there is habitually a police guard of any kind at the armory, or in the armory yard?

Answer. Nothing more than fire watchmen, as they are termed, in the armory service.

Question. How many of them are there?

Answer. Generally two or three. I think three is the usual number—one posted near the armory gate, the other two distributed about at equal distances throughout the whole length of the armory, the whole distance being, I suppose, three eighths of a mile in length.

Question. Do you call them fire watchmen?

Answer. Yes, sir. So that it may be fully understood by the gentlemen of the committee, I will say that the object of that watch is to pass through the shops after the workmen have left them, after working hours, to see that the fires in the forges and all necessary fires kept up in the workshops are put out, so that there may be no danger from fire, and also to prevent any individuals who might come in during the night to pilfer. They are generally not armed, however. They might be considered civic watchmen.

Question. They are only on duty at night?

Answer. Only at night.

Question. Are there watchmen in any of the other buildings except the armory buildings?

Answer. None other.

Question. Are there none at the arsenal?

Answer. None.

Question. Are there any at Hall's works?

Answer. We regard them as a portion of the armory. There are two watchmen there of similar character.

Question. Do you mean by armory, the place where the arms are manufactured?

Answer. We do; we designate the place of deposit of arms as the arsenal, which is a separate building, on the other side of the street; not in the armory inclosure.

Question. Is there any watch or police of any kind at night at the arsenal?

Answer. None.

Question. Are those four watchmen appointed by regulations made at the armory, or are they by any directions of the War Department?

Answer. I am not able to say, but I am inclined to think that it is required by the Ordnance Department. That is my opinion, but I have no positive information to that effect.

Question. Then these watchmen are the only watch, or police, or guard of any kind that are at the works?

Answer. Yes, sir.

Question. Is there any ceremonial of hoisting the flag there during the day?

Answer. None at present—only upon the visit of an officer of high command.

Question. Whose duty then is it to hoist the flag?

Answer. The assistant of the military storekeeper.

Question. Did you see any of the citizens who had been killed, after they were killed at the Ferry?

Answer. No, sir; I did not personally see any of them, I believe.

Question. Were you confined all day in the watch-house?

Answer. Not closely confined. I was kept within the armory yard under guard until probably the middle of the day. I cannot be precisely accurate as to the hour, but near about the middle of the day, when the report came that the Harper's Ferry bridge was in possession of a military force from some quarter—we did not know of course, being prisoners—but from some quarter of Virginia or Maryland, at that time ten of us were selected as hostages, and placed in close confinement in the engine-house.

Question. Did you remain there until you were rescued by the marines on the following morning?

Answer. Yes, sir; but under varied circumstances. I suppose other gentlemen who were confined have stated them.

By Mr. DAVIS:

Question. I should like to ask Mr. Ball, whether any of the hired men were subject to do duty at night as a guard; whether any of them except the watchmen were subject to do duty at night as a guard?

Answer. No, sir; none prior to the outbreak.

Question. Are the workmen hired by the piece or the day?

Answer. Generally the great majority of the men working at the Harper's Ferry armory work by the piece; but there are a number of men constantly employed by the day.

Question. Could they, under their contract as workmen, be employed as a guard day or night, instead of working. Are they so employed that they may be put to anything the superintendent may please, such as a guard with arms?

Answer. Any individual, I presume, who is working by the piece or by the day, could enter into a contract with the superintendent.

Question. I do not mean that; I mean under the present organization would they be required to do so?

Answer. Unquestionably not. Their labors cease with the ringing of the bell, and can only be called into requisition by another contract until the ringing of the bell the following morning.

Question. How long have you been master armorer?

Answer. At the time that I was captured I was master machinist of the armory; I am now master armorer, but prior to that time, and at that time, I was master machinist.

Question. How long have you been connected with the armory?

Answer. About twenty-five years.

A. M. BALL.

JANUARY 11, 1860.

LIND F. CURRIE sworn and examined.

By the CHAIRMAN:

Question. Will you state what your age is, and where you reside, and what is your occupation?

Answer. I am in my thirty-third year; I live about three miles and a half from Charlestown, Jefferson county, Virginia; I am a farmer there. In connection with my farming operations I have been also teaching school.

Question. Where was your school at the time of the invasion of Brown?

Answer. It was about half way, I think, between the house occupied by Brown, in Maryland, and the Ferry, probably about three miles from the Ferry.

Question. Will you state whether your school was in session on Monday the 17th of October last?

Answer. It was.

Question. How many pupils had you generally?

Answer. I think I averaged from twenty-five to thirty, probably; I think the full number was about thirty.

Question. Of both sexes?

Answer. Of both sexes.

Question. What were the ages of the children?

Answer. They varied from eight to fifteen or sixteen.

Question. Will you state whether an armed party appeared at the school-house on the morning of the 17th of October; and if so, who they were, what brought them there, what they brought with them, and what they did?

Answer. They came there on the morning of the 17th, I think, about ten o'clock; it was sometime after I had opened my school, and Cook seemed to be the leader of the party. There were three white men, Cook and Tidd, and the third I have heard since was Leeman; but I think there is no certainty about that.

Question. Were there any negroes with them?

Answer. Some negroes; I do not recollect the number exactly. There might have been five and might have been ten, but I cannot recollect very distinctly the exact number. There were not less than five, though, I know; Mr. Cook came there in company with Mr. Byrne.

Question. Did he come before the other men?

Answer. They all came together with the wagon with arms; Mr. Cook came in and demanded possession of the school-house.

Question. Were all the party armed?

Answer. They were all fully armed; Cook, I recollect, had a couple of revolvers sticking around his belt, and a large Bowie knife and a Sharp's rifle; I presume they were loaded.

Question. Had the negroes any arms?

Answer. These long pikes, nothing else.

Question. Was it such a pike as that you see now in the corner of the committee room?

Answer. Yes, sir; exactly.

Question. You say Cook came in the school-house; now go on with the narrative?

Answer. Yes, sir; he came in and demanded possession of the school-house. He said he was going to occupy it as a sort of depot for their arms; that they intended depositing their arms and implements of war there; and they brought them in. At the same time he did not want me to dismiss the school. He thought I had better keep on the school and we should not be interrupted. I told him I thought that would not answer. The children were then very much alarmed, and I could not do anything with them. They were not in a condition to engage in their usual duties, and it would be impossible to keep them there.

Question. Were the children alarmed by it?

Answer. Very much alarmed.

Question. What evidences did they give of alarm?

Answer. Their manner of acting, their expressions, and so on, indicated the greatest alarm, so much so that he tried to pacify them as much as he could, but it was impossible to do it, and I finally dismissed them.

Question. What was in the wagon?

Answer. There were long boxes containing probably a dozen Sharp's rifles, I should think.

Question. Do you remember how many boxes there were?

Answer. I do not know the number; there were a good many of them; the wagon was loaded; it was full.

Question. They brought them in and deposited them in the school-house?

Answer. Yes, sir.

Question. Was there anything in the wagon but these boxes?

Answer. They took out at the same time one very large black trunk and put it in the school-house; I think that was all except these boxes.

Question. Did they tell you anything about what their design or purpose was?

Answer. Yes, sir; Cook said their intention was to free the negroes; that they intended to adopt such measures as would effectually free them, though he said nothing about running them off or anything of that kind. He said this, too: that those slave-holders who would give up their slaves voluntarily would meet with protection, but those who refused to give them up would be quartered upon and their property confiscated, used in such ways as they might think proper; at least they would receive no protection from their organization or party. I distinctly recollect that he said that.

Question. Did he ask you if you were a slave-holder?

Answer. No, sir, he did not; but I am under the impression that he discovered it afterwards. I should have stated probably before, that after I was there awhile there was a little boy of a friend of mine going to my school, and I felt a special interest in him, and he was extremely alarmed, and I was fearful that bad consequences might follow if I could not get him home very soon or do something with him to get him out of that fix. I asked Cook if he would allow me to take him home; he said yes, he had no objection; and I took him home to his father's house, about half a mile from there; I was gone probably an hour.

Question. Did you leave these men at the school-house when you went away?

Answer. I left them all there with the wagon; their wagon was not unloaded when I left.

By Mr. Collamer:

Question. Had the children generally gone off then?

Answer. Yes, sir; most of them went before I left; I dismissed school and allowed the children to go, but I kept this little boy because I wished to take him home myself. There was no one going his road, and I felt rather a special interest in him. I would not have gone back, but that there was no way of getting out that I knew of. My road lay in that direction across the river. There were two other roads, one through the Ferry, but both were occupied. The Ferry at that time was occupied by these men, and I could not get through the Ferry. There was another road passing up by Brown's house, which would have led me some miles out of my route home, but I did not go that way because I presumed that also was occupied by these men.

By the CHAIRMAN:

Question. You did go back to the school-house?

Answer. Yes, sir.

Question. What did you find there?

Answer. I found, then, nobody there but Cook and one black man with this wagon, the load of arms stowed about in the school-house. I did not know the negroes, but they knew me I presume; they were Colonel Washington's negroes, and I lived but a mile from his house. I learned afterwards that they stated to Cook who I was, that I was a Virginian, a farmer and slave holder over there; and I noticed some slight change in his manner after I came back; he was rather cooler; but after I was there sometime he became rather more communicative, and spoke of a great many things.

Question. Did you remain there?

Answer. Yes, sir; until late in the evening.

Question. Did he detain you, or was it your voluntary act?

Answer. I felt as if I was detained; he did not tell me so in so many words, but, when I made motions to move about, he would rather get in my way and endeavor as if he would prevent me, and I scarcely knew what course to pursue. I asked his permission, however, to go towards night. I saw the sun was getting down, and I told him I was anxious to get home. He told me I might go, but exacted a promise that I would not reveal what I had seen going on there. I suppose between two and three o'clock, probably, in the afternoon, somewhere, the shots became very rapid and continuous; we could hear them from the Ferry; they were constantly firing, and I asked him, "Mr. Cook, what does that mean?" "Well," said he, "it simply means this: that those people down there are resisting our men, and we are shooting them down."

Question. When you got back, in what position or what duty apparently was the negro whom he retained with him?

Answer. He seemed to be there as an assistant in guarding those arms. Mr. Cook told me he was there under orders from Brown, and that he could not get away. His orders from Brown were to remain there and take care of that point and protect those arms.

Question. What was the negro doing, apparently?

Answer. He appeared to be an assistant of his; they were both sitting there watching.

Question. What arms had the negro?

Answer. Nothing but the pike.

Question. Did the negro recognize you there; did he speak to you?

Answer. No, sir; he did not recognize me; he evidently knew me, though he did not speak to me or make himself known. I did not know until afterwards that he was one of Colonel Washington's negroes.

Question. Have you seen him since?

Answer. No, sir; I have not.

Question. You were then allowed by Cook to go away on a promise that you would not reveal what you saw. What direction did you take to get home; did you go by the Ferry?

Answer. No, sir; I went down by a road directly leading to the river; I did not go by the Ferry; it was then occupied; I did not go to the Ferry at all that night; I went immediately home; there was nobody there but my mother and the negroes, and I was anxious to get home; I started the next morning, however, for the Ferry. I asked Cook, at the school-house, "with how many men did you commence this foray down there." He did not answer me directly, but said, "I do not know how many men are there now; there may be 5,000 or there may be 10,000 for aught I know." I believed it; I supposed it was all true; I had no idea that twenty-two men were going to attempt such a foray as that.

Question. Did he tell you anything about their expecting assistance, or where it was to come from?

Answer. Oh, yes, sir. I did not ask him; but I presumed it was to come from the north.

Question. What did he tell you about expecting assistance?

Mr. Collamer. Tell what he said.

Answer. He said these men were to be there. I did not ask him where they were to come from, and he did not say. I just formed that impression.

By Mr. Collamer:

Question. That is, when he told you there might be 5,000 there, those were the men you understood him to say he expected there?

Answer. Yes, sir. He was speaking, too, about different personages. Gerrit Smith and Fred Douglass he mentioned.

Question. What did he say of them?

Answer. He said they were interested in it, and knew of it. Those were the remarks he made. These, I think, were almost precisely the words he used: That Gerrit Smith knew of it, and was interested in it, and also Fred Douglass; and I asked him especially if Mr. Seward was concerned or interested in it, and I think he said he did not know.

By the Chairman:

Question. Was there any other conversation in the school-house besides what you have given us?

Answer. Yes, sir. Cook and myself were talking of the feeling entertained towards the south by the north generally. He said he had no doubt that the efforts would be strong now and unfailing in order to extirpate the institution of slavery from the entire land. I forget in what connection exactly he brought that in; but that was about the gist of what he was saying. He said, "We, as a little band, may perish in this attempt, but," said he, "there are thousands ready at all times to occupy our places, and to step into the breach." He said, further: "It is our design to use every effort to disseminate our sentiments in regard to the institution of slavery among your own people; we will scatter them among you in different ways; we will send our people among you as colporteurs and peddlers, and we will place them in your pulpits and schools; in different ways we will send our men among you, and by such means circulate our opinions and sentiments." Our conversation was long and varied. Those are the leading points that I recollect now.

By Mr. DOOLITTLE:

Question. On that subject of any other persons being concerned than those there, were these persons the only persons you recollect that were named—Fred Douglass and Gerrit Smith?

Answer. Those were the only persons he named as knowing of it and interested in it. He used those words.

Question. Those are the only two he named?

Answer. Yes, sir.

LIND F. CURRIE.

JANUARY 13, 1860.

ANDREW HUNTER sworn and examined.

By the CHAIRMAN:

Question. Will you please to state where you reside, and what your pursuit in life is?

Answer. I reside at Charlestown, Jefferson county, Virginia. I belong to the profession of the law.

Question. Will you state the distance of Charlestown from Harper's Ferry?

Answer. It is eight miles by the ordinary road, ten by the railroad.

Question. Will you state how soon after the attack made by Brown on Harper's Ferry you went to the Ferry?

Answer. The attack was made on Sunday evening, the 16th of October last, and continued Monday the 17th, and Brown was captured on Tuesday morning, the 18th, and I was at Harper's Ferry about three hours after his capture on Tuesday.

Question. Will you state whether you saw Brown at the Ferry; when you saw him; and under what circumstances you had an interview with him?

Answer. I saw him soon after I arrived there. He was lying wounded in one of the rooms of the superintendent's office, the prisoner Stevens alongside of him; and I went to visit him in company with Governor Wise, who had arrived there about the time I did, or a little after; and in the presence of a number of other gentlemen.

Question. Who was Stevens?

Answer. He was one of Brown's party who had been captured on Monday, and was held as a prisoner from about, as I was informed, I do not know personally, the middle of the day on Monday.

Question. Was Stevens wounded?

Answer. He was wounded very severely; but after the capture of the whole party, and the storming of the engine-house, he was brought over and joined to the other prisoners in this room with Brown. The other prisoners were in a different place, in the watch-house connected with the engine-house. These two were lying there, both severely wounded. It was supposed Stevens would not live over the night, and Brown appeared to be very severely wounded, but his wounds did not prove to be so dangerous.

Question. Now state whether you either held or heard a conversation held with Brown on the subject of his attack; under what circumstances the conversation was held; and what it was?

Answer. We had a very long interview with him. It continued two or three hours. I went in first, being introduced by the sentinel; saw Brown then, but for a moment; did not speak to him, except to inquire of his wounds; then retired, and conducted Governor Wise in, and told Brown who he was, and they passed salutations. Brown was lying down, with his face on the pallet where he was lying, and immediately the governor commenced a conversation with him, and, although not at first, very soon after, he distinctly told Brown that he did not desire to hear anything from him that he did not willingly, and in view of all the circumstances that surrounded him, feel disposed to communicate; that his case would not in any degree, and could not, be affected by anything he told. Brown immediately replied that he knew that very well, that he had never begged quarter, and he would not do so; that he had nothing, so far as he himself was concerned, to withhold; and seemed even desirous of making known what his plans and intentions were. I can hardly describe his manner. It struck me at the time as very singular that he should so freely enter into his plans immediately. He seemed very fond of talking. Very soon some particular reference was made to the object for which he came, when he referred to a pamphlet that he desired to have sent for to his baggage or papers, wherever they were. I did not understand at that time where the papers were, but learned afterwards that they had been gathered up, and were somewhere about the building where he was confined. I am not quite sure whether Colonel Robert E. Lee of the army, or some other person present, attended to the matter of getting this pamphlet. When brought, it proved to be a copy of the constitution of the provisional government. It was the same afterwards used in court, identified and proved, and also admitted by Brown on his trial.

Question. Have you that paper?

Answer. Yes, sir. I have it here; indorsed "Referred to by Brown, A. H." [It is also identified by the initials of the Chairman, and placed among the records of the committee.] That was the copy that was brought when sent for, as before stated, and he admitted it distinctly. The paper was shown to him, and he was inquired of as to the pencil marks on the back of it, "Owen Brown," and some other names, and he said it was a copy of his constitution for a provisional government, &c., under which he was acting. After being submitted to him, he requested Governor Wise to read it. He said he wanted the whole of it read, and remarked that he would find a large number of copies of it among his papers; that he had intended within the next fortnight to have published it at large and distributed those copies. We did find a number of them amongst his papers. The governor read two or three of the first and several of the latter articles.

Question. Did the governor read them aloud?

Answer. Yes; Brown called upon him to read them so that everybody could hear them. There were ten or a dozen in at the time. I remember that Col. Lewis W. Washington was present at the time, and Col. Robert E. Lee. Brown's attention was called to this forty-eighth article:

"Every officer, civil or military, connected with this organization, shall, before entering upon the duties of his office, make solemn oath or affirmation to abide by and support this provisional constitution and these ordinances. Also, every citizen and soldier, before being fully recognized as such, shall do the same."

He inquired of Brown specially and directly if all his men, and those who were connected with him, had taken this oath. He replied promptly that they had all taken it.

Brown's attention was drawn to divers other articles there, which were commented on after the Governor had finished reading what he did read. Brown was inquired of particularly where he intended to put his provisional government into operation. He rose partly up, and somewhat earnestly said, "Here, in Virginia, where I commenced operations." I think it was about that time the Governor desired me to make some memoranda of what was said, which I did, very briefly, but quite enough to recall what took place. He was inquired of what support he expected to enable him to accomplish this, having so small a number of men, or what number he expected to aid him; when he quite as promptly, and clearly, and distinctly replied to it: three thousand or five thousand, if he wanted them. There was a pause made there. I was struck with the reply, and I thought it was about to lead to some very important developments. I made a memorandum of it. Stevens was lying alongside of him wounded, and, as we supposed, mortally, although apparently not suffering. He was calm, but had his hands folded on his breast, and some one remarked that was the attitude dying men usually assume; but he seemed exempt from any acute suffering. Brown was suffering and complaining every now and then. On this reply being made by Brown, Stevens interposed and remarked he was not sure of any aid, but he only expected it; "you do not understand him." Brown immediately took it up, and said, "Yes, I merely expected it; I was not certain of any support." This modification of what he had said was evidently the result of the prompting of Stevens. The inquiry was pursued further, where he expected the support from, and he then replied that he expected the slaves and non-slaveholding whites to join him from all quarters. He was then inquired of how many arms he had brought there. The inquiry was how he expected to arm them, or how many he was prepared to arm. He said he was prepared to arm about 1,500, but not perfectly. Further inquiry was made on this point—I sometimes presenting the questions, but chiefly the governor—and he then replied he had 200 Sharp's rifles and about 200 revolver pistols, and had expected 1,500 spears, but the contractor had failed, and he had rceeived only about 950.

He was particularly inquired of—I do not remember by which of us—as to his intending to stampede slaves off, and he promptly and distinctly replied that that was not his purpose. He designed to put arms in their hands to defend themselves against their masters, and to maintain their position in Virginia and the South. That, in the first instance, he expected they and the non-slaveholding whites would flock to his standard as soon as he got a footing there, at Harper's Ferry; and as his strength increased, he would gradually enlarge the area under his control, furnishing a refuge for the slaves, and a rendezvous for all whites who were disposed to aid him, until eventually he overrun the whole South. That was his purpose, as distinctly stated on that occasion. If you desire it, I will now connect with it what took place afterwards in the jail bearing on that point.

The CHAIRMAN. I think that would be the best mode of doing it. I think you had best continue the narrative, so far as relates to Brown, touching this particular point.

WITNESS. When Brown was brought out to be sentenced, which, as well as I recollect, was about the 1st of November, in his speech in reply to the interrogatory whether he had anything to say why sentence should not be pronounced upon him, I was greatly surprised at his statement, so distinctly made, that his sole purpose in coming to Virginia was to run off slaves. Will you have a copy of that speech before you?

The CHAIRMAN. We have it only orally. It has been printed, but taken by a stenographer, I suppose?

WITNESS. It was taken down very accurately by a stenographer.

The CHAIRMAN. Give the substance of it as you recollect it.

Answer. He stated in substance, as I recollect, that his purpose in coming to Virginia was simply to stampede slaves, not to shed blood; that he had stampeded twelve slaves from Missouri without snapping a gun, and that he expected to do the same thing in Virginia, but only on a larger scale. I was immediately struck by the palpable inconsistency between that statement and what he had communicated to Governor Wise and myself at Harper's Ferry, just after his capture. I mentioned it probably to Governor Wise, who came up some time after on his second visit to Charlestown; and it appears, as I learned from Brown himself, and afterwards from the Governor, the Governor went to see him again, and they had a very kindly interview, for I was struck with the respect——

Question. Were you present?

Answer. No, sir; I was not present at that interview, but I learned of it from Brown himself. I was struck with the respect and courtesy they mutually had for each other. Brown was impressed with a high regard for Governor Wise, and the governor with an estimate of him, in which I at first participated. After this interview Brown wrote me a letter from the jail to my office.

Question. Have you got that letter with you?

Answer. No, sir; I can send you the manuscript. I regarded it as of very little importance, and handed it to one of our village papers to publish, for you will find from the letter Brown desired it to be published. I can state the contents.

The CHAIRMAN. We ought to have the original.

Answer. It was published in the "Spirit of Jefferson." I will send it to the committee. In that letter he refers to this interview, and attempts to correct an apparent "confliction," as he calls it, between what he stated on the occasion of receiving his sentence and what he had stated to Governor Wise and myself at Harper's Ferry. He goes back and takes the ground he occupied originally, and excuses himself for making the statement he did in the court, on the ground that he did not at that time expect to receive his sentence, and that he was confused and not prepared, and was misunderstood or blundered in stating what he intended. He also sent the jailor or some one else requesting me to come and see him. When I went there I found it was for the purpose of still further explaining and attempting to reconcile the conflict between his two statements. He then assured me that his statement originally made at Harper's Ferry was the correct one; that that was his purpose, and he desired me to vindicate his memory in that respect—to publish it or to make it known.

Question. Now, Mr. Hunter, go back to the scene at the Ferry.

Answer. One or two other provisions to be found in that constitution were read. The Governor read them, and asked him how he meant to carry them out. He stated that, in regard to all non-slaveholders, they were not to be interrupted, but to be protected, if they kept quiet; but that the property of all slaveholders was to be confiscated, and the proceeds applied as an indemnity for the trouble and expense of freeing the slaves and of carrying on the expedition.

Question. Did he speak of his having been anywhere else in the Southern States, except at Harper's Ferry, at any time?

Answer. He stated that he had been as far south as to the southern line of Virginia. He also stated distinctly that he had sent Cook, one of his men, to get information in regard to the relative numbers of slaves, and general information that might be useful to him in his intended attack upon Virginia, but that, in reference to Harper's Ferry, he had reconnoitered it himself, in person; he had intrusted it to no one. I mentioned to him the ingenious device Cook adopted; that he had been to my house and obtained a census of my family, living near Charlestown, on the road between Harper's Ferry and Charlestown, under the pretense of having been employed to decide a bet as to the relative number between the whites and the slaves, and I asked Brown, familiarly, if that device was of his contriving. He replied, no; that he had simply sent Cook for that purpose, and had left to his own discretion the mode of obtaining the information. I think this is all connected with this long conversation that occurs to me, of importance to your inquiry.

Question. Did he, at Harper's Ferry, speak of having emissaries in any of the Southern States?

Answer. I do not think he did, in that form. He spoke of having many friends in the slave States, and expecting large support from the slave States, but I do not think he ever referred to having any emissaries, except Cook.

Question. Will you state now, sir, whether you were employed by the State of Virginia to act as counsel to conduct the prosecution, in aid of the prosecuting attorney of the county?

Answer. I was employed by the Governor, while at Harper's Ferry, and, I believe, on the day after this interview, to conduct the prosecutions, and to attend to all matters connected with this affair in our community, generally, and also I was afterwards appointed by the judge to aid the regular prosecutor in prosecuting the prisoners.

Question. Did you discharge that office?

Answer. I did, sir.

Question. In the prosecution of Brown, and who else?

Answer. Cook, Coppic, Shields Green, and John Copeland, the two last being negroes. Copeland was captured on Monday, and brought up to jail. I saw him and had an interview with him that evening. He gave me a statement of the numbers, and it proved to be a very accurate statement. Copeland was a mulatto, a smart, intelligent fellow.

By Mr. DOOLITTLE:

Question. Did Brown state, in the course of the conversation which has been already referred to, where this provisional constitution was formed?

Answer. He said it was formed at a convention in Chatham, Canada; that he drew it himself, and that it was printed immediately afterwards, in the neighborhood; he thought at St. Catharine's.

By the CHAIRMAN:

Question. Did you bring these papers and deliver them to me?

Answer. Yes, sir.

[Among the papers brought by the witness is a bundle indorsed: "Papers in evidence in Brown's case. I wish these kept together and separate."]

Question. Will you state in what way these and any other papers of Brown came into your possession, and at what time?

Answer. On the day after we had this long interview with Brown, at Harper's Ferry, Governor Wise, then at the Ferry, apprised me that he had given directions to some one, who had gathered up the papers, to convey them to me.

Question. Who was the person?

Answer. I think it was Colonel Baylor, the colonel who commanded part of the day on Monday, the 17th. He was a colonel of the militia of Virginia. I had not seen the papers then. There was a large mass of them. The governor said he had directed them to be put into my keeping at once. I did not, however, receive the papers until the next day, Thursday, the 20th. The Chairman of this committee called on me at Charlestown, to learn if I had papers. I informed him what I have just now stated, that Governor Wise had directed them to be placed in my possession, but that they had not yet been received.

Question. State how they came, subsequently, into your possession.

Answer. They were subsequently placed in my possession, the bulk of them, by Colonel Mason, the Chairman of this committee.

Question. At what time?

Answer. I think on Thursday, the 20th.

Question. In what condition were they brought to you by me?

Answer. They were brought in a canvas sack. I do not remember whether they were sealed up or not. That was the bulk of the papers. Another portion of them were delivered to me by the Hon. Alexander R. Boteler, representative in Congress from that district; and still more of them, by direction of Governor Wise, had been procured from some military officers of the Baltimore troops, and, from some of the editors in Baltimore, were forwarded to me by the master of transportation of the Baltimore and Ohio Railroad Company, Mr. Smith. There was a schedule made of those that came from Baltimore, and they were placed in my possession upon condition that I would take care of them and return them. There was one paper, a military commission, issued to one of the parties by Brown, furnished me by Henry Davenport, Esq., an officer in the Jefferson Guard, who was engaged in the conflict at Harper's Ferry.

Question. Do you know whose commission it was?

Answer. I think it was Coppic's, but am not sure.

Question. Do you know how Davenport got possession of it.

Answer. I do not. I do not think he ever told me. It was proved in court that it bore Brown's signature.

Question. Will you state whether the papers which you have produced to the committee, and which are labeled as those that were given in evidence, were given in evidence, and how were they verified in court?

Answer. The papers labeled as "papers given in evidence," were produced upon the trial of Brown; about one half of them, as well as I recollect, without being able to discriminate which of them, were admitted by Brown. They were presented to his counsel, handed back to Brown, he having voluntarily proposed (to save us trouble) to admit them, as soon as they were shown to him; and, until we probably got through half, that process was resorted to, and we were getting along rather expeditiously with it, until at last I presented him a paper—I do not know whether it is here—an autobiography of Brown, or some other paper, and a question was made at once about its admissibility, and I then replied that we were prepared to prove them; and, as it was quite as summary a mode of reaching the end, I declined waiting for his information, and they were proved by the sheriff of the county, who had become familiar with his handwriting—all the rest contained in this bundle. Either the indorsements or the papers themselves were in his handwriting.

Question. What about the other papers, not indorsed as having been used on the trial?

Answer. [After examining the papers.] The paper indorsed "Journal of the provisional constitutional convention, held on Saturday, May 8, 1858," was found among the papers of Brown. I am very clear that Brown, on some occasion, recognized that paper as containing a true statement of the proceedings of the convention, but I do not recollect whether he did so at Harper's Ferry or subsequently. [The paper here referred to is identified by the initials of the Chairman being placed thereon.] Either Brown or Cook, I do not know which, informed me that the name of "Whipple," on the list of members of the convention, was an assumed name of Stevens.

Question. State the history of the eight letters which you produce, indorsed "Intercepted. A. H."

Answer. I will state, in explanation of the matter, that I am satisfied between two hundred and two hundred and fifty letters arrived at the post office for Brown, some of them directed to Harper's Ferry and forwarded to Charlestown. Of the eight letters now produced, seven of them were addressed to Brown at Charlestown, and one addressed "Smith & Son, Harper's Ferry, Virginia," which was afterwards forwarded to Charlestown. Seventy or eighty of these letters were intercepted. The others are now before the joint committee of the legislature, in Richmond, carried there by directions of the governor. I retained and examined all the letters before they were delivered. The postmaster put Brown's letters into my box, and I received them, until the writers found they were intercepted, and then they began to address them to the jailor alone, but he handed them to me. These were amongst the letters thus intercepted. I believe the jailor handed them all to me, for he handed me two or three offering to bribe him to aid in a rescue. No doubt, he was entirely faithful in delivering to me all that he received.

Question. Do you say the rest of the intercepted letters were sent to Richmond?

Answer. Yes, sir; and they are now in the hands of the committee of the legislature.

Question. There seems also to be a roll of maps, were these found among Brown's papers?

Answer. Yes, sir. I may state that Brown said he had been preparing for this, and for carrying out the purpose he attempted to accomplish on that occasion, ever since 1856.

Question. When did he state that; when he was at Harper's Ferry, or when he was in jail?

Answer. At Harper's Ferry.

Question. Can you state where these maps came from?

Answer. I do not know now from what source these maps came. I was engaged in the trials, and could not examine them for some time after, and am not able to tell certainly; but they were among the Brown papers, furnished to me in the manner before mentioned.

[The maps referred to are seven in number, and of the following States: Kentucky and Tennessee, Alabama, Mississippi, Louisiana, South Carolina, Florida, Georgia. Each map has pasted at its side a slip, evidently cut from the Census Report of 1850, showing the number and kind of inhabitants (whether free or slave, white or black, male or female) in each county of the State or States which it represents. On the maps of South Carolina, Alabama, Mississippi, and Georgia, there are also various ink marks, in the shape of crosses, at different points.]

The witness also identifies several bundles of papers, by writing on each package: "Found among the Brown papers. A. Hunter. January 13, 1860." The papers thus identified consist of ten separate bundles or packages, one memorandum book, and one roll of paper, on which is written what purports to be a declaration of liberty by the slaves.

The CHAIRMAN, [to the witness.] I do not know that you have any further information that would be useful in the inquiries committed to this committee. If you think of anything more, you can state it. The subjects-matter of inquiry referred to the committee are the facts attending the invasion at Harper's Ferry, and to get at, if we can, any persons connected with it in any way who were not present at the time of the invasion. That is about the substance of the inquiry.

The WITNESS. I am not able at this moment to recall anything else that it occurs to me would be of value to you.

By Mr. DAVIS:

Question. Have you any information in relation to the views which were taken by Brown of the manner of his trial?

Answer. Brown repeatedly admitted to me that his trial was fair, much fairer than he expected; he sent for me to write his will.

By the CHAIRMAN:

Question. Did you write it?

Answer. Yes, sir; about an hour and a half before his execution; he took leave of me very kindly, indeed; he thanked me for my kindness and attention. I should also say that I explained to him the reason why his trial came on so promptly; he was captured on Tuesday, and the regular semi-annual term of the court commenced on the Thursday following, and we would have no other term until spring. His trial, I thought, when into it, ought not to have taken three hours, for he admitted everything distinctly, and said to Governor Wise and myself that he was ready for the consequences at any time. The Governor asked him if he did not want to make preparation for the other world, and he said he had done that many years ago; he had nothing of that sort now to do.

ANDREW HUNTER.

The following is the note referred to by Mr. Hunter as having been received from John Brown after his sentence:

CHARLESTOWN, JEFFERSON COUNTY, VA.,
November 22, 1859.

DEAR SIR: I have just had my attention called to a seeming confliction between the statement I at first made to Governor Wise and that which I made at the time I received my sentence, regarding my intentions respecting the slaves we took *about the Ferry*. There need be no such confliction, and a few words of explanation will, I think, be quite sufficient. I had given Governor Wise a *full and particular* account of that, and when called in court to say whether I had anything further to urge, I was taken wholly by surprise, as I did not expect my sentence before the others. In the hurry of the moment,

I forgot much that I had before *intended to say*, and did *not* consider the full bearing of what *I then said.* I intended to convey this idea, that it was my object to place the slaves in a condition to defend their liberties, if they would, *without any bloodshed, but not* that I intended *to run them out of the slave States.* I was not *aware* of any such apparent confliction until my attention *was called* to it, and I do not suppose that a man in *my then circumstances* should be *superhuman* in respect to the *exact purport* of every word he might utter. What I said to Governor Wise was spoken with all the deliberation I was master of, *and was intended for truth;* and what I said in court was *equally intended for truth*, but required a more full explanation *than I then gave.* Please make such use of this as you think calculated to correct any *wrong* impressions I may have given.

Very respectfully, yours,

JOHN BROWN.

ANDREW HUNTER, Esq., *Present.*

STATE OF VIRGINIA, *Jefferson County, to wit:*

Andrew Hunter maketh oath and saith that the annexed letter from John Brown to him, dated on the 22d day of November, 1859, was received by him, he thinks, on the day of its date; that it is altogether in the handwriting of said Brown, and, in a subsequent conversation between Brown and affiant, said letter was referred to and fully recognized by Brown as having been written by him.

Sworn to before me, mayor of Charlestown, this 17th day of January, 1860.

THOMAS C. GREEN,
Mayor of Charlestown.

JANUARY 16, 1860.

WILLIAM F. M. ARNY affirmed and examined.

By the CHAIRMAN:

Question. Will you state to the committee where you reside?

Answer. I reside in Hyatt, Anderson county, Kansas.

Question. How long have you resided in Kansas?

Answer. I have resided there since the spring of 1857.

Question. Since what month in 1857, do you recollect?

Answer. My family, I think, arrived in Kansas in May, 1857. I was in Kansas myself previous to that.

Question. How long previous?

Answer. Nearly a year previous.

Question. Then you were there in the spring of 1856.

Answer. No, sir; not in the spring, but in the summer of 1856.

Question. Will you state what has been your occupation and pursuit since you have been in Kansas?

Answer. I have been connected with a railroad, and have been engaged in farming. I am president of a railroad company.

Question. Were you acquainted with John Brown, who was mixed up with the troubles and difficulties in Kansas, and who has been recently executed in Virginia?

Answer. I was. With reference to that, I should like to say a few words.

The CHAIRMAN. Certainly; any explanation you please.

The WITNESS. As a member and general agent of the National Kansas Aid Committee, I became acquainted with some matters pertaining to John Brown's operations in Kansas, which may throw some light on his operations at Harper's Ferry; but before I give testimony in regard to it, I would prefer to be sent home for my books and papers, so that I can give a full and accurate statement of the amount placed in my hands by individuals and committees, and how it was appropriated, so as to show definitely what was given to John Brown as far as I know, by whose order, and for what purpose. If this investigation is designed for a full exhibition of the relation of other parties to John Brown, it will be impossible for me to testify accurately, and so as not to do injustice to persons who are absent, unless I can have access to my memoranda, records, and books kept at the time, in which were recorded the events, to some extent, as they transpired.

Question. Will you state when your acquaintance with John Brown commenced?

Answer. Between twenty and twenty-two years ago.

Question. When did you first see him in Kansas; when was your acquaintance with him resumed there?

Answer. My first personal connection with him in Kansas was after the beginning of 1857. I saw him previous to that, however, but not in Kansas.

Question. Where did you last see him anywhere?

Answer. I have some minutes of dates. My memory is not very good as to dates, and that is the reason I wanted to get my books and papers. [After consulting a memorandum book.] I last saw him in the fall of 1858.

Question. Do you remember in what month, with any reasonable certainty?

Answer. No, sir. If I had my papers I could tell. During the summer and fall of 1858, I saw him some half dozen times.

Question. Where was he then; where did you see him?

Answer. I saw him in Lawrence, Kansas. I saw him at my own house once or twice; he spent a day or two with me there. I saw him down on the borders of Missouri while I was engaged in surveying a railroad there. I think these are all the places where I saw him in the Territory during that year.

Question. Did he tell you then, in any of those interviews, of any plans that he had in contemplation to make a descent on any of the Southern States with a view to incite insurrection amongst the slaves?

Answer. In my conversation with him, growing out of a communication that I was requested to make to him by Mr. Richard Realf, he said some things in regard to a plan that he had, but gave me no details in regard to it.

Question. What was the character of the plan? You may give that without going into details. What sort of a plan was it?

Mr. DOOLITTLE. Perhaps it would be better that the witness should state just what Brown said. His inference as to the character of the plan might mislead us.

The CHAIRMAN. Probably that would be the better form. The witness says Brown gave him some information as to a plan without going into details. Now, I want him to state the conversation as near as he can. [To the witness.] State what communication was made from Realf, and what conversation ensued upon it?

Mr. COLLAMER. And let us know when and where it was.

The CHAIRMAN, (to the Witness.) State what communication you made to Brown as coming from Realf, and what conversation ensued on it?

The WITNESS. In that statement shall I be allowed to give a full history of my acquaintance with Realf?

The CHAIRMAN. If you think it important in connection with your testimony, you will state what you please, and you will understand, Mr. Arny, that by the laws of the United States, nothing that a witness says here can ever be brought in judgment against him elsewhere.

The WITNESS. I have no apprehension on that score.

The CHAIRMAN. I did not know that, but wished to state what the law is.

The WITNESS. It may be as well to state here that in all my conversations with Brown I opposed any project of even carrying off slaves, and my connection with him was more interposing to prevent him from getting means to accomplish anything of that kind.

Question. Will you state what the communication from Realf was and what conversation ensued upon it with Brown?

Answer. Mr. Realf, previous to leaving Kansas, had communicated to me his intention to go to England to lecture upon the subject of Kansas and the West to raise means to go into business. I assisted him, as far as I was able, to go to England for that purpose.

Question. Assisted him with money?

Answer. I gave him some property from which he realized money, and I have his note now for the money. I have made a minute of the dates as far as I could recollect them. Sometime in the spring or the beginning of the summer of 1858, I saw Mr. Realf in New York; I expressed my surprise that he was not in England. He told me that instead of going to England, he had been to Iowa and Canada, engaged in some matters, and that as soon as he had attended to a matter of business that he was engaged in then in New York, he should start for England.

The CHAIRMAN. I do not think you stated when it was that Realf told you his purpose and wish of going to England, and your giving him that aid.

Answer. That was previous to the time; I think it was in the winter of 1857, or early in the spring of 1858; I think that was the time. If I had his note I could tell the exact day; that is with my papers. I then asked Realf what business he could have to detain him there in New York, instead of going to England, as was contemplated. He told me that there was an arrangement in which he was engaged by which they proposed to make slavery insecure in the slave States; but, says he, "this is a matter of secrecy, and I only mention it to you because I wish to get your advice in regard to a matter connected with it; you have shown your friendship for me by assisting me more than once, and therefore I want to ask your advice in the matter." I told him that I must know what the matter was first. He then stated that there was an arrangement in which this Brown was a leader, by which they proposed to make slavery insecure, and that it would have been carried out before that time if a man by the name of Forbes had not left them and threatened to betray them. Well, I immediately remarked to him, that I thought he had better carry out his first project of going to England, and have nothing to do with anything of that sort, and that if he wanted my advice on the subject I would advise him, just as I had in Kansas, to go to England, deliver his course of lectures, get some money, return to Kansas, and either get into some business there, or go on a farm and settle down and marry a young lady to whom he was paying attentions, and from whom, as I learned afterwards, he had borrowed money. Realf then said to me that if he had the matter then in hand disposed of, he would take my advice, but he did not know, he was somewhat conscientious in regard to his obligations to an association in Canada. He did not tell me what the association was, but he said he was under obligations to that association, and that he promised Brown to use the money that he realized to assist that association. Throughout the whole conversation, however, I urged upon him to let Brown alone and go to England. He then said to me, "If you will deliver a message for me to Brown when you meet him in Kansas, and do one favor for me when you are in Washington, so as to make that message perfect, I will go to England." I told him if it was nothing to compromise me in this matter, for I would have nothing to do with it, I would. He then told me that the object he had in view in remaining in New York was to find out what Forbes was at, and that he understood that Forbes was going or had gone to Washington to communicate to Governor Floyd what he knew with regard to the organization in Canada. I think it was the next day or the second day after, I shall not be certain which, I saw Realf again. He then told me he had received information that Forbes had gone on to Washington; that Forbes wanted to obtain money from parties in Washington, he did not mention their names, and to communicate their organization to the Secretary of War, and asked me, as I was coming to Washington, to look after Forbes and see if he had done so, and let Mr. Brown know it. I came on to Washington; I did not feel that it was my business to inquire into that matter; I attended to my business here and returned home to Kansas. I made no inquiries about that whatever; I attended to my own affairs and returned to the Territory.

Question. Did Realf refer you to any person particularly in Washington that you were to communicate with?

Answer. No, sir; no person.

Question. At what time was it that you were in Washington on the business you have alluded to?

Answer. I cannot be certain as to the time, but I can come somewhere near it.

By Mr. Collamer:

Question. Did you say it was in June you saw Realf in New York?

Answer. It was sometime early in the summer or spring; it may have been June; but I cannot say certainly what the date was. If I had my papers, I could give the date.

By the Chairman:

Question. Did you come from New York to Washington, directly?

Answer. Yes, sir; and from here I went to the Territory.

Question. When you were in Washington, did you hear or see anything of this Mr. Forbes?

Answer. I never saw the man in my life.

Question. Did you know whether he was in Washington or not? Did you hear anything about him here?

Answer. Not while I was here.

The Chairman. Well, you returned to Kansas. Now, you can go on.

The Witness. I think the first time I saw Brown after I got into the Territory was while I was engaged in surveying a road through Missouri and from Missouri to the Neosho valley, and during the time of the troubles in Linn county, Kansas. It was directly after what were termed the Marias-des-Cygnes murders, when Hamilton came over from Missouri and killed a lot of men, taking them from their farms in Linn county, on the borders of Linn. After speaking to Brown in regard to the position of things in the Territory at the time, I said to him that I understood that an organization was established in that county and neighborhood in which I was very glad to learn they had decided that no invasion should take place of the territory of the State of Missouri; I then remarked to him that I had seen Realf in New York. He was standing a little way off. He immediately came close up to me. He said, "I am glad to hear that, for I thought Realf was gone to England and had not attended to what we had appointed him." I then told him that Realf asked me to say to him that he did not know where to write to him, and that if I fell in with him to say to him that he had been engaged in the matter to which he was appointed, and that he could not learn definitely what Forbes was after, but that he had gone to Washington; that was about the substance of the message that Realf gave me. Brown then asked me if I knew who Forbes saw in Washington. I told him, no. Well, said he, you were in Washington. I said, yes. I then immediately remarked to Brown that I thought it very strange that he would be engaged in an enterprise such as Realf had intimated to me they were engaged in. "Why strange," said he. Why, said I, "Brown, when I knew you twenty years ago over in Virginia, you professed to be an abolitionist, and an abolitionist of that class that are termed non-resistants, and you refused to use arms in any shape or form; even when you were opposed, you would not resist, and you were considered ultra on that subject at that time; how is it that you have changed?" He then referred to the fact that he had sent his sons into the Territory of Kansas in 1853 or 1854 with a lot of blooded cattle and other stock with the intention of settling. Said he, "You know they did settle on the Pottawatomie, and that after they settled there, part of their cattle was stolen; they were notified to leave the country, and they had to sacrifice their cattle"——

The CHAIRMAN. I do not think it important to know Brown's reasons for this thing. We want to get at the facts.

The WITNESS. I am only detailing the conversation I had with him. There were three conversations with him; he was at my house two days.

The CHAIRMAN. The question put to you was whether Brown had communicated to you any plans he had in reference to inciting insurrections in the slave States. Whether he did or not, I am uninformed. But why he did it, or why he ceased to be a non-resistant, (if he did so cease,) are not exactly within the scope of the inquiry.

Mr. DOOLITTLE. Still, perhaps, as he has got into the conversation, he had better go through with it.

Mr. COLLAMER. If that is the conversation in which the communication was made, let us have it.

The CHAIRMAN. I have no objection, except that we are not inquiring into his motives for it in any way.

The WITNESS. He then, in addition, referred to the meetings in Missouri that had been held previously, and the resolutions that had been adopted by those meetings, and referred to the invasions of the Territory of Kansas in 1854 and 1855, the resolutions that had been passed in 1853, and the speeches of Mr. Atchison with regard to that matter——

Mr. FITCH. That kind of testimony, taken and published as part of the proceedings of this committee, conflicts directly with the refusal of the Senate to go into an investigation of Kansas affairs; and, if brought in on one side, must be brought in on the other. Brown alleges certain reasons, and, if his reasons form part of the testimony, it will be incumbent on those who do not think his alleged reasons to be the true ones, to hunt up others. If testimony of this kind is admitted on one side, it must be on the other.

Mr. DAVIS. It has no application either to the question or to the inquiry, that I see.

Mr. FITCH. We know Brown's alleged reasons; they have been stated repeatedly; but they have been contradicted. If statements of this kind are received, we shall have similar statements from the other side.

The CHAIRMAN. I took the liberty of suggesting just now that the matter which the witness was detailing was not pertinent to the subject submitted to the committee for inquiry, and is not within the scope of the question.

Mr. DAVIS. Certainly not.

The CHAIRMAN. But these two gentlemen (Messrs. Collamer and Doolittle) intimated a wish that he should continue to detail it. I think it is extraneous.

Mr. FITCH. I have no objection to it, if the committee will only decide that they will open the whole subject.

The CHAIRMAN. I protest against opening any subject connected with Kansas.

Mr. Doolittle. As I understood the question, it was to find out whether Brown communicated to the witness his design to make an attack on the slave States for the purpose of raising an insurrection among the slaves, and I understood the witness to state that, in answer to his communicating from Realf in New York a message to Brown bearing upon that project, it happened that Brown went into conversation, and in that conversation did disclose to him some plan on that subject.

Mr. Davis. The incentives which brought Brown to a proposition, if he made one, have nothing to do with the inquiry which we have before us.

The Chairman. Here is the exact state of the facts, and, by reading over the testimony, I think you will find it to be the case: the witness recalled it to Brown that he had always known him to be a non-resistant, although an abolitionist, to know why he had changed his character or purpose in reference to that rule of action in life, and he details now Brown's reasons for having done it. If there could be anything more impertinent to the subject of inquiry, I do not know it.

Mr. Collamer. When a man is asked whether Brown communicated to him any such purpose as is indicated, I know of no way in which that man legally can tell that he did, unless by stating what was said. You may detach it, and isolate it, and get out some one thing and suppress the rest, but legally you cannot. He is to tell it just as the man said it. I know of no other legal mode of doing it.

Mr. Davis. He may tell what was said about this particular subject.

The Chairman. I differ entirely with Judge Collamer. He might go into a detail of Brown's whole career in life, from his childhood up, and by his construction we must go into that.

Mr. Collamer. I think the conversation in which he made the communication in question, and all that was connected with that communication, should be given.

Mr. Davis. It is a mere question of what is connected with it, I suppose.

Mr. Collamer. There is always some difficulty as to that. We can never tell with perfect certainty what a witness will say.

The Chairman. I will put the question to the committee, whether the witness in his answer shall be confined to the scope of the question.

Mr. Collamer. Every one of us will vote affirmatively on that.

The Chairman. Then I will confine the witness to the scope of the question.

Mr. Collamer. The scope of the question is what the man said at that time in that connection.

Mr. Fitch. Instead of telling us of the communication of the plan, the witness seems to be giving a communication of Brown's reasons for a change of conduct on a particular subject, which involves an investigation of Kansas matters.

Mr. Collamer. I supposed the witness was coming to the point in the conversation at which Brown spoke of his plans, and in telling what his plans were he said why it was——

The Chairman. We do not want to know why it was, but what it was.

Mr. FITCH. But he is telling us Brown's remarks as to conduct in Kansas, which preceded his action elsewhere.

The CHAIRMAN. Then, unless overruled by the committee, I shall instruct the witness that in answering the question he will confine himself to what Brown told him—whether he had any plans to incite insurrection in the slave States; and, if he did tell him so, what they were; that his reasons for doing it, the motives that actuated him, are not within the scope of the question. [To the witness:] Confine yourself to the fact whether Brown did tell you that he had any plans to incite insurrection in any of the slave States, and, if he did, what those plans were.

Mr. COLLAMER. I presume you mean to include also plans for stampeding slaves?

The CHAIRMAN. No, sir. The question speaks for itself—plans to incite insurrection amongst the slaves in the slave States; and the witness will tell us, if he communicated them, what his plans of inciting insurrection were.

The WITNESS. Finding that Realf had intimated the matter to me, when I made that communication to him, the substance of what Mr. Brown said was that there was a proposition on foot to make slavery insecure in the slave States.

The CHAIRMAN. Was that what Brown told you?

The WITNESS. Yes, sir; and it is just the same language that Realf had used. He said they proposed to go somewhere and locate themselves, and then carry slaves off to Canada.

Mr. FITCH. Did Brown intimate that this was to be in a slaveholding State?

The WITNESS. He did not say.

Mr. DOOLITTLE. Allow me to suggest that it is not proper to have what he intimated, but what he said.

Mr. FITCH. That is what I meant—what he said.

The WITNESS. He did not say where it was to be. I will remark in regard to Mr. Brown, if I am allowed to do so, that he was a very secretive man, and that he would not communicate much to me, knowing that I was opposed to his plans, as I had expressed myself. Knowing that I was a peace man, he would not communicate, I suppose, to me as freely as he would have done to others.

The CHAIRMAN. That is your inference, Mr. Arny. You say this was in the summer or fall of 1858. Did he tell you at that time, or any other time, of any organization that he and others had entered into in Canada?

Answer. No, sir.

Question. Did he refer to any convention that they held, or proposed to hold, in Canada?

Answer. No, sir.

Question. Did he tell you of any persons or committees, in Kansas or elsewhere, who had been consulted, or who approved of his plans of "making slavery insecure in the slave States?"

Answer. He referred to a committee, and matter of action of a committee, at which I was present at the time that action was taken, and of which I was a member and the general agent. He conversed with regard to that matter. He mentioned also the fact of applications having been made to parties in the East, and the result of those applications.

Question. What was the committee, and what was the action of that committee, when you were present, to which he referred?

Answer. That committee was the General National Kansas Aid Committee, organized at Buffalo, in the State of New York, in July, 1856. That committee was composed of one member from each of the States.

Question. All the States?

Answer. All the free States. It grew out of the fact that the State of Missouri had invaded Kansas.

The CHAIRMAN. We do not go into that inquiry.

The WITNESS. You asked in reference to the committee. I wanted to give its history.

The CHAIRMAN. I only wanted to know what the committee was, and who it consisted of. Can you state that?

Answer. I cannot recollect all the members of the committee. There was Mr. B. B. Newton, of Vermont; Thaddeus Hyatt, of New York; John H. Tweedy, of Wisconsin; Governor Reeder, of Kansas. Abram Lincoln, of Illinois, was appointed at Buffalo on that committee, but when he was notified that he was appointed, he declined the appointment. He was then the elector for the State, and took the position that this was a matter that had nothing to do with politics, and therefore he did not wish to interfere. I was appointed in Mr. Lincoln's place for Illinois. I then lived in Illinois. I think John W. Wright was the member from Indiana.

Question. Do you know who was there from Massachusetts?

Answer. If I remember right, Dr. Howe was appointed; but so far as Massachusetts was concerned, they refused to coöperate with the committee, or have anything to do with it. They worked "on their own hook," as they say out West.

Question. Who was there from Connecticut?

Answer. W. H. Russell, of New Haven.

Question. Who from Rhode Island?

Answer. I do not recollect who was from Rhode Island.

Question. Who from New Hampshire?

Answer. I forget who was from New Hampshire.

Question. Ohio?

Answer. I forget. Some gentleman living in Cleveland. I do not recollect his name though.

Question. From Iowa?

Answer. I think Mr. Clark—I do not now know his first name—of Iowa City.

Question. From Michigan?

Answer. S. S. Bernard.

Question. Who from Pennsylvania and New Jersey?

Answer. I will not be certain as to the name of the member from Pennsylvania, but I think his name was Grier, and he lived at Pittsburg. I do not recollect who came from New Jersey.

Question. You have said, as I understand, that Brown, in speaking of his plan, referred to some action taken by that committee, and that you were present when that action was taken. Now state what the action was that was taken, to which Brown referred?

Answer. Mr. Brown applied to that committee——

Mr. COLLAMER. State when and where.

Answer. He applied sometime in the month of January, 1857, at a meeting of that committee, at the Astor House, in New York, for some arms, and other assistance, to organize a company of men and equip them, to repel the invasion in Kansas by Missourians. The action of the committee, to be short about it, was a refusal at that time to let him have the arms, and an agreement to let him have some other matters that would assist him in a peaceable settlement in the Territory of Kansas.

By Mr. COLLAMER:

Question. What were they?

Answer. Clothing and other things.

By the CHAIRMAN:

Question. What do you mean by a peaceable settlement?

Answer. That he should settle there in the Territory of Kansas, and not go out. The policy of this committee was a policy of self-defense, and opposed to any attempt to invade the States or any Territory.

By Mr. DAVIS:

Question. Do you mean clothing for himself or for others?

Answer. Clothing for himself and others.

Question. For distribution?

Answer. Yes, sir; for persons that he was acquainted with and that he represented were in a destitute condition.

By the CHAIRMAN:

Question. Was that the action that Brown referred to?

Answer. Yes, sir; he referred to that action, and I think I can give the exact words, after speaking of that action.

By Mr. DOOLITTLE:

Question. Was this an application for the arms which were in Iowa at that time?

Answer. Yes, sir.

By the CHAIRMAN:

Question. You say that Brown told you of some plans that he had to arm a party of men for the purpose of stampeding slaves from the slave States, and in telling you of that plan he referred to the action taken by this committee at Buffalo, and you now tell us what that action was. That action was an application of Brown for certain arms for certain purposes, and a refusal of the committee to let him have those arms for those purposes, and a proposition by that committee to aid him in a peaceable settlement. I want to know how Brown referred to that action in connection with any plan of running off slaves. What had that to do with running off slaves?

Answer. In my conversation with him, I was finding fault with him for proposing a plan of that kind; and, in finding fault with him in regard to that matter, he classed me with the abolitionists, with the republican party, and with this committee, and he said we were all alike a set of cravens, that we all refused to give him assistance to carry the war into Africa.

Question. You spoke of applications he had made to some people in the East. What applications did he speak of there that he had made?

Answer. After his application to the committee at the Astor House, in New York, in 1857, and a refusal by all who were there to do anything for him at that time, and after the settlement of the ownership of these arms, which was made afterwards at the same meeting, he then addressed a letter, a copy of which I have and can furnish to the committee if I have my papers, to the people of the States.

Question. What States?

Answer. Of the free States, asking for aid.

By Mr. DOOLITTLE:

Question. Was that letter published at the time?

Answer. It was published at the time.

By Mr. DAVIS:

Question. How was it published?

Answer. In various newspapers. I have a copy of it filed with my papers, as published in the New York Tribune.

By the CHAIRMAN:

Question. Well, sir; do you know of any person in any of the States to whom Brown made application for aid to carry out the plans that he had communicated to you?

Answer. I do not know of any persons in any State who aided him; and I do not suppose that Mr. Brown would have let me know of any, because he knew I was opposed to it.

Question. Now, as to the arms you have referred to, over which, as I understand, this committee to which you belonged had control, what arms were they? What was their character?

Answer. In order to a definite understanding of the answer to that question, probably it would be proper for me to give a detailed account of those arms and what I know about them.

The CHAIRMAN. Well, we do not want any details, because they consume time. What I want to get at is, first, what was the description of arms. Were they guns, pistols, swords, or what, and how many; what was their quantity?

Answer. There were said to be in the boxes 400 Sharp's rifles.

Question. Well, what other arms?

Answer. In that lot there were no other arms. In another lot, which was forwarded to Chicago a few days afterwards, there were 200 revolvers in boxes.

Question. What period are you speaking of now when you say such was the quantity of arms at that period?

Answer. That was in the spring of 1856, at the time when the

Missouri river was closed and the citizens of Missouri had stopped citizens of Illinois from passing across the State into Kansas.

Question. Where were those 400 Sharp's rifles at that time?

Answer. They were shipped from Massachusetts to Chicago.

Question. In the spring of 1856?

Answer. Yes, sir.

Question. Then as to the 200 revolving pistols, where were they?

Answer. They were sent about the same time from Massachusetts, but that was previous to the organization of any national aid committee or society.

Question. Were both parcels shipped from Massachusetts to Chicago?

Answer. I believe they both were. At that time I had the bills of lading, and probably may have them now among my papers.

Question. Do you know by whom they were shipped from Massachusetts?

Answer. I do not recollect.

Question. Would your papers enable you to tell that?

Answer. I cannot say; I think probably they would.

By Mr. DOOLITTLE:

Question. Were they shipped to your care?

Answer. No, sir; I do not know to whose care they were shipped, but they were placed in my hands after they got to Chicago, with instructions how to proceed with them, by a committee which was organized previous to the national Kansas committee.

JANUARY 17, 1860.

Mr. ARNY's examination resumed.

By the CHAIRMAN:

Question. I wish you to state who composed the committee that placed those arms in your hands, and where they resided?

Answer. Probably I should change the phraseology of my last answer yesterday, and instead of saying that they were placed in my control by a committee, I should say they were placed in my hands by a person who was not a member of any committee, but who was representing the meeting held at Chicago previous to the organization of the national Kansas committee, that raised twenty thousand dollars from the citizens of Chicago by subscription.

Question. Who raised the money, the person you speak of or the committee he represented?

Answer. It was done at one meeting in one night.

Question. By a committee?

Answer. By the citizens of Chicago; there was no committee appointed that I recollect; but, so far as that is concerned, I cannot recollect whether there was a committee appointed or not without my papers. This money was subscribed by the citizens of Chicago in one night. I would just suggest here that probably you would have understood this matter better if you had allowed me to give a history of those societies and their organization.

The CHAIRMAN. Well, sir, give as briefly as you can the history of the committees, their organization, and their object—all the committees that were connected with Brown in reference to supplying him with arms or with money, &c., or the parties with whom Brown was connected.

The WITNESS. I do not know that any of the committees were connected with Brown's supply of arms, but what I mean is the committees that had these arms that I speak of.

The CHAIRMAN. Well, go on with them, assuming that they were the same arms subsequently brought here.

The WITNESS. I stated, I believe, yesterday that previous to the invasion of Kansas by Missourians, in May, 1856, there was no committee or society organized in the North to aid in any matter connected with Kansas; previous to that time the Emigrant Aid Company was incorporated; the first committee that I know of was appointed in Illinois.

By Mr. COLLAMER:

Question. Did that Emigrant Aid Company ever have anything to do with any arms that you know of?

Answer. I went myself in behalf of the national Kansas committee to Boston to confer with them after they were formed.

Mr. COLLAMER. He seems to have broken the thread of your story, Mr. Mason.

The CHAIRMAN. It is not my story, it is his; he says he thought we should have a better knowledge of the history and destination of these arms by a history of the committees connected with them; I thought it was pertinent.

Mr. COLLAMER. I have no objection to his giving it.

The WITNESS. According to the question Mr. Collamer asks me——

Mr. COLLAMER. I will withdraw that, and let Mr. Mason proceed in his own way.

The CHAIRMAN. You were speaking of going to Boston.

The WITNESS. After the national Kansas committee, of which I spoke yesterday, was organized, in July, 1856, I went to see whether the Emigrant Aid Company would coöperate with them in supplying the destitute in the Territory and assisting them; I had a conference with one or two members of that company; they called a meeting of the company, but I was not present at it; they gave me their answer the next day; their answer was that their organization was a financial one, that they had organized to establish mills and assist in the——

The CHAIRMAN. I do not think it necessary to detain the committee by details that really are not interesting to them, and are not before them. What I wanted to know, and what I understood you were about to state, was any history of the committees that had any connection with those arms. As to supplying provisions or anything else to Kansas, we have nothing to do with that.

The WITNESS. I wanted, in my statement, to show what committees there were, and what connection those committees had with these arms; that is what I desired to do.

The CHAIRMAN. So I understand; but I do not see that this diverging into the objects of committees to send provisions or supplies of clothing to destitute people in Kansas has anything to do with it.

The WITNESS. I was going on to remark that I knew of an application being made to that committee afterwards for arms, but that was refused in the same way on that ground that they were a financial company and not a committee.

Question. By whom was the application made?

Answer. By Mr. Brown; the first committee that was organized that I know of was a committee organized in Illinois, at a meeting called for the purpose of taking into consideration what should be done in consequence of the closing of the Missouri river, and the refusal to allow citizens of McLean county to pass through Missouri into Kansas

Question. You speak of the closing of the river. What closed it?

Answer. The citizens of Missouri had declared that no person should pass up that river. They had stopped persons who were going up the river and sent them back.

The CHAIRMAN. I thought you might have meant closed by ice. I really did not know.

The WITNESS. No, sir. This meeting appointed a committee of three, of which I was made the financial agent and secretary.

The CHAIRMAN. That was a committee of Illinois?

The WITNESS. Yes, sir; and the first committee that I know of that was organized at all. It was in Illinois. A few days after, I cannot recollect the exact date, a meeting was held at Chicago, and $20,000 was raised there. That $20,000 was subscribed. I was authorized to go up and confer with the committee in Chicago in order to coöperate together so as to appropriate all the money that should be raised in Illinois judiciously, and my instruction from the other members of the committee was to avoid, if possible, the use of that money which we had raised in our part of Illinois, and to induce the others to use theirs if possible for the feeding and clothing of persons and assisting them in defending themselves in the Territory, and to open that river if it could be done without causing any disturbance. Then there was the national committee, organized in July, 1856, at Buffalo, New York. I have already stated all I can recollect about it. If I had my papers I could tell more.

Mr. COLLAMER. State its purpose.

The WITNESS. Its object was to assist in relieving Kansas of the invasion that then existed in the Territory, and supplying the destitute in the Territory with provisions which they required. It was reported to them that they were in a starving condition. After that time, a society was organized in each State, auxiliary to that society.

The CHAIRMAN. Each of the free States?

The WITNESS. Yes, sir, each of the free States. Massachusetts refused to coöperate with the others. I have, however, at home a report from officers of that society showing how much they contributed, but I cannot recollect exactly the amount; it was some forty or fifty thousand dollars. I was appointed to go and confer with that society and with the Emigrant Aid Company, and they both refused to coöperate with us or have anything to do with us—one because their company was a financial one. This was the state of things at the time when those arms were sent to Chicago and came into my hands; at least, I should say the two hundred rifles. Probably I had better tell here what became of the other two hundred.

The CHAIRMAN. Two hundred of the four hundred Sharp's rifles, I suppose you mean.

The WITNESS. Yes, sir. Only two hundred actually came into my possession, but I know what became of the others.

Mr. DOOLITTLE. Tell that. Let us have the whole story.

The WITNESS. One hundred were started up the Missouri river, not of those I had, but part of the four hundred. One hundred were started up the Missouri river on a steamboat. When that steamboat arrived at Lexington the boat was stopped; the man who had them in charge was, I think, taken off the boat; at any rate all that he had in his charge were taken away from him, and that one hundred rifles were held at Lexington.

The CHAIRMAN. Where is Lexington?

The WITNESS. In Missouri, on the Missouri river. Another hundred were taken up the Missouri river shorly after that in another boat, and they were stopped, I think, at Waverley, in Missouri, just below Lexington. Those two hundred Sharp's rifles remained there until the latter part of 1856, or the beginning of 1857. I was traveling in Missouri purchasing provisions for the——

The CHAIRMAN. Never mind that. Have you any reason to believe that the two hundred guns of which you are now speaking, ever came into the possession of Brown?

The WITNESS. Not those two hundred.

The CHAIRMAN. Then I do not see that we have anything to do with them.

The WITNESS. I want to tell the whole story.

The CHAIRMAN. But we do not want the whole story, because it is not before us. You have traced two hundred rifles as sent up the Missouri river and stopped there. You say those arms never came into the possession of Brown. Then we have nothing to do with them.

Mr. DOOLITTLE. [Showing the witness Sharp's rifle carbine produced by Mr. Kitzmiller.] Is that what you mean by Sharp's rifle?

The WITNESS. Yes, sir.

The CHAIRMAN. What about the other two hundred Sharp's rifles?

The WITNESS. I cannot tell about the other two hundred definitely unless I have my papers, but I think they were sent to Iowa City.

Mr. COLLAMER. Did you send them there?

The WITNESS. Yes, sir, I shipped them. I think it was to Iowa City. If I remember aright a report came from Iowa City——

Mr. COLLAMER. You say you sent those rifles in boxes to Iowa City. Did you mark the boxes yourself?

Answer. I believe I did.

Mr. COLLAMER. [Showing the witness the box lid heretofore proved by Mr. Kitzmiller, and marked "T. B. Eldridge, Mount Pleasant, Iowa."] Is that your mark?

Answer. Yes, sir. If you will let me get on a little further, I will tell you about it. A report came that the warehouse in which the arms were stored in Iowa City had been broken open and the arms taken. I went immediately to Iowa City, and then, I think, if I remember aright, the arms were shipped from Iowa City to Mount Pleasant, and arrangements made there for them. Mr. Eldridge was the agent at Mount Pleasant. Mount Pleasant is beyond Iowa City, in the State of Iowa. They were directed to him, but they were stored in Iowa City, in the warehouse; there held in reserve. Now, probably, it would be well for me to tell why they were stored there, and state what transpired.

The CHAIRMAN. Do I understand you to say that when you shipped them from Chicago to Iowa City, you addressed them to Mr. Eldridge, in Mount Pleasant?

Answer. I do not know whether I addressed them to Eldridge in Chicago, to go to Mount Pleasant, or whether I addressed them in Iowa City.

Mr. DOOLITTLE. You do not know whether you put that mark on at Chicago or at Iowa City?

The WITNESS. Exactly. I do not recollect.

The CHAIRMAN. I do not think that is important; but I understand you to say that you did ship them from Chicago to Iowa City, and afterwards, hearing that for some reason they were supposed to be insecure there, you went up to Iowa City and shipped them to Mount Pleasant?

Answer. Yes, sir. They were sent, at any rate, from Chicago to Mount Pleasant, to the care of Mr. Eldridge, who was staying there for the time, as the agent of the committee.

Question. Do you know what became of them after they were sent to Mount Pleasant?

Answer. Yes, sir. I went afterwards to Mount Pleasant. Wagons were hired there, and they were sent to Tabor.

Question. Where is Tabor?

Answer. On the borders of Nebraska, in Frémont county, Iowa.

Question. In whose charge were they at Tabor?

Answer. I do not know, sir. Without I had my papers I could not tell. I do not recollect. They were placed in the hands of teamsters to haul there. Now, do you wish me to tell what I know with regard to them afterwards?

The CHAIRMAN. Yes; but I wish you would say in what year and month they were sent to Tabor.

Answer. It was in 1856, but the exact month I cannot recollect without my books.

The CHAIRMAN. Now, anything you know of them after they arrived at Tabor, or the disposition made of them afterwards.

The WITNESS. Probably I had better go back. After I had gone into the Territory of Kansas, and found the destitution of the people and the condition of things, after I had come down the Missouri river and found the condition of things in Missouri, being engaged in purchasing provisions there, and after these arms were at Tabor, and I had been somewhat in Iowa, I became fully convinced that a civil war was pending. The river was closed; no person could go into Kansas without being stopped, and these arms were on the borders of Nebraska, and citizens of Iowa were threatening to go into Missouri, if Missouri invaded Kansas again. Twenty-eight hundred Missouri men were at Lawrence just previously to that, or about that time. I came right on here to Washington——

The CHAIRMAN. What has all that got to do with the subsequent disposition of the guns?

The WITNESS. It has something to do with it.

The CHAIRMAN. We do not want to know the reasons that actuated you or anybody. All I want to get at is in whose charge these guns remained while they were at Tabor, and, when they went from Tabor, in whose charge they were then.

Answer. I do not recollect who the man was. I sent twice to him messages not to let them go into the Territory, by messengers, but I do not recollect his name.

Question. Did they remain at Tabor?

Answer. They remained at Tabor until Mr. Brown came to New York, and, in the application that I alluded to yesterday, he asked the national committee for these guns. That was in 1857, at the Astor House, and that committee refused him.

Mr. DOOLITTLE. I do not want any long story from Mr. Arny, but at that point, in speaking of the guns, in the way they were there, he speaks of the country being in a disturbed condition and civil war threatening, and he speaks of coming to Washington, and just about that time, as I understand the fact to be, President Pierce appointed Mr. Geary as governor of the Territory. He went out there, and put an end to the civil war, and the arms lay without being used by anybody.

The CHAIRMAN. What have we got to do with that? It is perfectly immaterial for what reasons they made this disposition of the arms—immaterial to our inquiry, why they did not use them there. I only want to know where they remained, and how they came subsequently into Brown's possession.

Mr. DOOLITTLE. So far as I look at it, it has this bearing: to show what the intention of the men who originally parted with their money, which bought these guns, was; what the intention of the great mass of the men who paid this $20,000 was. It was on the impulse of the moment, when they supposed a conflict was imminent, that the money was contributed and arms sent there, and Governor Geary going out there and putting an end, as he did, to the conflict, the arms remained there.

The CHAIRMAN. That refers entirely to the motives that governed those men in reference to the affairs of Kansas, with which we have nothing to do.

Mr. DOOLITTLE. I think we have so much to do with these arms as to show the fact that the men who contributed the money to buy them did not do it for the purpose of invading Virginia. That is what I want to show.

The CHAIRMAN. That cannot be questioned, I suppose; but if they allowed them to be used afterwards for that purpose, it is a different thing.

Mr. DOOLITTLE. I admit that, so far as regards those concerned in it.

The CHAIRMAN. We have nothing in the world to do with matters transacted in Kansas.

Mr. DOOLITTLE. Not to go into details.

The CHAIRMAN. We have nothing to do with the inquiry of how these arms were sent there, except to trace the arms in their various progress, and then to show ultimately, if we can, how they came into the possession of Brown. Why they were sent to Kansas, we have nothing to do with.

Mr. COLLAMER. It seems to me we have. I think that to rebut the presumption of their having been got up by the people for the occasion for which Brown used them, it becomes material to show that they were intended for an entirely different purpose, at the time they were purchased.

The CHAIRMAN. I think that is clearly shown, for surely they would never have sent arms out to Iowa that were intended to be used at Harper's Ferry.

Mr. COLLAMER. I should think so.

The CHAIRMAN. Probably this question would solve the whole: Why were the arms not sent to Kansas?

The WITNESS. I was going on to tell why they were not sent on to Kansas, when you interrupted me. As I said, I came to Washington——

Mr. DOOLITTLE. Understand, we do not want you to go into details. Just state the reason in substance.

Mr. COLLAMER. Were they not carried into Kansas because the difficulties there were ended?

The WITNESS. That is the reason—because the difficulties were ended there.

By the CHAIRMAN:

Question. Did those arms ever leave Tabor, as far as you know, before Brown got hold of them?

Answer. Brown told me, in January, 1857, in New York, that the rifles were at Tabor.

The CHAIRMAN. You mean those 200?

Answer. Yes, sir. He said he wanted to get them, and he applied to that committee for them. They refused his application, on the ground that Governor Geary had gone into the Territory and disbanded the 2,800 men; that the Missouri river was open; and that there was no necessity for the arms going into the Territory.

Question. At the time you deposited these rifles at Tabor, whose property did you consider them; who had the right to control them and dispose of them?

Answer. There was a controversy gotten up by some persons from Massachusetts—their names I cannot recollect without having my memoranda here—who claimed that they had sent those rifles to Chicago, and that they had the control of them. The national Kansas committee claimed that they had the control of the rifles, they having been sent by contributions from Massachusetts for the committee to transport; or, at least, they had fallen into their hands after the two committees had given up to the national committee. I, as general agent, had the control of them at Tabor, and had sent orders twice up there, not to let any person have them without an order from an officer of the national committee. We wanted to keep them there, and not have them go anywhere or be misused. That controversy was settled by the national committee agreeing that the Massachusetts people should have them and have the control of them.

Question. Then, as far as you were informed, the ultimate control of them rested with those Massachusetts people?

Answer. Yes, sir.

Question. Can you tell who those Massachusetts people were?

Answer. I cannot recollect the names of them. I know there were a number of persons, but I cannot recollect, unless I have my papers, who were the persons to whom they were given, or who was the officer of that committee.

Question. Can you tell us the name of the Massachusetts committee? I do not mean the names of the people.

Answer. I think it was the New England Kansas Aid Society, but I cannot tell distinctly.

By Mr. DOOLITTLE:

Question. Was it that, or the Massachusetts Kansas Aid Society?

Answer. I cannot tell unless I have my books.

By the CHAIRMAN:

Question. Do you know the name of any one or more persons who were members of that society?

Answer. Well, there were three or four societies there, county societies and other organizations, and they are so mixed up that I would decline giving any names, unless I could give them definitely and distinctly. I consider it unsafe to implicate persons in any way unless I could do it definitely and distinctly.

The CHAIRMAN. I understood you to say, Mr. Arny, that these arms were entirely given up by the national Kansas committee upon a counter-claim to a society in New England, the name of which you have given, as far as you recollect.

The WITNESS. I did not say they were given up to that society. I stated that some of the members of that society were there, and that there were other persons there from New England, and I cannot recollect how they were disposed of. They were given to the control of the persons there, however.

Question. Where do you mean by "there?"

Answer. In Massachusetts. It was agreed, in other words, that the national Kansas committee would have no further control of them. They might dispose of them as they saw proper, as they came from there.

The CHAIRMAN. Well, sir, whether it is agreeable to you or not, I request you to give the names of any one or more persons who were members of that society that you refer to as the New England Kansas Aid Society, or the Massachusetts Kansas Aid Society. I want the name or names of any person or persons who were members of this society that you have spoken of as the New England Kansas Aid Society, if that be its right name; whether the arms were left under their control or not, so far as you can recollect?

Answer. There is but one name that I can recollect distinctly, and

I do not remember the first name of that man; that is a gentleman named Clifford, who, I think, signs this report that I spoke of that I have, giving an account of $40,000 that was collected.

Question. Where does he reside?

Answer. In Massachusetts; but whereabouts I do not recollect. I never saw him in Massachusetts.

The CHAIRMAN. I do not want to prosecute the examination any further on that branch of it. Other gentlemen can inquire about it.

By Mr. COLLAMER:

Question. After you had thus given up the control of the arms to those from whom you supposed they had come, do you know anything of the disposition of them after that time?

Answer. I do not know anything about it.

By the CHAIRMAN:

Question. Well now, sir, you said in a former part of your evidence that there were 200 revolver pistols also sent to Chicago, and left under your charge. Can you tell us what became of them?

Answer. Yes, sir. I cannot tell you into whose hands they went unless I had my books and papers; but I can tell you where they went to. They were taken into Kansas. That was previous to or about the time the river was closed. The distribution of them there I cannot tell anything about unless I had my books.

Question. When you parted with them at Chicago, where were they sent to then, and by whose order were they sent?

Answer. They were sent to Mount Pleasant, and, by wagon, taken into the Territory afterwards.

By. Mr. COLLAMER:

Question. Were they sent with the rifles?

Answer. No, sir; but in another conveyance.

By the CHAIRMAN:

Question. Were those arms—the rifles and pistols—both in the manufacturers' original boxes when they were under your charge?

Answer. Yes, sir.

Question. Did you ever learn from Brown in what way he got the control of those arms subsequently?

Answer. No, sir.

Question. Did the conversation with Brown in your house, of which you spoke yesterday, relate to his ulterior views in reference to operating against the institution of slavery anywhere?

Answer. Yes, sir. It related to his intention and a difference of opinion with the people of the North in regard to that matter.

Question. State the conversation, as near as you can come at it, in substance.

The WITNESS. Do you want me to give the whole conversation, or just his language as far as I can recollect it?

The CHAIRMAN. Of course, his language, as far as you recollect it; but what I want is all the conversation he had with you, or you with him, on the subject of his ulterior views in reference to operating against the institution of slavery, and, if you please, any differences of opinion between him and those with whom he was associated or acquainted in that same business. State the conversation. If you give anything irrelevant, we can stop you.

The WITNESS. The conversation was one somewhat similar to the previous conversation with regard to our difference of views upon the subject of his going to the slave States. In the course of his remarks, in the commencement of one of his conversations, when he remained with me a day and over a night, he said that he believed the only way to abolish slavery was to situate a company of men somewhere in the mountains in the slave States, and to assist slaves in escaping, so as to make the system of slavery insecure—make that species of property insecure. I told him that I thought he was doing an injury to the whole country in pursuing that course; that it was contrary to his former views on the subject; that I did not suppose he could get any person to assist him in it; that I felt satisfied his good friend Gerritt Smith would not assist him, because Gerritt Smith had placed in our hands ten thousand dollars, and when be placed that money in our hands he made it an especial condition that every dollar of it should go for food and medicine and not for matters of war; he professed to be a peace man. I told him that I knew he was acquainted with Dr. Howe, and I did not suppose Dr. Howe would do anything of that sort, and no republican would. His answer was, that he disliked the do-nothing policy of the abolitionists East, and he said they would never effect anything by their milk-and-water principles; as to the republicans, he thought they were of no account, for they were opposed to carrying the war into Africa; they were opposed to meddling with slavery in the States where it existed. He said his doctrine was to free the slaves by the sword. I then again asked him how he reconciled his opinions then with his peace principles that he held when I first knew him in Virginia, more than twenty years ago. He said that the aggressions of slavery, the murders and robberies perpetrated upon himself and members of his family, the violation of the laws by Atchison and others in Kansas, in 1853, and from that time down to the murders on the Marias-des-Cygnes, convinced him that peace was but an empty word, and he repeated that the republican party and the abolitionists were cravens. I then told him that I could not give him any advice or assistance for that. He had at that time asked me to take charge of some papers that he had in his possession, but I declined doing so.

The CHAIRMAN. That was in the fall of 1858?

Answer. Yes, sir.

Question. I saw that you read the last answer?

Answer. I read it because I sat down and made it out.

Question. I saw that you read the answers which you gave from John Brown, from a book in your hand. Will you please to state when those answers were written down?

Answer. They were written in the course of the past week. I wanted to refresh my own memory, and I prepared them so that I could give, as near as I could recollect, the exact conversation that I had with him, as I supposed that would be asked of me, and I felt it was important that I should give it.

The CHAIRMAN. I only wanted to know when they were written. Mr. Arny, can you tell the committee at what time these rifles or pistols came under Brown's control?

Answer. No, sir.

By Mr. COLLAMER:

Question. You think these pistols were not at Tabor with the rifles?

Answer. I am certain they were not, because I saw them afterwards in the Territory. The pistol here I have looked at. It is a different pistol altogether from those shipped to me; they were Colt's pistols.

Question. You said yesterday your committee furnished clothing. What do you know about that?

Answer. The day he applied in New York for these arms and they refused him, he left displeased, and said he would appeal to the public and get what he wanted. The letter I alluded to yesterday was published afterwards. The next day, I think, he came back and met the committee again, and then represented to them that there were persons who had been with him and who were suffering, and that he needed clothing for certain purposes, and that if they would not give him anything else, to give him clothing. They then gave me an order to furnish him with clothing. I then received instructions from him, and a correspondence was carried on between him and myself, and probably the letters were found among his papers for aught I know, in regard to that matter. After I went to St. Louis, he gave me instructions to box his clothing and ship it to Nebraska City, as he could not get up the Missouri river, and he was going into the Territory; he wanted the clothing at Nebraska City.

The CHAIRMAN. Was that in 1857?

Answer. The beginning of 1857, after the river was opened by Governor Geary.

Question. Will you state now what that clothing consisted of generally?

Answer. It consisted principally of boots, shoes, a few caps, and I think some little female clothing. In all the clothing that was contributed to Kansas there was very little female clothing. It was principally male clothing, but I gave him some for the women as well as for the men, if I recollect aright.

Question. Were there coats, and waistcoats, and pantaloons, as part of the clothing?

Answer. Yes, sir; all kinds and all varieties.

The CHAIRMAN. As near as you can estimate it, not in money but in amount, was there clothing for fifty, or one hundred, or two hundred people?

The WITNESS. Less than one hundred. I do not think it would have clothed more than sixty persons. With regard to the other funds, if you desire me, I will state——

Question. What other funds?

Answer. The amount that was expended by me as agent.

The CHAIRMAN. Oh, no.

Mr. Doolittle. Will you state the amount of moneys expended by you in Kansas as the agent of the various Kansas aid societies, and for what purposes?

The Chairman. I object to that question, as irrelevant. It is submitted to the committee whether it is a relevant question or not.

The question was overruled, Messrs. Davis and Fitch concurring with the Chairman.

W. F. M. ARNY.

January 21, 1860.

Richard Realf sworn and examined.

By the Chairman:

Question. Will you state to the committee of what country you are a native, and what your age is?

Answer. I am a native of England. I was born in the year 1834. I shall therefore be twenty-six next June.

Question. When did you first come to this country?

Answer. In 1854.

Question. Are your parents living now in England?

Answer. They are.

Question. Will you state what was the occupation in life of your father?

Answer. At the time I left England my father was filling the position which he now fills, namely, an officer of the English rural police.

Question. To what occupation had he been bred?

Answer. My father was a blacksmith at one time. That trade he learned himself. He was a peasant, which means an agricultural laborer.

Question. Will you state what brought you to the United States in 1854?

Answer. I had been a *protégé* of Lady Noell Byron, widow of Lord Byron. I had disagreed with Lady Noell Byron, on account of some private matters, which it is not necessary to explain here, but which rendered me desirous of finding some other place in which to dwell. Moreover, my instincts were democratic and republican, or, at least, anti-monarchical. Therefore I came to America.

Question. Had you any acquaintance in this country when you came over?

Answer. No, sir; no personal acquaintance.

Question. Will you say whether you formed the acquaintance of John Brown, who was recently executed in Virginia for murder and treason?

Answer. Yes, sir; I did form his acquaintance.

Question. When?

Answer. In the year 1857. I cannot say whether it was the last day of November or the first of December, but within two or three days of that time.

Question. Will you state what brought you to his acquaintance, and where it was?

Answer. I was residing in the city of Lawrence, Kansas, as a correspondent of the Illinois State Gazette, edited by Messrs. Bailhace & Baker. I had been, and was, a radical abolitionist. In November, 1857, John Edwin Cook, also recently executed in Virginia, came to my boarding-house, in Lawrence, bringing me an invitation from John Brown to visit him at a place called Tabor, in Iowa. There I met John Brown.

Question. You went with Cook?

Answer. I went with John E. Cook.

Question. Did Brown then make known to you the object of the invitation to come and see him?

Answer. John Brown made known to a certain, but not to any definite and detailed degree, his intentions. He stated that he purposed to make an incursion into the Southern States, somewhere in the mountainous region of the Blue Ridge and the Alleghenies.

Question. What was the plan and purpose of the incursion, or did he develop it?

Answer. At Tabor, in Iowa, no place was named.

Question. What were the character and object of the incursion? Did he tell that?

Answer. To liberate the slaves.

Question. Did he disclose how he proposed to effect it?

Answer. Not at that time.

Question. Did you enter into any arrangements or engagements with him in reference to it?

Answer. Yes, sir.

Question. State what they were.

Answer. I agreed to accompany him.

Question. Did you remain under his control or guidance? What subsequent disposition did you make of yourself, or did he make of you, after that interview at Tabor?

Answer. I will tell you. From Tabor, where I myself first met John Brown and the majority of the persons forming the white part of his company in Virginia, we passed across the State of Iowa, until we reached Cedar county, in that State. We started in December, 1857. It was about the end of December, 1857, or the beginning of January, 1858, when we reached Cedar county, the journey thus consuming about a month of time. We stopped at a village called Springdale, in that county, where, in a settlement principally composed of Quakers, we remained.

Question. Did John Brown accompany you there?

Answer. John Brown accompanied us thither, but, whilst we ourselves remained there, John Brown went on East.

Question. Now, will you state who composed the company that Brown had assembled there, distinguishing between the whites and blacks, if there were any blacks?

Answer. Myself, Mr. Kagi, Mr. Cook, Mr. Stevens, Mr. Tidd, Mr. Leeman, Mr. Moffet, and Mr. Parsons, all these being whites, and Mr. Richard Richardson, a colored man, whom I met with Brown, at Tabor. These composed our company.

Question. How long did you remain at Springdale?

Answer. From the month—whether it be, I cannot now remember, the latter part of December, 1857, or the beginning of January, 1858, but from that time up until about the last week in April, a period of nearly three months.

Question. What was your occupation while you were there?

Answer. We were being drilled a part of the time, and receiving military lessons under Mr. Stevens. A part of the time I was lecturing.

Question. Did Brown provide for the support of the company while you were there?

Answer. Brown provided for the support of the company whilst we were there in this way: upon reaching there he, finding himself unable to dispose of the mules and wagons with which he transported us across the State, and unable to get the price he desired for them, left us there to board, the property named to belong to the man who kept us, a price having been agreed upon between himself and Mr. Brown.

Question. Whom did you board with?

Answer. With a Mr. Maxom.

Question. Did he keep a tavern?

Answer. No, sir; a private farm-house.

Question. You remained there, you say, until the following April?

Answer. Yes, sir.

Question. Will you inform the committee whether, during your residence there or at any time subsequent to Brown's inviting you to join that party, you heard of a man or made the acquaintance of a man named Forbes?

Answer. I never made the acquaintance of Colonel Forbes. I have heard of such a man.

Question. Will you say whether it was expected that he should be your military instructor? I mean anything you learned from Brown on the subject.

Answer. Yes, sir. You did not ask me the question, but I may as well state the fact that during our passage across Iowa, Brown's plan in regard to an incursion into Virginia gradually manifested itself. It was a matter of discussion between us as to the possibility of effecting a successful insurrection in the mountains, some arguing that it was, some that it was not; myself thinking, and still thinking, that a mountainous country is a very fine country for an insurrection, in which I am borne out by historic evidence which it is not necessary to state now.

Question. Brown's plans, then, were to make an incursion somewhere into the mountainous regions of Virginia?

Answer. Yes, sir.

Question. Did he say when he expected to effect it?

Answer. In that spring.

Question. Will you state whether the military training that he proposed for you and the company, had a reference to that incursion?

Answer. It was my belief that it had.

Question. Did he give you, in the course of conversation, any outlines or plans as to how he proposed to effect it—the mode of doing it?

Answer. Not during our residence in Iowa.

Question. You say Brown left you there. When did he return?

Answer. Brown returned a day or two before the period at which we left, namely, the last week in April, 1858.

Question. Did he inform you or the company, in conversation, how he had been occupied during the period of his absence?

Answer. No, sir; and here I ought to say, which you have also omitted to ask in regard to Colonel Forbes, that whereas we expected Colonel Forbes to be our military instructor, yet, in consequence of a disagreement between himself and John Brown, the latter wrote us from the East that Forbes would not become our military instructor, and that we should not expect him.

Question. Do you remember the point in the East he wrote from?

Answer. I do not. He used to write to his son Owen, one of the deceased persons, and in stating the number of persons comprising our company, I accidentally omitted his son. Owen was with us.

Question. Did Brown have much correspondence with his son while he was absent.

Answer. No, sir; the correspondence was very rare.

Mr. Collamer. In stating what was said by Brown, I desire the witness, as much as possible to give exactly what Brown himself said—the words used.

The Chairman. Exactly. It is desirable, of course, that you should give, if you can, the exact language; or if you cannot do that, give the substance of any communication from Brown.

The Witness. I will endeavor to do so.

Question. What was the next movement made by the company and Brown after his return in April?

Answer. The next movement after his arrival was an immediate departure from Iowa into Canada, *via* Chicago and Detroit.

Question. You remained at Springdale, you say, January, February, and March, something more than three months?

Answer. Yes, sir.

Question. Were the objects of your assembling there made known to the people around, in any way?

Answer. Not by myself; I cannot tell whether by others.

Question. Could you not learn something of it from conversations?

Answer. I am inclined to think that the people knew nothing at all of our movements for the reason that by some we were suspected to be Mormon emissaries.

Question. Did you not divest yourselves of that suspicion.

Answer. No, sir.

Question. Can you inform the committee whether there was any person or persons in that neighborhood who did know of the object of your assembling and your future plans?

Answer. I do believe that John Brown had given a man named Townsend, I cannot remember his first name, a member of the Society of Friends, some indirect and indefinite hints of his plan. I do also think that from the nature of a conversation which a Mr. Varny, also residing in the immediate neighborhood, and being also a Quaker, held with myself, that some one must have given him. some hints in regard to the same matter; but neither of those people were evidently, from the tone of their conversation, possessed of any definite information in regard to the matter.

Question. How were your military trainings conducted? Where were they conducted?

Answer. Principally in a field behind the house of Mr. Maxom; it being generally understood in the place where we were boarding, in the vicinity and round about, that we were thus studying military tactics and being thus drilled in order to return to Kansas and prosecute our endeavors to make Kansas a free State.

Question. That was the first idea?

Answer. That was the general understanding.

Question. Had you arms?

Answer. Yes, sir. John E. Cook had his own private arms. We had our private arms. I had my pair of Colt's revolvers.

Question. Did Brown furnish you with any arms?

Answer. No, sir, not myself, ever.

Question. I mean any of his company?

Answer. Not to my knowledge, because I suppose you will remember that I met the people comprising this company gathered together at Tabor. All of these people had been engaged in Kansas warfare. Everybody at that period in Kansas went armed, and the inference is that they were well armed before they met John Brown. Indeed, I am certain of that matter, because, in a greater or less degree, all of them had been engaged in the Kansas troubles.

Question. I only wanted to know whether Brown had furnished you any arms for the purpose of training.

Answer. No, sir.

Question. What part of Canada did you stop at?

Answer. We stopped at a town called Chatham, in Canada West.

By Mr. Collamer:

Question. What time did you get there?

Answer. It must have been about April 28 or 29, 1858, I think; or perhaps the 1st or 2d of May. I cannot remember within two or three days. I recollect it was at that time, because the convention, to which we shall come presently, was held on the 10th of May; and we were there a sufficient time to allow John Brown to write letters, about which I shall, doubtless, be asked.

By the Chairman:

Question. Will you state who of the company that you had at Springdale, accompanied John Brown to Chatham?

Answer. All of the company whom I named as having gone to Springdale and two others: a young man named George B. Gill, who resided at Springdale, who had learned of our plans, from whom I do not know, but I suppose from John Brown, inasmuch as he never manifested any desire to accompany us anywhere until the return of John Brown; and another young man, named Stewart Taylor, the latter of whom was killed at Harper's Ferry, and the former of whom, so far as I have been able to learn, was not present at the incursion.

Question. Where did Stewart Taylor come from?

Answer. I do not know.

Question. Did this man Richardson, the negro, go with you to Chatham?

Answer. Yes, sir.

Question. Was Brown's intercourse with the negro of a character to show that he treated him as an equal and an associate?

Answer. It certainly was. To prove it, I will simply state that, having to wait twelve hours at Chicago, in order to make railroad connection from Chicago to Detroit, and to Canada, we necessarily had to breakfast and dine. We went into one of the hotels in order to breakfast. We took this colored man, Richardson, to table with us. The keeper of the hotel explained to us that it could not be allowed. We did not eat our breakfast. We went to another hotel, where we could take a colored man with us and sit down to breakfast.

Question. Where you could enjoy your rights, I suppose?

Answer. Yes, sir.

Question. Will you state in what way the expenses of your transportation were defrayed?

Answer. They were defrayed by John Brown.

Question. What was done on your arrival at Chatham?

Answer. Upon our arrival in Chatham, Canada West, we boarded at a hotel kept by a colored man, (I do not remember his name,) whence written (not printed) circulars were sent to certain persons east and west, for Chicago is west of Canada, inviting their attendance at a quiet convention of the friends of freedom, to be held on the day named, namely, May 10, 1858.

Question. Did you remain there during the intermediate time between the last of April and the 10th of May; or was the convention held earlier?

Answer. There were two conventions. The constitutional convention was held two days previous to the election of the officers. The constitution had been adopted, and then the election of the officers was held. I had forgotten that before. The constitutional convention was on the 8th of May, 1858.

The CHAIRMAN here submits to the witness the papers heretofore produced by Andrew Hunter, and purporting to be the minutes or "Journal of the Provisional Constitutional Convention," and of the convention to elect officers, signed respectively by "J. H. Kagi," as "secretary of the convention," and asks the following

Question. Do you know the handwriting of these papers?

Answer. I do; it is the handwriting of John Henry Kagi.

[The papers are identified by the chairman placing his initials thereon.]

Question. It is stated in these minutes that "on motion of Mr. Delany, Mr. Brown then proceeded to state the object of the convention at length." Did you know this "Mr. Delany?"

Answer. Yes, sir; he was a colored doctor, residing in Chatham, Canada West.

Question. Do you mean a negro when you say "colored?"

Answer. Yes, sir.

Question. Who was the presiding officer of this conveution?

Answer. A man named Munroe—a preacher.

Question. Where did he come from?
Answer. I believe the city of Detroit?

By Mr. COLLAMER:

Question. Was he a colored man?
Answer. Yes, sir; a mulatto.

By the CHAIRMAN:

Question. Do you recollect Brown's speech, which, it is said in these minutes "developed the plan?"

Answer. I cannot remember his speech. I can remember certain salient points and leading ideas in his speech.

Question. He did make a speech?

Answer. Yes, sir.

Question. Of course you cannot remember the speech; but will you state as briefly but as exactly as you can, what he did state to be the object in view of this constitution and all that?

Answer. John Brown, on rising, stated that for twenty or thirty years the idea had possessed him like a passion of giving liberty to the slaves. He stated immediately thereafter, that he made a journey to England in 1851, in which year he took to the international exhibition at London, samples of wool from Ohio, during which period he made a tour upon the European continent, inspecting all fortifications, and especially all earth-work forts which he could find, with a view, as he stated, of applying the knowledge thus gained, with modifications and inventions of his own, to such a mountain warfare as he thereafter spoke upon in the United States. John Brown stated, moreover, that he had not been indebted to anybody for the suggestion of this plan; that it arose spontaneously in his own mind; that through a series of from twenty to thirty years it had gradually formed and developed itself into shape and plan. He stated that he had read all the books upon insurrectionary warfare which he could lay his hands upon—the Roman warfare; the successful opposition of the Spanish chieftains during the period when Spain was a Roman province; how with ten thousand men divided and subdivided into small companies, acting simultaneously, yet separately, they withstood the whole consolidated power of the Roman empire through a number of years. In addition to this, he said he had become very familiar with the successful warfare waged by Schamyl, the Circassian chief, against the Russians; he had posted himself in relation to the wars of Toussaint L'Overture; he had become thoroughly acquainted with the wars in Hayti and the islands round about; and from all these things he had drawn the conclusion, believing, as he stated there he did believe, and as we all (if I may judge from myself) believed, that upon the first intimation of a plan formed for the liberation of the slaves, they would immediately rise all over the Southern States. He supposed that they would come into the mountains to join him, where he purposed to work, and that by flocking to his standard they would enable him (by making the line of mountains which cuts diagonally through Maryland and Virginia down through the Southern States into Tennessee and Alabama, the base of his operations) to act upon the plantations on the plains lying on each side of that range of mountains, and that we should be able to establish ourselves in the fastnesses, and if any hostile action (as would be) were taken against us, either by the militia of the separate States, or by the armies of the United States, we purposed to defeat first the militia, and next, if it were possible, the troops of the United States, and then organize the freed blacks under this provisional constitution, which would carve out for the locality of its jurisdiction all that mountainous region in which the blacks were to be established, and in which they were to be taught the useful and mechanical arts, and to be instructed in all the business of life. Schools were also to be established, and so on. That was it.

Question. Did he develop in that plan where he expected to get aid or assistance; who were to be his soldiers?

Answer. The negroes were to constitute the soldiers. John Brown expected that all the free negroes in the Northern States would immediately flock to his standard. He expected that all the slaves in the Southern States would do the same. He believed, too, that as many of the free negroes in Canada as could accompany him, would do so.

Question. Was anything said in his developments of his expectations and resources after he got into the slave States, of any division of sentiment between the slaveholders and non-slaveholders?

Answer. The slaveholders were to be taken as hostages, if they refused to let their slaves go. It is a mistake to suppose that they were to be killed; they were not to be. They were to be held as hostages for the safe treatment of any prisoners of John Brown's who might fall into the hands of hostile parties.

Question. As to the non-slaveholders; was there anything said about them?

Answer. All the non-slaveholders were to be protected. Those who would not join the organization of John Brown, but who would not oppose it, were to be protected; but those who did oppose it, were to be treated as the slaveholders themselves.

By Mr. Davis:

Question. Where did he expect in the first instance to get his resources of money and arms?

Answer. John Brown expected that——

Mr. Collamer. Did he say that? We are talking now of what he said in his speech.

Mr. Davis. What he stated.

Answer. John Brown did not make any explicit or definite statement in his speech at all as regarded where the money was to come from.

Mr. Fitch. I do not understand that the witness is limited to that speech.

The Chairman. No, sir.

Mr. Fitch. The understanding was that he was to state to the committee any information derived from Brown himself at any time.

The Chairman. It was to prevent confusion of what he did derive from Brown and from other sources, that I put the questions as I did.

Mr. Collamer. But I suppose what he is telling us now is what Brown stated in that speech on that occasion.

The WITNESS. I have been stating what Brown said in that speech, all this being a part thereof.

Mr. DAVIS. So I understood, and that is the reason I asked the question I did.

The WITNESS. It is not yet quite all of that speech.

Mr. DAVIS. I did not wish to break the chain.

The CHAIRMAN. Go on and give us all you can recollect of Brown's exposition on that occasion.

Answer. Thus, John Brown said that he believed, a successful incursion could be made; that it could be successfully maintained; that the several slave States could be forced (from the position in which they found themselves) to recognize the freedom of those who had been slaves within the respective limits of those States; that immediately such recognitions were made, then the places of all the officers elected under this provisional constitution became vacant, and new elections were to be made. Moreover, no salaries were to be paid to the office-holders under this constitution. It was purely out of that which we supposed to be philanthropy—love for the slave. Moreover, it is a mistake to suppose, as Cook in his confession has stated—and I now get away from John Brown's speech—that at the period of that convention the people present took an oath to support that constitution. They did no such thing. This Dr. Delany of whom I have spoken, proposed, immediately the convention was organized, that an oath should be taken by all who were present, not to divulge any of the proceedings that might transpire; whereupon John Brown rose and stated his objections to such an oath. He had himself conscientious scruples against taking an oath, and all he requested was a promise that any person who should thereafter divulge any of the proceedings that might transpire, agreed to forfeit the protection which that organization could extend over him.

Mr. DAVIS. If the witness has concluded his recollection in relation to what Brown stated——

The WITNESS. No, sir; I have not. John Brown stated in that convention, in the speech he made, that there were a great number of rich people all over the free States who, he doubted not, would assist him. He stated that he had some rich friends in the free States who had assisted him, and who had promised further to assist him, but John Brown did not disclose their names, being too profound and sagacious a man to do so.

Question. Did he say, do you recollect, that the friends to whom he referred had promised aid, or that he expected it only?

Answer. That they had assisted him in some degree; that they had promised to assist him further.

By Mr. COLLAMER:

Question. Did he state that those people understood this—his plan?

Answer. No, sir; he did not state so explicitly, but that was the idea which he conveyed to us. In order to render that answer intelligible, I should say that John Brown had, from the time he went to Kansas, devoted his whole being, mental, moral, and physical, all that he had and was, to the extinction of slavery. He stated that he only went to Kansas in order to gain a footing for the furtherance of this matter. He stated that explicitly and emphatically.

Question. That that was his private purpose?

Answer. Yes, sir; that that was his private purpose; and he stated that, having left his wife and children and home, these friends had assisted him to prosecute his designs against slavery in Kansas first, and next generally in his enterprises in the cause of freedom.

By the CHAIRMAN:

Question. Have you gone through with your recollections of Brown's exposition to the convention?

Answer. I have, except that if any questions should be asked me in regard thereto, they might suggest certain things to me which I cannot now remember without those questions. I have stated as much as I can, of my own recollection, remember.

Question. Will you tell us this: was there any person belonging to Canada in that convention who took any part in the discussion of John Brown's plan, after his exposition?

Answer. Yes, sir; Dr. Delany was one of the prominent disputants, or debaters.

Question. Will you state, as far as you can recollect, anything that fell from Delany showing a coincidence of purpose with John Brown?

Answer. The whole tenor of Dr. Delany's speeches was to convey the idea to John Brown that he might rely upon all the colored people in Canada to assist him.

By Mr. DAVIS:

Question. Were there any Canadians other than negroes?

Answer. No, sir; not one.

By the CHAIRMAN:

Question. Have you any reason to know whether the purposes of the convention, or the purposes ultimately disclosed in the convention, were known to the white people around you there in Chatham?

Answer. I am confident that they were not.

Question. Was the convention held in the presence of an audience or in secret?

Answer. The convention was held with closed doors, all other persons present excepting Brown's original party being colored men.

Question. And Canadian negroes?

Answer. Yes, sir, Canadian negroes.

Question. You have stated that in traveling from Tabor across Iowa to Springdale, you were about a month engaged in it, and that John Brown conducted the expedition and defrayed the expenses, and that he left you then, and left his mules, &c., in pledge for the expenses of the party. Did he tell you or the company of the object of his going eastward?

Answer. Yes, sir. He had two purposes in going to the East; one to secure the services of Colonel Forbes, and bring him on, in order to instruct us. Another purpose was to secure funds.

Question. How do you mean "to secure funds?"

Answer. To secure funds to enable him to prosecute his business.

Question. How was he to get them?

Answer. I do not know; he did not state. It was to collect funds. Here I ought to state, inasmuch as it may be of use during this examination, that John Brown was a man who would never state more than it was absolutely necessary for him to do. No one of his most intimate associates, and I was one of the most intimate, was possessed of more than barely sufficient information to enable Brown to attach such companion to him; and none of us were cognizant of more than the general plan of his design until the time we reached Chatham, Canada West.

By Mr. DAVIS:

Question. Have you, from Brown or other sources, any means of informing us where the money and arms were expected to be obtained?

Answer. No, sir; I have not, except to say this—and I am glad that the question is put—that a certain number of arms had been placed in the hands of John Brown by Dr. Howe, or which it was *supposed* had thus been placed, by Dr. Howe, of Boston. Dr. Howe was the Massachusetts representative of the national Kansas committee, a committee which received contributions and made collections to be applied to the assistance of the free State settlers in Kansas during the troubles in that Territory. Afterwards, on account of disagreement, the Massachusetts committee withdrew from the national committee, and had received back a certain quantity of arms which it, Massachusetts, had purchased and thrown into the general granary, so to speak.

Mr. COLLAMER. Where were those arms, do you know?

The WITNESS. They had been at Tabor, in Iowa.

Mr. DAVIS, (to the witness.) You were going on to say something. What was it?

The WITNESS. Dr. Howe, as the representative of Massachusetts, immediately following the disagreement, withdrew the control of those arms from the national committee, and had therefore himself control over them.

The CHAIRMAN. But the arms, I understand, still remained at Tabor.

The WITNESS. I do not know whether they did or not. I cannot tell, inasmuch as when I reached Tabor John Brown had made all his arrangements for immediate passage across Iowa.

Mr. DAVIS. The witness was interrupted in what he was going on to state. I desire him to continue it.

The WITNESS. I do not *know* that Dr. Howe placed those arms in John Brown's possession, but I *supposed* so, for a reason which I will explain immediately. Within a day or two following the convention at Chatham, John Brown said to me that he had received a copy of a letter written by Senator Henry Wilson, of Massachusetts, from Washington city, to Dr. Howe, of Boston. Brown then stated to me that Colonel Forbes, maddened by the failure to receive money from John Brown, as had been agreed on according to Forbes's statement, and exasperated by the dreadful condition in which his family were, or in which he claimed that they were, in Paris, had threatened to make disclosures of Brown's plan, unless Brown forwarded money to him. Forbes was cognizant of Brown's plan, for the reason that at one period he had agreed, as I learned, to head the expedition; but a rupture occurring between him and Brown, he, being possessed of Brown's plans, threatened to divulge them, and did divulge them, or so much of them as was necessary to put people on the alert. He divulged them, as I say, to Senator Wilson, in this city.

Mr. COLLAMER. That is what Brown told you.

The WITNESS. Yes, sir; that is what Brown told me. To explain it a little more, I should perhaps say that Brown had written to us whilst we were at Springdale, that Forbes and himself had disagreed. On the occasion of which I have just spoken, at Chatham, Brown said to me that Colonel Forbes, maddened by the non-receipt of moneys which he had expected to receive, had threatened to divulge Brown's plans, and had done so by coming to Washington, and stating to Senator Henry Wilson, of Massachusetts, that Brown had a purpose in view of effecting an insurrection in the Southern States. Senator Wilson, immediately upon receipt of the news, said that he did not think any man, or any company of men, could be wild enough and mad enough to do such a thing; but knowing the character of John Brown, and supposing——

The CHAIRMAN. Are you giving this as what Brown told you?

The WITNESS. I have given that which Brown said to me, and now I am making a statement in regard to what Henry Wilson said.

Mr. COLLAMER. What Brown told you Mr Wilson said?

The WITNESS. What Brown told me he said. Thus, then: Forbes has made this revelation to Wilson, whether definite and in detail I do not know, but he had made a revelation of that kind. Immediately upon receipt thereof, Senator Wilson sat down and wrote to Dr. Howe that, understanding or supposing that arms belonging to the Massachusetts committee, which Howe had withdrawn from the national committee, had been placed by his, Howe's, hands in care of John Brown, he, Wilson, requested him, Howe, to withdraw from John Brown's hands all command over those arms, lest in a moment of madness, he might possibly put into operation such a scheme. This letter was written by Senator Wilson to Dr. Howe, of Massachusetts. All along, I say Dr. Howe, but I cannot swear that it was Dr. Howe; but if it was not he, it was Sanborn. Whilst I have one thought out of ten that it might be Sanborn, I have nine out of ten that it was Howe. It was one of those two men, and Howe I believe.

Mr. DOOLITTLE. I think there was one sentence you did not finish when you were interrupted by another question. You began a sentence, stating that Mr. Wilson said that he did not think any man or any company of men could be found to go into such a scheme. Please finish it.

The WITNESS. But lest they should be mad enough to do it, he Wilson, requested him, Howe, to withdraw from Brown's hands those arms, so as to place it out of his power to do the thing. A copy of this letter, thus written by Wilson to Howe, was forwarded by Howe to Brown, at Chatham, and in compliance with the request made to Howe by Wilson, he did withdraw those arms from Brown; that is, he made a requisition on Brown to deliver them up, stating that he withdrew from him the *carte blanche*, or power of attorney, or whatever it was he had over them. Whether or not he afterwards reinstated Brown in the possession of those arms, I cannot say. That is so much as relates to that matter.

By the CHAIRMAN:

Question. You have spoken of the contents of the copy of a letter from Wilson to Howe; will you state how you derived a knowledge of those contents?

Answer. John Brown read those letters to me.

Question. Howe's letter to him, and Wilson's letter to Howe?

Answer. Yes, sir.

Mr. DAVIS. Did the letter of Senator Wilson disclose the fact that Forbes was enraged?

Answer. Only that Forbes had made such a statement to Wilson.

The CHAIRMAN. You have stated to us, as I understand, that Brown read to you the copy of Wilson's letter to Howe, which he alleged Howe had sent to him. Now, will you give to the committee, as nearly as your memory will allow, the contents of Wilson's letter to Howe.

The WITNESS. I can but remember the things of which I have spoken in regard to it, the contents of his letter being that Forbes had made such a revelation to him, Wilson.

The CHAIRMAN. What revelation?

The WITNESS. A revelation that John Brown proposed to commit an incursion on the Southern States. I stated before that I did not know whether Forbes gave any definite or detailed information in regard to the plan or not; because, if he did so, Wilson did not state it.

The CHAIRMAN. We do not want your inferences, but we desire you to state, as nearly as you can, the contents of the letter from Wilson to Howe, and the request which you say was contained in it.

The WITNESS. The request was based upon the statement made by Forbes to Wilson, and Wilson either knowing or supposing, I cannot tell which——

The CHAIRMAN. We do not want anything about that. Did the letter itself say what statement Forbes had made?

The WITNESS. I cannot tell whether it ran in so many words or not, but it said that John Brown had designs against the Southern States, calculated to effect a rupture between the free and the slave States, and in order to stop it he wrote.

By Mr. DAVIS:

Question. Did Brown's knowledge of Forbes's intention to divulge his secret come from the copy of the letter received by him from Dr. Howe, as having been sent to Dr. Howe by Senator Wilson, or did he know it anterior to that?

Answer. He knew previously to that, that Forbes had threatened to do these things, in several letters.

Question. And now he was made aware that he had done it?

Answer. Yes, sir. Now, he was made aware that Forbes had done so.

By the CHAIRMAN:

Question. Do you know whether Brown remained in possession of the arms spoken of by Senator Wilson and Dr. Howe, or whether he afterwards got them into his possession?

Answer. I do not know; for the reason that a very short time following the receipt of that letter by John Brown, I left the party, and have since had no connection with them.

Mr. COLLAMER. What was the occasion of your leaving the party? For what ostensible purpose did you leave?

The WITNESS. I will tell you.

The CHAIRMAN. Before that, I want to ask what became of the members of the convention when they adjourned.

The WITNESS. The answer to that will include the answer to the other question.

The CHAIRMAN. After the convention adjourned, what became of those members of the convention that had been with you under military drill at Springdale, including yourself?

Answer. Immediately following the adjournment of the convention, a portion of the original company went from Chatham, in Canada, to Cleveland, in Ohio, in the United States.

Question. Who went there?

Answer. I cannot now remember all the party who went there; but I know that Cook was one who went; I know that Stevens was one who went; that Tidd was another; that G. B. Gill was another; that Stewart Taylor was another; that Owen Brown was another; and I think they were all.

Question. Were you with them?

Answer. No, sir; but very shortly afterwards, myself, the colored man Richard Richardson, and another colored man, whose name I cannot recollect, residing in Canada, and who had agreed to accompany us, went from Chatham to Cleveland. In addition to these persons, I now remember that Mr. Leeman, one of the persons killed at Harper's Ferry, went with me, too. Our departure, by which I mean the departure of those who were with me, as contradistinguished from those that went before, was about two weeks later than the departure of the first company.

Question. Then you remained at Chatham for two weeks after the adjournment?

Answer. About that time.

Question. Then you went to Cleveland?

Answer. We went to Cleveland. Now, I ought to say here that those persons comprising the first party who went from Chatham to Cleveland did not remain in the city. They went out into the surrounding country and procured work, John Brown's means being so limited that he could not pay their board. I have not stated what John Brown did yet. He went East, leaving me to go on to Cleveland, and there await the receipt of letters from him from the East, and his own return from that quarter. John Brown went East. He went to North Elba, where his family resided. He wrote to me from North Elba that he would shortly return. Afterwards he went to Boston. He again wrote me from Boston that he had been delayed, but would shortly return. None of John Brown's letters to me, of which I think I received during my stay in Cleveland three, contained over four lines; therefrom you may judge how much John Brown allowed his people to be cognizant of his plans.

Question. Have you preserved those letters?

Answer. No, sir; I destroyed them a long time ago. Well, John Brown returned to Cleveland from the East in the beginning of June, 1858, having, perhaps, been absent East a month from his departure from Chatham, Canada West. On his returning to Cleveland, those of our company who had been out in the country procuring work returned to Cleveland to the hotel where John Brown came, and where I was boarding. I ought, however, now that I remember it, to state that John H. Kagi did not go there to Cleveland with the first party or with myself; but he went to a town called Hamilton, in Canada West, and there, being (among his other accomplishments, for he was a very accomplished man) a practical printer, he privately superintended the printing of the constitution adopted at the convention. Kagi reached Cleveland a few days previous to the arrival of Brown from the East. We were all united there, consequently, once again. John Brown arrived from the East. John Brown had not procured money. He had probably about $300 altogether. He had not enough to pay the necessary expenses for the printing of the copies of the constitution in Canada. He had barely enough to give those who accompanied him a sufficient amount of money to enable them to return back to their different places of abode. Mr. Kagi, John Brown, and Mr. Tidd went back to Kansas. John E. Cook received his quantum of the money. I do not know whither he went. Stewart Taylor received his, and went to Ann Arbor, Michigan. G. B. Gill and Mr. Stevens returned to Springdale, Iowa, the brother of Mr. Gill residing there, and Mr. Stevens having formed some connections which induced him to return. I was to go on to New York city.

Question. Did you go by direction of anybody?

Answer. I went——

Question. What sent you there, or who sent you there?

Answer. John Brown sent me to New York city for this purpose: Knowing that Forbes had made these revelations about which I have spoken, and knowing, too, that it incapacitated him for the time being from prosecuting this plan, he desired me to go on to New York, somehow or other procure an introduction to Forbes; and he being an Englishman and I being an Englishman, he thought we might presently establish mutual good relations; that by ingratiating myself into his esteem, I might ultimately be able to possess myself, acting for Brown, of that obnoxious correspondence held by Forbes, written by Brown to him, in which Brown had developed his plans. For that purpose, I went on to New York, and I ought, in justice to myself to say, that I went with the intention of securing that correspondence; for at that period, though I had not been at all satisfied with the condition of the negroes in Canada, I was still an abolitionist, and I went to New York city purposing to possess myself of this correspondence. I arrived in New York city——

The Chairman. Stop a moment. What were you to do with the correspondence, if you got it?

Answer. Return it to John Brown, so that when Forbes was called upon, (as Brown supposed would be the case,) to substantiate his statements, he should not have the means of doing so. I went to New York. In New York city, I met, for the first time, with a book called "Limitations of Human Responsibility," written by Dr. Wayland, a philosophic author. I had thought a great deal about human responsibility and my own responsibility, perhaps, indeed, a little too much; but I had never thought anything in regard to the limits of it, and that book taught me that there were certain things which I might thoroughly believe myself, but which I had no right to enforce *nolens volens* on my neighbor, and it set me pondering on a new train of ideas. I did not see Colonel Forbes in New York city. I cannot recollect whether I made any attempt to see him or not. What I know is, that I did not see him. I met in New York city with Judge Arny, examined before your committee the other day, with Thaddeus Hyatt, a mutual friend of ours. To Judge Arny I made a statement of Brown's purpose; not, however, in detailed terms, but I said to him that Brown had in view a project of liberating the slaves in the South. I stated the same to Thaddeus Hyatt. Because the lapse of time is so great, and because I have had so many things passing through my brains since, I have forgotten whether I held any conversation with those men beyond making that simple revelation. I know that I went to England; I know that Judge Arny strongly advised me, instead of connecting myself with any such wild movement, to get married, which he thought would most effectually quiet me. I went to England. Cook, in his confession, states that I went to England for the purpose of procuring assistance for John Brown. I did not. I went to England; I wanted to see my father and my mother. I was home-sick. I did very probably say, indeed I know I have often said to Cook, during my acquaintance with him, that England would be the proper place in which to raise money for abolition purposes. I do not know how Brown became cognizant of my departure for England, or Cook either, except in this wise: Arny, knowing I was going to England, I having consulted him in regard to it, and he having advised me, and assisted me to do so, I suppose that on his return to Kansas, he must have told Brown and Kagi, and the rest of them who were there. I saw a statement in a paper, I do not remember what paper, but sometime ago, I saw a statement that the internal evidence of the letters of Brown and his friends plainly revealed the fact that, though they could trace my departure for England, they could not learn anything of me or my movements since. That, therefore, is evidence that I was not collecting money for them in England, or that if I did, they did not get it; which, so far as implicating me is concerned, amounts to about the same thing. Well, I went to England——

Mr. Collamer. Now, stop. There is no use of pursuing this any further, unless the witness had further connection with Brown. Had you any further connection with Brown?

Answer. No, sir; I knew nothing at all about him.

Mr. Davis. Let the witness proceed, because it has been alleged that he went to England to lecture for the purpose of raising money. The best way in which he can satisfy not only the committee, but others, in relation to what he went there for, is to tell his story.

Mr. COLLAMER. It has nothing to do with this inquiry before the committee, but I shall not interpose.

The CHAIRMAN. Let us have the whole ground.

Mr. COLLAMER. Very well, if you desire it.

The WITNESS. I went to England. I lectured in England. I lectured, among other things, on temperance—principally on that subject. Among other things, too, I lectured on the literature, liberty, &c., of the United States. I was an abolitionist at the time, too. I never, during the period of my sojourn in England, collected, or endeavored to collect, a single cent of money for any purpose whatever. I was paid for lecturing; and "the laborer is worthy of his hire," and I put that money in my pocket. Then I went to France. As I stated just now, I had witnessed a great discrepancy between the actual condition of the negroes in Canada and the statements which I had read in regard to their condition in Canada——

Mr. DOOLITTLE. One word in relation to that. I have no objection to its going down as far as he wants to exculpate himself from any allegation that he has collected money and misapplied it. Any personal explanation I have no objection to; but then, to lumber up the record with giving his peculiar views about one thing or another which does appear on our investigation, seems to me to be improper.

The WITNESS. No, sir; but I will not be one minute longer, if you will permit me.

Mr. COLLAMER. That might lead to considerable inquiry, and perhaps cross-examination on that point, if you desire to go into it.

The CHAIRMAN. I agree we have nothing to do with his mission to England.

Mr. COLLAMER. Or his return to America, and going to New Orleans, and from thence to Texas, &c.

Mr. DAVIS. I have no desire to go beyond the subject before us.

Mr. COLLAMER. The subject is John Brown and his foray.

The WITNESS. I have finished in regard to my connection with John Brown. I never wrote him a single letter; never received a single letter from him; never had, directly or indirectly, any acquaintance or connection, in the most remote degree, with the party after my departure from Cleveland.

The CHAIRMAN. You have said that in New York you revealed to Arny and to Thaddeus Hyatt what you learned from Brown were his plans as to incursions into the Southern States.

The WITNESS. Not as a detailed plan; but a broad statement, that he did purpose to put into operation a movement having for its object the liberation of the slaves.

Question. Did you tell, either to Arny or Hyatt, your mission to New York—what brought you there?

Answer. I cannot remember whether I did or not, it being such a period of time removed. I will not say I did not I will say it is possible, nay, probable, that I did tell them what my mission there was.

Question. But you never did see Colonel Forbes?

Answer. I never saw Colonel Forbes, to my knowledge, in my life.

Question. Or had any communication with him?

Answer. No, sir.

Question. Now, I will put this general question: Did you go to

England with any view to collect funds for the purpose of carrying on any abolition schemes in the United States?

Answer. No, sir.

Question. You did collect no funds for that purpose?

Answer. I collected none.

Question. Will you tell us when you returned to the United States?

Answer. I returned to the United States, leaving Havre on the 2d of March, 1859, and arriving in New Orleans the 17th of April, the same year.

Question. What brought you back to the United States?

Answer. My desire to return.

Question. And since your arrival, tell us where you have spent the intermediate time?

Answer. I spent part of my time in New Orleans. Now I ought to say, in justice to myself, that part of my mission in England was in order to procure the consent of my father and mother to join the Catholic Church. They would not give it to me. Coming back, I immediately joined the Catholic Church without their consent. I purposed to become a Jesuit priest——

The CHAIRMAN. I do not want to know anything about that.

The WITNESS. But you asked me——

The CHAIRMAN. I asked you for your reason for coming back to the United States.

The WITNESS. And where I had been, and what I had been doing since I came back.

The CHAIRMAN. But it does not follow that you should tell us what your plans and pursuits in private life were. I only want to know what points you have been at in the United States since your return?

The WITNESS. Well, sir, New Orleans for one. In New Orleans it was proposed to establish a new Democratic paper, the "Delta" having, as they thought, written itself out. Mr. Semmes, now attorney general of the State, had spoken to some friends of mine——

The CHAIRMAN. We do not want that. My question simply was, at what parts of the United States you had been since your return to this country?

Answer. New Orleans, Mobile, and Austin, in Texas.

Question. Had you any purposes in view, at either of those places, connected with your former views in reference to the abolition of slavery?

Answer. No, sir; but I had in view the purpose of investigating the condition of slavery for myself.

By Mr. DAVIS:

Question. Were you secretary of state of the proposed government to be established by John Brown?

Answer. I was.

Question. Did you receive and preserve, or was he the depositary of the correspondence which was held with the friends of such a movement on the part of John Brown?

Answer. I was not. John Brown was.

Question. Were you the organ of any correspondence as secretary of state?

Answer. No, sir.

Question. Were the letters written and the answers received in relation to funds, exhibited to you?

Answer. No, sir; for this reason: that but a period of a week or two elapsed between my nomination and election as secretary of state and the disbanding of the whole party, John Brown being in the mean while absent.

Question. Did you, from your relation to John Brown and to this organization, know the names of persons who were relied upon to furnish money, or who did furnish money?

Answer. Not any other names save those of Dr. Howe, whom Brown mentioned, F. B. Sanborn, whom Brown mentioned, and Gerritt Smith, whom Brown also named.

Question. How did he mention them? as having given or being expected to give money?

Answer. That Gerritt Smith had given Brown money; that he had assisted Brown from the time when he first went to Kansas, and had promised to assist him further in his enterprises against slavery; whether or not in this particular movement against the South I cannot say, but I suppose that was the understanding.

Mr. Doolittle. The supposition ought not to go down.

Mr. Davis. I think the impression made upon his mind, considering the position he occupied, is legitimate.

The answer was allowed to remain as given by the witness.

The Witness. Here I may as well state, once for all, that I do not believe John Brown would intrust to any man, no matter how intimate his friendship might be, more than barely sufficient of his schemes to secure his coöperation and support.

Mr. Collamer. You spoke of Brown having received aid from wealthy people at the North. Did that relate to Kansas?

Answer. He said he had received aid from those wealthy people from the time he went to Kansas, and that they had promised to assist him in any enterprises which he might undertake against slavery and in behalf of freedom. That was it; a general promise of assistance—he having left his farm, wife, home, friends, everything.

Mr. Collamer. I wish to know whether your position as secretary of state, as it is said, furnished you with any information on which you could found a supposition, more than you had when you were not secretary of state?

Answer. No, sir. I should like to say this: Gerritt Smith having been, as I learned from John Brown, one of the persons who had principally supplied him with means, and John Brown having stated that Gerritt Smith had promised to assist him in any enterprises he might undertake for the furtherance of freedom, that he would enable him to prosecute all such movements—on that statement of Brown I based my supposition.

Mr. Collamer. And on that only?

Answer. On that only.

Mr. Davis. The question, however, was, whether your position enabled you to form a supposition?

Answer. My position did not; because, before I became secretary of state I possessed that information; and after I was secretary of state

I possessed no more. That information, therefore, was the cause of my supposition, which I not only had *after* I was secretary of state, but *before* it.

Mr. DAVIS. I ask whether, as secretary of state, the witness was not put in more confidential relations with John Brown than he was before?

Answer. No, sir; for the simple reason that, before there was any opportunity of establishing any confidential relations, the whole affair was broken up.

By Mr. COLLAMER:

Question. Did Brown at any time suggest to you that he had disclosed to Gerritt Smith the purpose which you know he entertained?

Answer. Never, sir.

By the CHAIRMAN:

Question. Did you learn from Brown, at any period of your intercourse with him, and up to the latest period, when he proposed to carry his plans into execution in the Southern States?

Answer. John Brown had purposed, immediately upon his return from the East, in June, 1858, to endeavor to put them into operation then. On account of the failure to receive money, as also on account of the revelations Forbes had made, the matter could not proceed. Nothing was to be done, or could be done, Brown said, until I had secured the correspondence to which I have alluded. I did not secure that correspondence, and therefore I supposed the matter could not go on.

The CHAIRMAN submitted to the witness a paper marked with the chairman's initials and indorsed "members of the convention," (produced by Andrew Hunter,) asking the following:

Question. Will you be good enough to state what knowledge you have of this paper on which your name appears?

Answer. That is my name, in my own writing. This paper is the one appended to the constitution. All of the persons signing this paper agreed to accept the constitution, and to devote themselves to the furtherance of the purposes for which the constitution was established. The name occurring first is the name of the president of the convention, William Charles Munroe. He was a mulatto. The next is G. I. Reynolds. I cannot remember him; he was not a white man, however. Then there is a name I cannot read; it looks like J. C. Grant; I do not remember him. There were a good many negroes there; and in a convention of two days it would be difficult to remember all their names. The next is A. J. Smith; I remember him as a Canadian negro. The next is James M. Jones; I do not know him; he was not a white man, however. Then comes the name of G. B. Gill, a white man, of whom I have already spoken. The next is M. F. Bailey, a negro. W. Lambert was a negro. S. Hinton was a negro. C. W. Moffett was one of our original party. J. J. Jackson I do not know; he must have been a negro. Then comes ——— Anderson, the christian name I cannot make out; he was the colored man of whom I spoke as having come with us from Canada. The next name is Alfred Whipper; I do not remember him. James M. Bell was a mulatto residing in Chatham. William H. Leeman was one of our original party. Alfred M. Ellsworth was a colored man living in Windsor, a village in Canada, opposite Detroit. John E. Cook and Stewart Taylor I have already spoken of as belonging to our company. Charles W. Purnell must have been a colored man. Then comes George Akins, his x mark; Akins was a negro. Robison Alexander was a negro. Then comes my own name, Richard Realf. Thomas F. Cary was a negro. Richard Richardson was the negro who accompanied us from Iowa. I taught him to write. L. F. Parsons was one of our company. Thomas M. Kinnard was a negro. M. H. Delany was the colored doctor of whom I spoke. Robert Van Vraiken must have been a negro. Thomas W. Stringer was a negro. Charles P. Tidd was a white man, one of our original party. John A. Thomas was a negro. C. Whipple is the next; that was the name by which Stevens was called. J. D. Shadd was editor of a paper in Canada—a mulatto. Robert Newman was a negro. Owen Brown was the son of John Brown. Then comes old John Brown's signature. J. H. Harris was a colored man. Charles Smith was a colored man. Simon Fisher was a colored man. Stephen Dutton was a colored man. Isaac Holden was a colored man. Giles Chitman was a negro. Thomas Hickerson was a colored man. John Launcel was a colored man, and so was James Smith. John H. Kagi, secretary of the convention, was one of the original party.

Mr. DAVIS. Do you know whether these negroes, or any part of them, were runaway negroes?

Answer. I have no knowledge as to that.

The CHAIRMAN exhibits to the witness a paper purporting to be a list of those men who were with Brown at Harper's Ferry, and asks this:

Question. Can you state the age of John Brown at that time?

Answer. No, sir, except that I suppose him to have been almost 60 years of age.

Question. What was the age of Owen Brown, as nearly as you can tell?

Answer. Owen Brown was about 29 or 30.

Question. Of Watson Brown?

Answer. Watson Brown was not one of the original party, and I never knew him.

Question. Of Oliver Brown?

Answer. I never knew him.

Question. Of Aaron D. Stevens?

Answer. I did not know Stevens's Christian name. His age was 28. He was 27 at the time of the convention.

Question. Of Albert Hazlett?

Answer. I never knew him.

Question. Of John H. Kagi?

Answer. Twenty-three at the time of the convention.

Question. Of Edwin Coppic?

Answer. I think I met him once or twice in Iowa, but never had any speaking acquaintance with him. He must have been about 18.

Question. Of Barclay Coppic?

Answer. I do not know him. He must have been a brother of the other, I suppose.

Question. May you not have confounded the two Coppics?

Answer. I may have done so.

Question. What was the age of Charles P. Tidd?

Answer. About twenty-five or twenty-six—near the age of Stevens.

Question. Are you speaking now of their ages at the time of the convention?

Answer. Yes, sir.

Question. What was the age of William H. Leeman?

Answer. Not more than eighteen at the time of the convention.

Question. Of Francis J. Meriam?

Answer. I never knew him.

Question. Of William Thompson?

Answer. I never knew him.

Question. Of Dolphin Thompson?

Answer. I never knew him.

Question. Of Jeremiah Anderson?

Answer. A stranger to me.

Question. Of Stewart Taylor?

Answer. About nineteen at the time of the convention.

Question. Of John E. Cook?

Answer. Probably between twenty-three and twenty-four at the time of the convention.

Question. Now, as to the negroes with John Brown. What was the age of Shields Green?

Answer. I never knew him.

Question. John Copeland?

Answer. A stranger to me.

Question. ——— Anderson?

Answer. That must have been the negro who accompanied us down from Chatham to Cleveland. He was about 24 or 25 years old.

Question. Newby?

Answer. I never knew him. Indeed, I knew no others, save those two negroes, Anderson and Richardson, who afterwards returned from Cleveland to Canada.

Question. The remaining negro with Brown was named Leary; did you know him?

Answer. I did not know him. I will give you my own age at that time. At the time of the convention I was not quite 24 years old.

Question. Can you state whether the signatures to the paper, which you say was appended to the constitution, are the original signatures of those who made them.

Answer. I saw the persons sign this document, and do testify thereto. In those cases where "his mark" follows the name, the mark was made by the person whose name appears, the writing having been done by Mr. Kagi.

Question. Were you present when the paper was signed?

Answer. I was.

Question. Was it signed before the convention dispersed?

Answer. Yes; before the convention dispersed, after the adoption of the constitution.

Question. You have spoken of three persons who you there learned from Brown had supplied him with money. Do you know of any other persons with whom Brown was in communication upon the subject of getting money?

Answer. I understood that a clergyman, whose name is Thomas Wentworth Higginson, who, I believe, resides at Worcester, Massachusetts, was an intimate friend of John Brown, and that he, as were these other men, was one of those who supplied him with funds to enable him to prosecute his movements in behalf of freedom in Kansas, and who had given him a general promise to assist him in whatever enterprises he might undertake.

Question. Can you recollect any others?

Answer. I cannot.

Question. Can you rememember the names of any persons, in any of the States, with whom Brown, during your acquaintance with him, was in correspondence?

Answer. No, sir. I do not believe that Brown was in correspondence with more than half-a-dozen people during my connection with him; for you must remember, that during our passage across Iowa, occupying a month, in which we camped out every night and walked across the plains every day, he could have no correspondence then. Immediately after we reached Springdale, in Iowa, he went on East. I could not be cognizant of his correspondence then, he being absent. Immediately on his return to Springdale, we departed for Canada, and on our passage thither we could not do anything in the way of correspondence. Just after we arrived there, the convention was held, and there was no chance for correspondence at that time. After the convention was disbanded, I left for New York city.

Mr. Collamer. But you had been to Cleveland.

The Witness. Yes; I went from Chatham to Cleveland, and from Cleveland to New York.

Mr. Doolittle. I understood the witness to state that he, in general terms, communicated to Arny and to Hyatt, of New York, what he supposed was the general purpose of Brown—to produce an insurrection, or do something upon the South somewhere.

The Witness. Yes, sir.

Mr. Doolittle. Then I ask you this question, are those the only two persons to whom you ever communicated any such thing, aside from those who went with you from Iowa to Canada, and those you met there?

Answer. No, sir; there is one other. His name is Charles Carroll Yeaton, a young gentleman, formerly a very intimate friend of mine, but not an abolitionist.

Mr. Doolittle. Where does he reside?

Answer. He resides now in New York. He was a junior partner in a banking and brokerage house in Wall street.

Mr. Doolittle. Did you ever communicate to any other person, or have any conversation with any other person, in relation to this programme, except those three?

Answer. No, sir; except as follows: during the time when we were in Iowa, and when it was thoroughly expected that, immediately on leaving Canada, we should go down into the South, I wrote a letter on some private matters to a lady, hinting therein that very probably she would hear of us again, and, perhaps, in the Southern States; but I never told her anything in regard to the plan. Those three persons are the only ones to whom I ever communicated anything about it.

Mr. FITCH. Did John Brown admit to you, or state to you, that Forbes was fully cognizant of his plans, as far as he had formed them?

Answer. Yes, sir; because Forbes at one period purposed to conduct the movement.

The CHAIRMAN. As you understood from Brown?

The WITNESS. As I understood from Brown; and you will permit me to say, that in any question of veracity arising between Forbes and Brown, I should, without hesitation, decide for Brown.

RICHARD REALF.

JANUARY 23, 1860.

WILLIAM H. D. CALLENDER sworn and examined:

By the CHAIRMAN:

Question. Will you state whether you were the cashier of the State Bank at Hartford, in Connecticut, in June, 1857?

Answer. Yes, sir; I was.

Question. Are you still the cashier?

Answer. I am.

Question. In my summons I asked you to bring us a statement of the account of John Brown, if he had one. Had he an account in your bank?

Answer. He had not.

Question. Had he the control of any funds in your bank in June, 1857?

Answer. No, sir.

Question. Had you any acquaintance with John Brown, personally?

Answer. I knew John Brown in 1846 and 1847; I boarded with him in Springfield, when he was connected with Simon Perkins, of Ohio, wool-dealer; Perkins & Brown was the firm.

Question. Did you know him in 1856 or 1857?

Answer. Yes, sir.

Question. Here is an order drawn by John Brown, dated the 22d of June, 1857, upon Colonel H. Forbes, at New York city, in these words—(by reading that you will probably know what we desire your testimony to)—"Sir, if you have drawn on W. H. D. Callender, Esq., cashier, at Hartford, Connecticut, for $600, or any part of that amount, and are not prepared to come on and join me at once, you will please pay over to Joseph Bryant, Esq., who is my agent, $600, or whatever amount you have so drawn." The indorsement on it is, "I did not present this to the colonel, as I presumed it would be of no use, and then he is, I am persuaded, acting in good faith. (Signed,) Joseph Bryant." Have you any knowledge of the $600 that Brown assumed the right to draw from this bank?

Answer. He drew upon me individually for that amount?
Question. Brown?
Answer. No, sir; Forbes. I had instructions from Mr. Brown to pay him $600; that was about the 1st of April, 1857; the two drafts I have with me.
The CHAIRMAN. Let us see them.
[The witness produced two drafts, which are identified by placing the initials of the chairman thereon. These drafts are in the following words and figures, respectively:

"No. —. $400. NEW YORK, *April* 27, 1857.
"At sight, pay to the order of Ketchum, Howe & Co., four hundred dollars, value received, and charge the same to account of
(Signed) HUGH FORBES.
"W. H. D. CALLENDER, Esq.
"*Hartford, Connecticut.*"

Indorsed: "Cr. our account,
KETCHUM, HOWE & Co."

"No. —. $200. NEW YORK, *April* 29, 1857.
"Pay to the order of Ketchum, Howe & Co., two hundred dollars, value received, and charge the same to account of
(Signed) HUGH FORBES.
"W. H. D. CALLENDER, Esq.,
"*Hartford, Connecticut.*"]

Question. Will you state by what authority Forbes drew these drafts on you?
Answer. On the strength of what Mr. Brown told me. He said that Mr. Forbes might draw upon me for $600; that was about the 1st of April, 1857; these drafts soon afterwards came on, and I paid them.
Question. Did Brown furnish you with the money to pay the drafts?
Answer. He furnished me, I think, with $400, which came from Springfield, Massachusetts.
Question. Will you tell us all about it?
Answer. I will give you the whole story: $400 was sent to me, I think, from Springfield, by Mr. Brown; I raised for him, I think, $600 in Hartford, or rather part of it was sent to me and part given by various parties in Hartford and its vicinity. Brown himself had the balance of the money after $600 was applied for these drafts. I handed it over to him.
Question. Do you say that Brown furnished you with $400?
Answer. He sent me $400 from Springfield, Massachusetts, towards paying for these drafts.
Question. Did he send you that before the drafts were drawn, or afterwards?
Answer. Before they were drawn.
Question. And then you raised for him $600 in Hartford?
Answer. About that, I should think.

Question. Will you state who contributed the money?

Answer. I cannot recollect; I have destroyed everything about it a long time ago; Brown came there and lectured there; he was therefore some time in that vicinity.

By Mr. COLLAMER:

Question. What time did Brown come there?

Answer. Not far from the 1st of April, 1857.

Question. Now, tell us what took place?

Answer. He stated to me that he had seen a great deal of trouble in Kansas, and was anxious to vindicate the rights of the free-State men, and he appealed to the sympathies of northern men to give him aid. I told him I had no objection to his using my name, and he published an appeal, in which my name was introduced, that contributions might be handed to me. He gave a lecture there, and also in Collinsville, I think.

By the CHAIRMAN:

Question. What was the subject of his lecture?

Answer. In regard to his troubles in Kansas and the troubles of the free-State men.

Question. And it resulted in contributions amounting to some $600?

Answer. About $600, if I recollect aright.

Question. Were they contributions collected at the time from the audience at the lecture, or contributions sent in subsequently?

Answer. They were sent in afterwards, some few days; I think during the week afterwards.

Question. Were they contributions brought to you by the contributors, or by Brown?

Answer. By the contributors.

Question. Can you give the names of any of the contributors?

Answer. I cannot at present. One I remember particularly, who was strongly enlisted in his behalf, James M. Bunce, who is now dead.

Question. It would be desirable to learn the names of any of those persons who contributed, if you can recollect them with safety to your memory, of course?

Answer. I cannot recollect them.

Question. Was no list kept of them?

Answer. No, sir; but it was all settled with Brown. Every paper of every kind was destroyed, and these two drafts were all that I retained in regard to the matter.

Question. The money was afterwards paid to Forbes upon these drafts?

Answer. Yes, sir.

Question. Were the payments ever questioned by Brown afterwards?

Answer. Not at all.

Question. Did you ever see Forbes?

Answer. I never saw him.

By Mr. COLLAMER:

Question. When Brown told you to answer Forbes's drafts, what did you tell Brown about knowing Forbes's handwriting?

Answer. I told him I wanted a signature to recognize it.
Question. Did he furnish you one?
Answer. Yes, sir.
Question. Have you got it now?
Answer. Yes, sir.
Mr. COLLAMER. Let us see it.
[The paper was produced.]
Mr. COLLAMER. This was a paper that Brown furnished to you to show the handwriting of Forbes, and you kept it for that purpose.
Answer. Yes, sir, I did; and filed it away with those drafts.

By the CHAIRMAN:

Question. Have you any further knowledge of any pecuniary or other matters with which John Brown was connected in your State or neighborhood?
Answer. Not at all, sir; nothing since the payment of these drafts.
Question. Do you know the firm in whose favor these drafts were made, Ketchum, Howe & Co., New York?
Answer. Yes, sir; they are private bankers in New York.
Question. Will you give the names of the persons composing the firm?
Answer. Morris Ketchum, Edmund G. Howe, and Thomas Belknap, jr.
Question. Where is their house in New York?
Answer. Exchange Place.

By Mr. DAVIS:

Question. I wish to know whether the application of Brown for funds, either in his lecture or otherwise, indicated any purpose than that of a political character in Kansas?
Answer. Not at all, sir. All his movements since then are something I know nothing about.

By the CHAIRMAN:

Question. Did you pay the balance of the money of which you speak to Brown himself, or to his order?
Answer. To Brown himself. I sympathized with him. I gave him a Sharp's rifle myself, and went to Colt's to get a pistol for him. I believe I remarked at the time he ought to use it with care. It was a very good gun. I recollect that distinctly. He said, "I shall not use it unless I am called on to defend myself."

W. H. D. CALLENDER.

BENJAMIN B. NEWTON sworn and examined.

By the CHAIRMAN:

Question. Will you please to state where you reside?
Answer. At St. Albans, Vermont.
Question. Were you at any time acquainted with John Brown, who was recently executed under the laws of Virginia for offenses against that State?
Answer. Yes, sir.

Question. State when and where you made his acquaintance?

Answer. In the city of New York, in January, 1857.

Question. Did you see him afterwards?

Answer. No, sir, not after that time; I saw him daily for perhaps ten days about that time.

Question. You did not see him subsequently in 1857, or subsequent years?

Answer. No, sir.

Question. Had you any communication with him by correspondence or otherwise?

Answer. No, sir.

Question. Do you know of any arms, consisting of rifles and revolving pistols, that were under his control in the year 1857 or 1858?

Answer. I know nothing about that, except what he told me at that time, and what I learned at that time; I suppose the same arms that he had control of at that time were those he had in 1858.

Question. State what Brown told you at that time?

Answer. Brown told me that he had some 200 Sharp's rifles, or had an order for some 200 Sharp's rifles, and some revolvers that were then at Tabor, Iowa.

Question. Did he state the number of revolvers?

Answer. No, sir; I do not remember that he stated accurately the number.

Question. When and where was this statement made to you?

Answer. In January, 1857, at New York.

Question. Did he tell you from whom he got the order?

Answer. Yes, sir.

Question. Who was it?

Answer. It was from the Massachusetts State Kansas committee.

Question. Do you know who was the organ of that committee, or who represented it in giving that order?

Answer. The secretary of the Massachusetts State committee was present, and told me that such an order had been given; I cannot say whether he gave it, or whether the chairman of the committee gave it.

Question. You mean that the secretary was present at the conversation with Brown?

Answer. That was another source of my information; he was not present at the conversation.

Question. State that, if you please?

Answer. The secretary of the Massachusetts State committee told me also that Mr. Brown had an order for those rifles.

Question. Will you give the name of that secretary?

Answer. Sanborn.

Question. His first name?

Answer. I do not remember his first name.

Question. Do you know where he resided?

Answer. Concord, Massachusetts.

Question. Will you state what was your business at that time in New York, in January, 1857? I want to know whether you came there as the member of any committee.

Answer. I went there as a member of the national Kansas committee, as we called ourselves; Governor Reeder's committee it was sometimes called.

Question. What did that committee consist of? How were they composed?

Answer. They were composed of one individual from each State; whether they were all full or not, I cannot say; they were not all present, of course.

Question. Each State of the United States, comprising all the States?

Answer. I do not know how that was? I suppose not, of course; I presume it was the Northern States.

By Mr. COLLAMER:

Question. Was that the same committee of which Arny was agent?

Answer. Yes, sir; he was agent of that committee.

By the CHAIRMAN:

Question. What was the object of your assembling in New York?

Answer. Our object was general; we had a great deal of business; one special object was to hear reports from our agents, and to consult about future action and effort; we had a good deal of business of one sort and another which was incomplete.

Question. What was the object of the committee? What business was intrusted to them? What were they created for?

Answer. Chiefly an organ of communication between the Northern States and Kansas—agents, so to speak, for the Northern States.

Question. Had these arms that Brown referred to been at any time under the control of that national committee?

Answer. No, sir.

Question. Do you know from what source they were derived; where they were purchased?

Answer. No, sir.

Question. Was there any member of this Massachusetts State committee in New York at that time, except the secretary, as far as you know?

Answer. No, sir. The secretary of the Massachusetts State committee was there in place of the member of the national committee from Massachusetts. The secretary of the Massachusetts State committee was not a member of our committee, except as taking the place of some other gentleman of that committee.

Question. He was, then, a constituent member of the Kansas committee from Massachusetts?

Answer. He was, at that time. Other States were represented in the same way. Ohio was, I remember.

Question. I think you stated that you did not know in what way these arms had been purchased?

Answer. No, sir.

Question. What led Brown to inform you that they were under his control?

Answer. He was asking aid of us, and of course he told us what means he had got in his hands; what encouragement he had to offer us to assist him.

Question. Was he asking aid in money or in arms, or in both?
Answer. In money and clothing.
Question. Was any money furnished him by your committee?
Answer. No, sir; none furnished, although a vote was taken that we would furnish him with a small amount; but I believe he never got it. I have been told so, at least.
Question. Is that committee of which you speak—the national committee—still in existence?
Answer. Yes, sir. I mean to say it is in existence, because at that time it contemplated a full and complete report, which has never yet been made. I suppose probably we shall have another meeting; but I do not know that we shall. At that time we thought we should.
Question. When was your last meeting of that committee?
Answer. That was the last one.
Question. In January, 1857?
Answer. Yes, sir.
Question. Can you state what amount of money was furnished by the constituents of that committee in the different States, and expended in any way in Kansas?
Answer. I cannot, accurately.
Question. Did you say you did not see John Brown after January, 1857, and had no communication with him?
Answer. Yes, sir.
Question. Have you any other information that would throw light upon the inquiries of this committee relating to Brown's operations in 1858 and 1859, what he was engaged in, or what his plans were during those years?
Answer. That is a general inquiry. I had no communication with him of any kind directly. I, of course, heard about him, and supposed I knew where he was during 1857. I was, during nearly the whole of that year, in Kansas myself. I heard of his being in Kansas, but did not happen to meet him. I left the Territory in June, 1858, and I heard nothing from him until I learned from the papers of his being at Harper's Ferry.
Question. Did you whilst in the Territory, or otherwise, at any time in 1857 or 1858, hear that Brown or any of his accomplices projected a plan of exciting insurrection amongst the slaves in the South?
Answer. No, sir. I did not hear of his being in the Territory but once. He came in for a very few days, and then left, and from the papers afterwards I learned that he entered the Territory several months after I left it.
Question. My question was confined to any plans of inciting insurrection in the Southern States. Did you hear of any plans by Brown or his accomplices to go into any of the slave States?
Answer. Not at all. He was before us for a specific purpose, and there was nothing said in regard to anything further.
Question. I mean while you were in Kansas, in 1858.
Answer. Yes, sir.

By Mr. COLLAMER:

Question. You say he applied to you for clothing. Did you furnish him any?

Answer. Yes, sir.

Question. Did you deliver it to him, or give him an order for it?

Answer. We gave him an order for it.

Question. Was it in Kansas?

Answer. A part of it was in Kansas, a part of it in Illinois. The river closed before it all reached the Territory.

Question. What did he state were the uses of the money and the clothing he wanted from you?

Answer. He wanted to provide a supply for a company. He wanted to organize a company.

Question. For what purpose?

Answer. To defend the people of Kansas against invasions.

By the CHAIRMAN:

Question. That was the object of his application to your committee for the aid he mentioned?

Answer. Yes, sir.

Question. And you promised him a supply of money, but did not afterwards furnish it, but furnished him a supply of clothing?

Answer. Yes, sir. The money was contingent on the ability of the committee to furnish it. The amount was small, but I heard he never received it, because we never had it to give.

Question. Can you tell the amount of clothing you furnished?

Answer. I cannot; but I presume he had all he wished, because we had a large supply of it.

Question. Where was the clothing at the time?

Answer. There was a very large amount of it in the State of Illinois, and a large amount had gone into the Territory, which had not been distributed. It went in very late. I do not know where he got his.

By Mr. FITCH:

Question. Did either Mr. Brown or Mr. Sanborn intimate any use to which those arms might probably be applied outside of the Territory?

Answer. Oh no, sir. This was three years ago, you will remember.

Mr. FITCH. But we have it in testimony here that he had this thing contemplated for many years.

The WITNESS. That might be. Mr. Brown did not tell us his secrets very much, if he had any. He was very reserved on such matters. He never intimated to us anything about it.

Question. Are you unable to say whether Brown had possession finally of those arms at Harper's Ferry by virtue of this transfer from Mr. Sanborn?

Answer. I did not see the order, but from the way he asked aid of us, I should presume he had them in such a way that he had control of them.

Question. Unlimited?

Answer. Yes, sir. Mr. Brown was not willing to be under Jim

Lane, or Governor Robinson, or anybody. He would not take anything if it was conditioned. He said to us: "You know me, and I know you; I should like aid; I have been in Kansas, and was poor, and not able to aid myself or any one else, and I want aid; but I do not want it if I am to go here and there, where anybody orders me to go;" and I should infer, of course, although I did not see the order, that it was in such a form that he would have control of them. Indeed, as committees, the money was raised in such a way that when we had given it to an individual or agent in Kansas, it passed out of our hands, and our responsibility ended. It was given for a particular purpose; it must be intrusted to somebody, and when placed in the hands of an agent our responsibility in regard to it was ended, as we thought.

BENJAMIN B. NEWTON.

CHARLES BLAIR sworn and examined.

By the CHAIRMAN:

Question. Will you state where you reside, and what is your occupation?

Answer. I reside at Collinsville, Connecticut; I am a blacksmith by trade—a forger.

Question. Did you know the late John Brown who was recently executed under the laws of Virginia?

Answer. I did.

Question. Will you state when you made his acquaintance, and under what circumstances?

Answer. I made his acquaintance in the early part of 1857, if I mistake not, in the latter part of February or fore part of March. He came to our place, Collinsville, as I supposed, to visit connections who lived in our town. He himself was born, as I have understood, in Torringford, ten miles from there, and some of his relatives lived in a town five miles from our village. He spoke in a public hall one evening—perhaps by invitation of some of the community, but I do not know how that was—and gave an account of some of his experience in Kansas, and at the close of the meeting made an appeal to the audience. After stating the wants of many of the free settlers in Kansas, their privations and need of clothing, &c., he made an appeal for aid for the purpose of furnishing them the necessaries of life, as he declared. I think there was no collection taken up for him at that time. I do not know that I spoke with him that night, but on the following morning, if I mistake not, he was exhibiting to a number of gentlemen who happened to be collected together in a druggist's store some weapons which he claimed to have taken from Captain Pate in Kansas. Among them was a two-edged dirk, with a blade about eight inches long, and he remarked that if he had a lot of those things to attach to poles about six feet long, they would be a capital weapon of defense for the settlers of Kansas to keep in their log cabins to defend themselves against any sudden attack that might be made on them. He turned to me, knowing, I suppose, that I was engaged in edge-tool making, and asked me what I would make them for; what it would cost to make five hundred or a thousand of those things, as he described them. I replied, without much consideration, that I would make him five hundred of them for a dollar and a quarter a piece; or if he wanted a thousand of them, I thought they might be made for a dollar a piece. I did not wish to commit myself then and there without further investigation, but it was my impression that they might be made for a dollar a piece. He simply remarked that he would want them made. I thought no more about it until a few days afterwards. I did not really suppose he meant it then. I will endeavor to state the circumstances as correctly as I can, though three years have transpired, and I may find it necessary to refer to some of his letters to quicken my memory in regard to the matter. I have several of his letters with me here. I think he went to Springfield, Massachusetts, before a bargain was made between us; at any rate, the result was that I made a contract with him. From the tenor of this letter, [producing a letter from John Brown to Charles Blair, dated Springfield, Massachusetts, 23d March, 1857,] I think, he ordered me to make a dozen as samples, and I had forwarded them to Springfield before receiving this letter?

Question. Will you be good enough to look at that weapon in the corner of the room [referring to the pike produced and identified by A. M. Kitzmiller] and see whether that is according to the sample that you furnished?

Answer. That is nearly like it. The first dozen that I made as samples had wrought-iron ferules, rivetted through and blacked. When he came to make the contract, he wrote it to have malleable ferules cast solid, and a guard to be of malleable iron. That was all the difference.

Question. Was your contract to furnish the handles as well as the weapons?

Answer. Yes, sir.

Question. Did you, in your samples, furnish handles as well as weapons?

Answer. Yes, sir. After seeing the sample he made a slight alteration. One was to have a screw put in, as the one here has, so that they could be unshipped in case of necessity. To go back a little; when it became apparent to me that he was in earnest about having them made, I began to demur a little, doubting whether he was able to pay me, and I said to him, "Mr. Brown, I am a laboring man, and, if I engage in this contract with you, I shall want to know how I am going to get my pay." He said, "That is all right. It is just that you should, and I will make it perfectly secure to you; I will give you one half the money, that is $500, within ten days; I will pay you the balance within thirty days, and give you ninety days to complete the contract." That would carry it to somewhere near the 1st of July, 1857. Before making any move in the matter, I waited to receive the first installment.

Mr. Collamer. Was there a written contract?

Answer. Yes, sir; he drew up a contract in writing himself.

The Chairman. Have you got that contract?

Answer. I have not. It was a short contract, written on half a page of paper, perhaps; simply stating what the terms of it were.

Question. How many was the contract for?

Answer. A thousand.

Question. At what price?

Answer. One dollar each; pretty good, stiff pay; and hence, when I made the offer to make them for a dollar, it occurred to me as a matter of course that he would demur to the price, and it would fall through. He paid me $350 within ten days. This advancement was made in the latter part of March, 1857. I then went and purchased my materials. I went to a handle-maker in Massachusetts and engaged him to make a thousand handles. I purchased the steel for the blades and set a man forging them out, and he forged out perhaps five hundred of them. In the beginning of April I received another letter from him, stating that he was then unable to pay the balance of the money; that he had not the funds, but hoped to have them soon. [Letter produced, dated Springfield, Massachusetts, April 2, 1857, addressed by John Brown to Charles Blair.] Soon afterwards I received another letter sending me a draft for $200, making altogether $550, fifty dollars more than he promised to give me as the first instalment. [Letter produced, dated Springfield, Massachusetts, April 25, 1857, addressed by John Brown to Charles Blair.]

Question. This letter says, "If you do not hurry out but 500 of those articles it may, perhaps, be as well, until you hear again;" did you construe that as a revocation of the order for the remaining 500?

Answer. I did not. The thirty days, I think, must have expired at the time that letter was written; and it alludes to the fact that $200 did not come until after the expiration of the first ten days. He explained to me in a letter, which I have lost, why the $200 of the $550 had not been paid me until after the expiration of the ten days. The last time I saw him before that, he inquired of me whether he could get two or three heavy wagons built in that vicinity, to be done in a short time, and I remarked to him that I had a friend who was engaged in the manufacture of heavy wagons, who lived at Colebrook, and that if he chose I would write to him and see if he could furnish any. I had done so, but this man, whose name was Parsons, wrote me that he could not furnish them in the time required, and of course nothing further was done about it. That explains the allusion in the letter to my Colebrook friend. Shortly afterwards, in May, I received a letter from Brown, saying that I need not hurry out the first 500 until the handles were properly seasoned, nor the remainder till I heard from him. [Letter produced, dated Cannistota, New York, May 14, 1857, addressed by John Brown to Charles Blair.] I at that time contemplated a journey into Iowa. About the time he left our place he said to me that he was going back to Kansas. I told him I had never made a journey west, and that I contemplated going into Iowa, and should be happy of his company. That explains part of his letter. In regard to the rest of it, the handles were in a green state, and I wrote him that unless they were seasoned, when the blades came to be put in, they would shrink away and all become loose; and if he was not in any particular hurry he had better let them remain and become seasoned. I worked on perhaps until several days after the expiration of the thirty days in which the second installment was to come, but, receiving no further funds from Mr. Brown, I stopped the thing right where it was, determining that I would not run any risk in the matter. I just laid it aside, and there it lay, the work in an unfinished state, the handles stored away in the store-house, the steel which I had purchased stored away in boxes, the few blades which I had forged were laid away. Thus it was until last June; nothing more was done.

Question. Did you hear anything more from Brown?

Answer. I will read to you all the letters I received from him during the time; that is, all I have preserved, and I think they embrace all I received.

Mr. FITCH. Had you in the mean time sent him the five hundred?

Answer. Not one; I never finished any of them. It is possible that I have lost two or three of the letters; but, if I mistake not, the next letter I received from him was dated at Rochester, in February, 1858, nearly a year after the contract I made with him. [Letter produced, dated Rochester, New York, February 10, 1858, addressed by John Brown to Charles Blair.] What he meant by saying in this letter that he was again in the United States I did not know, for I did not know that he had been out; but, since the Harper's Ferry affair, I have learned what that means. In answer to that, I wrote to him—I have no copy of that letter, and I must give you my best impression of it—that immediately after the expiration of the thirty days I dropped the thing; that I had never finished any of the articles, and, of course, had none to forward; that I considered the contract at an end, and had other business to attend to. That was the substance of my letter. After that, I received another letter from him, dated at Philadelphia. [Letter produced, signed John Brown, and addressed to Charles Blair, dated Philadelphia, Pennsylvania, March 11, 1858.] Nothing more was heard from Mr. Brown at all, in any way or shape, until on the 3d day of June, 1859, the old man appeared at my door, unexpectedly of course, and said to me, "I have been unable, sir, to fulfill my contract with you up to this time; I have met with various disappointments; now I am able to do so." I say it was the 3d day of June, because the receipt that I gave him, which I presume you have, bears date June 4. That is the only thing I have to remind me of the date.

The CHAIRMAN exhibits to the witness (from among the papers proved by Andrew Hunter as having been produced at John Brown's trial) a paper in these words: "Received, Collinsville, June 4, 1859, of John Brown, on contract of 1857, $150. Charles Blair." And asks: Is that the paper to which you refer?

Answer. Yes, sir. The evening before, he came in by the train, made his appearance about six o'clock, and stated to me that he was now able to fulfill his contract with me. I remarked, "Mr. Brown, the contract I consider forfeited, and I am differently situated from what I was then; it will be exceedingly inconvenient for me to do any more with it; I have business now of a different kind; my men are fully employed on other work; and I do not see how I can do it." "Well," said he, "I want to make you perfectly good in this matter, I do not want you to lose a cent." I said "I shall not lose anything; I was careful in the first place not to exceed the amount of money I had in my hands, and I shall lose nothing if I drop it right here.'' I said to him, however, that he might take the steel and the handles just as they were, and I would pass receipts with him. "No," said he, "I do not want to do that; they are not good for anything as they are." At that point I remarked, "What good can they be if they are finished; Kansas matters are all settled, and of what earthly use can they be to you now?" "Well," he replied, "that they might be of some use if they were finished up, that he could dispose of them in some way, but as they were, they were good for nothing." I then said to him, "I will receive of you the remaining $450, if you have it and wish to pay it to me, and if I can find a man anywhere in the vicinity that is accustomed to doing such work who will finish up the work, I will do so, provided I can do it and come within the means, and it will not be much trouble to me, because I am very busy and have not time to attend to it; but in case I do not succeed in finding a man to do it, I will refund you this $450." Said he, "That is all right, and I will agree to it." A short conversation passed on that day, and he left me with that understanding, but paid me no money then. He went to the hotel and stayed over night, and in the morning, about seven o'clock, he came again and told me that he was about to start for New York, and that he would pay me $150 then, and would send me from New York on the following day, or from Troy, within a day or two from that time, $300 more. I said "very well." He took out his pocket book and paid me fifty dollars in bills and a one hundred dollar check, and I gave him the receipt which has been shown; I scratched it off in a hurry. He hurried to the cars and went off, as I supposed, for New York. A few days after that, four or five perhaps, I received a letter inclosing a draft for $300. [Letter produced, dated Troy, New York, 7th June, 1859, addressed to Charles Blair by John Brown.] The letter that I wrote in answer to that, has appeared in the public prints, and I presume you have it.

The CHAIRMAN exhibits to the witness a letter dated Collinsville, Connecticut, June 10, 1859, addressed to "Friend Brown," and signed Charles Blair, being one of the papers proved by Andrew Hunter, as having been produced at John Brown's trial, and asks: Is that it?

Answer. That is the reply I made to that letter. In regard to the time, I think he spoke to me something about liking to have them finished up as soon as possible, and that was the reason of my saying that man could not finish them up any sooner. In the month of July, I was absent on a business tour at the West, and during my absence a letter was received from John Brown, requesting me, when those goods were finished up—if I remember aright, the term "goods" was used, as in all his writing—to forward them to Chambersburg, in Pennsylvania, to J. Smith & Sons, at the same time requesting me to give him the price of axes, hatchets, broadaxes, and picks. That letter my son received, and I have not got it with me. My son replied to it, in my absence, telling him where he could find the price of those articles, which we were making; that I was absent, and probably, when I got home, I would write him. It is that letter, I presume, which caused the subpena for me to be directed to "Charles H. Blair, *alias* Charles Blair." Charles H. Blair is my son. Soon after I arrived home, I received a letter, in an entirely different handwriting, from Chambersburg. [Letter produced, signed J. Smith & Sons, addressed to Charles Blair, and dated Chambersburg, Franklin county, Pennsylvania, August 24, 1859.] A few days subsequent to that, I received another letter from J. Smith & Sons, requesting me to forward the "freight," when ready. In my reply, which I have also seen in the papers, I made use of the term "freight" because they had used the term, and said it had not been forwarded, but would be in a few days.

The CHAIRMAN exhibits to the witness a letter, dated Collinsville, Connecticut, August 27, 1859, addressed to Messrs. J. Smith & Sons, and signed Charles Blair, being one of the papers proved by Andrew Hunter as having been produced at John Brown's trial, and asks: Is that your reply, of which you now speak, to the letter signed J. Smith & Sons?

Answer. Yes, sir; that is my reply to that letter. I do not know that I said, if I did not I will here say, that I went out of town and got a man by the name of Hart to finish up this work for me. Mr. Hart was an acquaintance of mine, whom I had formerly known, and I knew him to be engaged in edge-tool manufacturing, a competent man to do it, and I submitted the whole thing to him. I received one other letter, which I cannot find, before a letter dated September 15, which I shall presently produce, simply saying to me that, when I sent the goods to Mr. Brown, I should send them to the care of Oakes & Cauffman. I presume that, when I marked the goods, I left that letter at the manufactory of the man who finished them. When they were done, I saw that the blades were tied up in boxes, and the handles in bundles. I simply marked them according to the directions. Then the next letter I received was dated September 15, acknowledging the receipt of the goods. [Letter produced, dated Chambersburg, Pennsylvania, Thursday, September 15, 1859, addressed to Charles H. Blair, and signed J. Smith & Sons.] That is the whole story, I believe.

Question. Will you state of what wood those handles were made?

Answer. Ash timber.

Question. Was that Brown's selection or yours?

Answer. My impression is that it was his selection. It is a common timber that we use for fork handles. In the course of the conversation I had with him, he spoke of the handle being made like a fork handle, about the size of a hay-fork handle, and of the same material.

Question. Did he prescribe the length?

Answer. The contract, I believe, was that they were to be six feet or six feet and a half long. I am not positive which.

Question. And the form of the weapon he showed from a weapon that he alleged he had taken from Pate, but to be accommodated to that fashioned pike?

Answer. Yes, sir; and I will show you the difference. The dirk had a ridge in the middle and was beveled each way, and was not as wide as this by about one fourth. His direction was to have it made two inches wide, if I mistake not, and a trifle longer than the blade he showed me. That had a guard shorter than this, and had a neat handle. It was an expensive weapon.

Question. Can you get a copy of that contract or the contract itself?

Answer. No, sir; I could not lay my hands on it when I came away; it has been lost.

Question. Did the contract prescribe minutely the mode and fashion and material of which the weapon was to be made?

Answer. It did not describe the blade, but simply that the ferules and guard were to be made of solid malleable iron and a screw through the shank and the ferules; that, I believe, was the description; it described the length of the handle.

Question. What is the blade there made of?

Answer. Of cast steel.

Question. You said they were put in boxes—by whose direction were they put up separately from the handles when they were sent on?

Answer. By Brown's direction, in one of the letters I read to you; they were tied up in bundles of about twenty or twenty-five in a bundle.

Question. Do you recollect in the address that you gave them to J. Smith & Sons to the care of Oakes & Cauffman, whether they were described as fork-handles?

Answer. They were marked fork-handles; I do not think that was Brown's direction; it was my own; I did not know what else to call them. They were properly fork-handles, and I so marked them.

Question. How could you say they were properly fork-handles when they were intended for a weapon?

Answer. Because they were just about the length and size; that is all.

By Mr. Collamer:

Question. Did you go to a fork-handle maker to get them?

Answer. Yes, sir; I ought to say, perhaps, that they are rather smaller than he ordered. They are much smaller than they were when they were green. They have been made three years and they have shrunk some.

By the Chairman:

Question. Was the whole affair furnished by that man or yourself, including the screw which connects the handle and the blade?

Answer. We furnished them all, although the ferules and the screw were made at New Haven. The malleable iron was made by a firm in New Haven.

Question. But the whole thing was furnished, so that nothing was required but to put them together?

Answer. Yes, sir; that was in accordance with the contract.

Question. What was the whole number furnished, did you say?

Answer. The contract was for a thousand, but I think there were nine hundred and fifty-four sent.

Question. Were they all sent at one shipment?

Answer. Yes, sir; all at once.

Question. I think you said it was in June, 1859, that he came back to your place of business?

Answer. Yes, sir.

Question. Did he tell you nothing in reference to what use he proposed then to make of them, except what you have already spoken of?

Answer. Not another syllable; I have stated precisely the language that he used: "I think that they might be useful if finished up, but they were good for nothing as they were." That, I think, is all he said about them. The idea I got, when he first spoke of them, was that he was going to sell them to the people in Kansas, and I think he made use of this expression, that he wanted them for the poor settlers in Kansas who were not able to purchase fire-arms; that they needed some weapon of defense to keep in their cabins, and such a thing would be useful to them.

Question. When you were required afterwards to send them to Chambersburg, Pennsylvania, did any conversation arise as to the reason of their change of destiny?

Answer. Not at all; I got the impression that they were on their way to Ohio or to the West; I never know where Chambersburg was at all, and having always had in my mind the idea that they were first originally destined for the West, I did not know but that he might send them, and that Oakes & Cauffman, as I supposed, were forwarding merchants, and J. Smith & Sons, I supposed, were a *bona fide* firm. Since further developments have come out, it appears who J. Smith & Sons were, but I certainly knew nothing about it at that time.

By Mr. Collamer:

Question. You say Mr. Brown made you these payments? Did he make them in money?

Answer. Part was in money and part was in a draft or check.

Question. In 1857, it would seem, he sent you a check on New York?

Answer. A draft on New York for $200.

Question. Was that of his own drawing? Did he sign it?

Answer. I think not; I think it was a draft drawn by one of the banks in Springfield on a bank in New York, payable to bearer, or it might have been payable to Brown, I cannot remember that; that draft came to me through a man by the name of Rust, living in the same town where I am. In writing to him Brown inclosed this draft, and requested him to hand it to me.

Question. Now, when you come to 1859, and what he sent you from Troy, what did he send you?

Answer. A draft, according to my statement, for $300. If I remember right, I cannot say positively, but it is my impression, it was a draft drawn by the cashier of one of the banks of Troy, payable to me.

Question. Did you receive from him any other checks or drafts of any kind towards these payments?

Answer. When I received the $150, for which I gave him that receipt, dated the 4th of June, 1859, he gave me, as part of that, one check drawn by Gerritt Smith for $100. The rest was in bank bills of the Springfield or Boston banks. I cannot say which.

Question. Where was that check of Gerritt Smith upon?

Answer. Upon one of the Albany banks, if I remember aright. I think it was a check made payable to John Brown, or bearer, or perhaps Brown's name was not contained in it; but I remember it distinctly, because it occurred to me at once that Gerritt Smith was a prominent man, here was his check for $100, and I supposed him to be good for it. I was inclined to be more particular about the check than I was about the drafts. I knew the drafts must be good, having been drawn by the cashier of a bank.

By the CHAIRMAN:

Question. The money was all received on them; they were all paid?

Answer. Yes, sir; as far as I know. I never heard anything from them. They were checked for me by Mr. Norton, who is treasurer of our saving's bank.

CHARLES BLAIR.

JANUARY 26 1860.

JAMES JACKSON affirmed and examined.

By the CHAIRMAN:

Question. Will you state where you reside?

Answer. In Boston.

Question. Were you acquainted with a man named Francis J. Meriam, of Boston?

Answer. I was.

Question. Are you connected with him?

Answer. I am.

Question. State what the connection is, if you please?

Answer. I am his uncle.

Question. Where is Francis J. Meriam's residence?

Answer. In Boston. Do you mean his present residence?

The CHAIRMAN. I mean his usual place of abode.

Answer. In Boston.

Question. Will you state whether, on the 14th day of last October, or on what day about that time, you sent to him, then at Baltimore, from Boston, a sum of money in gold or otherwise?

Answer. I cannot give the date.

The CHAIRMAN. As near to it as you can come.

Answer. I have no idea hardly of the date.

Question. Can you give the month?

Answer. I think it was in October.

Question. Cannot you speak of the probable time in reference to the outbreak at Harper's Ferry?

Answer. I have no distinct recollection of the date; hardly of the month. I think it was October.

Mr. DOOLITTLE. You can state whether it was before or after the outbreak.

Answer. It was before.

By the CHAIRMAN:

Question. State any facts within your knowledge in reference to you or any body else sending to him a sum of money from Boston by Adam's express, to Baltimore, at the time you speak of; state the facts?

Answer. Mr. Meriam sent a telegraphic dispatch to me to send on $600 in gold. I sent it by the express.

Question. Addressed to him, where?

Answer. I think it was at Baltimore.

Question. Will you state where that money came from, whence it was derived?

Answer. I raised the money. It was his own money.

Question. How did you raise it?

Answer. I think I got it from my father; he handing me a check on one of the Boston banks for the amount.

Question. How did you obtain it from your father? If the money belonged to Meriam, what need had you to resort to your father for it?

Answer. I think some property of Mr. Meriam's came into my hands, and I think I lent a portion of it to my father, and when this telegraphic dispatch came I called on him for the amount.

Question. How long before you sent that money had you seen Meriam?

Answer. I think it was two weeks; thereabouts.

Question. Where did you see him then?

Answer. In Boston.

Question. Did he tell you that he would have occasion for that sum, or any other sum of money in a short time?

Answer. He did not.

Question. Had you any knowledge of his wanting the money until you received the telegraph?

Answer. None.

Question. Did you know where he was before he sent you the telegraph?

Answer. I cannot say, positively.

Question. Give us your nearest recollection.

Answer. I think I did.

Question. State it, if you please.

Answer. I think that he was at Chambersburg.

Question. Did you know that John Brown, who was afterwards hung in Jefferson county, Virginia, was in the neighborhood of Chambersburg?

Answer. I did not.

Question. Did you know John Brown?

Answer. I did not.

Question. Did you never see him?

Answer. I never saw him.

Question. Did you know that Meriam had any connection with him in any way?

Answer. I did not.

Question. Did you never hear Meriam speak of him?

Answer. I think not.

Question. Do you know what use he made of the money after he got it?

Answer. I do not.
Question. Have you seen Meriam since?
Answer. Yes.
Question. Where did you see him?
Answer. In Boston.
Question. When?
Answer. Perhaps six weeks since—from four to six.
Question. Did he tell you what use he made of the money when he got it?
Answer. I think not.
Question. Was any reference made to your having sent him the money?
Answer. I do not remember.
Question. Did you obtain any receipt for it?
The WITNESS. From him?
The CHAIRMAN. From him.
Answer. I did not.
Question. Have you no evidence that you sent it at his direction?
Answer. Not from him.
Question. Then from whom have you it?
Answer. From the express office.
Question. What did you get from the express office as a voucher for the payment?
Answer. A receipt.
Question. Was the money in gold?
Answer. In gold.
Question. How was it put up?
Answer. I think it was taken from me at the express office loose.
Question. Do you know a man named Lewis Hayden, in Boston?
Answer. I do.
Question. Is he white or black?
Answer. Black.
Question. What is his business there?
Answer. I think his business is in the State House. I think he is one of the runners in the secretary of state's office; but I am not sure.
Question. Was there any acquaintance between him and Francis J. Meriam that you know of?
Answer. Yes.
Question. Are you aware that Hayden had control of any money belonging to Meriam?
Answer. I think I am.
Question. Give us the facts—your knowledge of the matter.
Answer. I understood that he had a small amount of money—five dollars, I think.
Question. You heard that he had a small amount of money, five dollars; whom did you learn that from?
Answer. I think from Meriam.
Question. Did you hear that from Meriam after his return, or before he went away?
Answer. I think afterwards.

Question. Do you know of Hayden's receiving a telegraph from Meriam shortly after you sent him that $600?

Answer. I do not remember.

Question. If you had heard shortly after you sent that money, about the time of the outbreak at Harper's Ferry, that Hayden had received a telegraph from him, would you recollect it?

Answer. I think I should.

Question. Did not Hayden tell you that he had received a telegraph from Meriam about money?

Answer. He did not.

Question. Do you know of any other persons in Boston who had the control or possession of money belonging to Meriam, except Hayden?

Answer. I do not remember of any.

Question. Did you never hear of Meriam's having any connection with the John Brown who has been spoken of?

Answer. Never.

Question. When Meriam came back to Boston, did he tell you what use he had made of the money you had sent him?

Answer. I think not.

Question. How long did he remain in Boston after he came back?

Answer. Two or three days, perhaps.

Question. Where did he stay during that time?

Answer. He stayed at Dr. Thayer's.

Question. What is Thayer's first name?

Answer. I think it is David.

Question. In what part of Boston does he reside?

Answer. In Beach street.

Question. What is his profession?

Answer. A physician.

Question. Is Meriam related to him?

Answer. Not at all.

Question. Did Meriam keep himself concealed when he came there?

Answer. Yes.

Question. Did you know his reason for it?

Answer. I supposed he feared being arrested.

Question. Why did you suppose so?

Answer. From his keeping himself concealed.

Question. What was he afraid of being arrested for; what had he done?

Answer. He had participated in the Harper's Ferry matter.

Question. How did you learn that?

Answer. From his statement.

Question. Well, now, give us his statement if you please; tell us what he told you, and say when it was?

Answer. During the time I have spoken of, when he was at Dr. Thayer's, he told me he was, with a few others, stationed at some distance from Harper's Ferry to guard arms that were there.

Question. What arms?

Answer. Arms that were to be used, if necessary, I suppose.

Mr. Davis. Perhaps the witness does not understand that he is permitted to go on and tell his narrative.

The CHAIRMAN. [To the witness.] I ask you to give the whole statement he made to you.

The WITNESS. What I have just stated is about all that I remember of any consequence, bearing directly on the point.

Question. Did you have any communication with him whilst he was at Chambersburg?

Answer. I think I did.

Question. State what it was, if you please?

Answer. It was in reference to the money that he requested me to send him. He requested me to send it immediately. I told him that when he ordered money, he must give me a little time to raise it for him. That was about the substance of my letter to him.

Mr. DOOLITTLE. Was the request to you by letter, from him at Chambersburg?

Answer. Not from him; but when I sent the letter, I requested him to give me a little time to raise it.

The CHAIRMAN. You said that he ordered the money by telegraph; was that telegraph sent to you from Chambersburg?

The WITNESS. I think, from Baltimore.

The CHAIRMAN. I understood you to say that you had some communication with him whilst he was at Chambersburg in reference to money?

The WITNESS. I should have said at Baltimore.

Question. Did you have any communication with him whilst he was at Chambersburg?

Answer. I do not remember that I did.

Question. Then how did you know that he had been at Chambersburg?

Answer. I think he wrote a letter to me when he was at Chambersburg.

Question. What were the contents of the letter?

Answer. He said that he should not return for, perhaps, some time; that we need not look for him to return home soon. That was about the substance of the letter.

Question. Did he tell you what he was doing there, what took him there?

Answer. Not at all.

Question. How long had he left Boston before he sent to you for that money from Baltimore?

Answer. Perhaps two weeks. These matters I do not remember distinctly the exact dates of.

Question. Did you say that you had no knowledge of his being in any way connected with Brown?

Answer. I did say so.

Question. Had you any knowledge of Brown's plans or purpose to make an attack on Harper's Ferry?

Answer. Nothing.

Question. Did you know of any contributions being taken up for Brown, in Boston, during that year, 1859?

Answer. I did not.

Question. What is Meriam's business or pursuit in life?

Answer. He has not been in any business for some five or six years? He has been in Europe part of the time.

Question. What is his age?

Answer. Twenty-two.

Question. Are his parents living?

Answer. His mother is.

Question. Where does she live?

Answer. In Boston.

Question. You say that Meriam left Boston from four to six weeks ago?

Answer. I think it was thereabouts, perhaps less.

Question. Where did he go when he left Boston?

Answer. He went to Canada.

Question. To what part of Canada.

Answer. I do not remember.

Question. Do you know where he is now?

Answer. I do not.

Question. Have you heard anything from him whilst he has been in Canada?

Answer. I have.

Question. What did you hear?

Answer. He asked me to send him money there.

Question. Did he tell you what his plans were—where he was going to, what he was going to do?

Answer. He did not.

Question. Did you send him any money?

Answer. I did.

Question. How much?

Answer. One hundred dollars.

Question. What is the value of his whole property; what are his means in life?

Answer. Six or eight hundred dollars.

Question. Do you mean that that is what he is worth now?

Answer. At present.

Question. Have you any knowledge of this negro, Lewis Hayden, collecting any money for Meriam?

Answer. Not any.

Question. Any knowledge of his money transactions with Meriam, except what you have spoken of?

Answer. That is all.

Question. No knowledge of Meriam's telegraphing to him about the time you sent the money?

Answer. Not any.

Question. Did you know from Meriam of any persons who were advising or counseling or aiding Meriam in his connection with Brown?

Answer. He gave me no names, of any one that I knew of, who advised him. He gave me the names of some who were with him.

Question. With him where?

Answer. Wherever he was stationed.

Question. Some of those who were with him of Brown's party, do you mean?

Answer. Of Brown's party.

Question. Have you any knowledge, whether derived from Meriam or otherwise, of any persons in Boston, or anywhere in New England, who were in communication or advising Meriam as to this Brown affair?

Answer. No.

Question. Or who was aiding Meriam in any way?

Answer. No one.

Question. Do you know of anybody who gave any money, directly or indirectly, to Brown?

Answer. No one.

By Mr. COLLAMER:

Question. How long has that young man's father been dead? I suppose his property came from his father?

Answer. It came from his father, who has been dead ten years, or perhaps fifteen years.

By Mr. DOOLITTLE:

Question. You say he had been in Europe. How long was he in Europe?

Answer. About a year.

Question. What business was he brought up to—any profession or business?

Answer. I think in the English goods business.

Question. What is your business or profession?

Answer. I am a florist.

Question. Have you acted as attorney, or agent, or guardian for him, or anything of that kind?

Answer. I have.

By the CHAIRMAN:

Question. What were you—his guardian, appointed by law?

Answer. Appointed by law.

Question. Did he not inform you of what his pursuits were, or what he was engaged in?

Answer. He did not.

By Mr. DOOLITTLE;

Question. As I understood you, the six hundred dollars was the greater part, or pretty much all of the money that now belongs to him?

Answer. That did belong to him at that time.

Question. Has he property in expectancy?

Answer. I think not.

By the CHAIRMAN:

Question. Do you know a man named James Redpath?

Answer. I do.

Question. Do you know of Meriam having any acquaintance or connection with him?

Answer. I do.

Question. What was their acquaintance and connection?

Answer. He went off with Redpath, I think, a year since, perhaps more, to Hayti to learn the condition of things there. He went with him as interpreter of the French language.

Question. Was he paid by Mr. Redpath for the service?

Answer. I do not know.

Question. What condition of things did he want to learn there? Do you know what interested him in Hayti?

Answer. Principally the condition of the colored population.

Question. The condition of the negroes?

Answer. Yes, sir.

Question. What interest had he in learning the condition of the negroes? What was his object, so far as you derived it from him?

Answer. I presume his object was to see how it bore in relation to the character of the negroes in this country.

Question. Do you know where Redpath is now?

Answer. I do not know.

Question. Have you seen him latterly?

Answer. I have.

Question. When?

Answer. I saw him five or six days since.

Question. Where did you see him then?

Answer. In Boston.

Question. What part of Boston was he in—in the streets or in a house?

Answer. He was in the Emigrant Aid rooms.

Question. The room of the Emigrant Aid Society?

Answer. Yes, sir.

Question. Where is that room?

Answer. In Winter street.

Question. What sort of a building is it in—a dwelling house or a public house?

Answer. A public house.

Question. Do you mean a hotel or tavern?

Answer. No.

Question. What is the purpose of the building?

Answer. Various purposes. It is a large building.

Question. Was it night or day when you saw him?

Answer. Daytime.

Question. Do you know whether he is in Boston now?

Answer. I do not.

The CHAIRMAN exhibits to the witness from among the papers proved by Andrew Hunter to have been produced at John Brown's trial, a letter, dated Boston, December 23, 1858, signed Francis J. Meriam, and asks: Is this letter in Mr. Meriam's writing?

Answer. I should think it was.

Question. The whole body of it, as well as the signature?

Answer. I should think it was; it looks like his.

Question. When did you make Redpath's acquaintance?
Answer. Just previous to their leaving for Hayti.
Question. Was this man Meriam at any time in Kansas, that you know?
Answer. I think he was not.
Question. Are you aware of his having been in Kansas in the winter of 1858–59?
Answer. I never knew him to be in Kansas.
Question. Did you never hear of his having been there?
Answer. Never.
The CHAIRMAN. I asked the question because this letter seems to imply that he had been there.

By Mr. FITCH:

Question. Who was present at the Emigrant Aid Society's room, besides Redpath, on the occasion when you recently saw him there?
Answer. I do not remember.
The CHAIRMAN. You were there.
The WITNESS. I was there.
Mr. FITCH. What was the character of the business which convened those who were there? What was the object of the meeting?
Answer. I merely called in to see Redpath on a matter of business.

By the CHAIRMAN:

Question. How did you know Redpath was there?
Answer. He usually calls there.
Question. Will you state what your business was with Redpath?
Answer. Relative to a note which is due to Meriam.
Question. From Redpath to Meriam?
Answer. Yes.
Question. What is the amount of it?
Answer. Five hundred dollars.
Question. Do you know what that money was loaned for by Meriam to Redpath?
Answer. I do not.

JAMES JACKSON.

JANUARY 30, 1860.

SAMUEL CHILTON sworn and examined:

By the CHAIRMAN:

Question. Will you please to state where you reside and what your occupation is?
Answer. I reside in this city; I am a lawyer, professionally and practically.
Question. Will you state whether you were counsel for John Brown, who was recently executed in Virginia, under the laws of that State?
Answer. I was.

Question. Were you counsel on his trial on an indictment for treason and murder?

Answer. Yes, sir.

Question. Did you attend his trial?

Answer. I did.

Question. Will you state by whom you were employed as counsel, and the circumstances under which you were employed?

Answer. It was on Friday, after his trial had been commenced, as I was informed—I had been absent the week preceding, and got some indulgence at the hands of the circuit court here, which was then sitting, and I had just returned and was engaged in business there—when Judge Montgomery Blair, of this city, came to the court-room and said he desired to see me, and I went out with him into one of the rooms; he told me that he had received a letter from a gentleman by the name of John A. Andrew, of Boston, whom he represented to be one of the leading lawyers of the Boston bar; in which letter he said that he desired Mr. Blair to go up to Charlestown and appear in Brown's case on behalf of his relatives and friends; that was the statement contained in the letter.

Question. Of whose relatives and friends?

Answer. Brown's; and the letter said if he could not go to employ suitable counsel at his discretion. He told me that he could not possibly go, and did not think he was the proper person to go and defend him, and asked if I would undertake his defense. I told him that I should have a little time to consider it; that I had just returned home, and I could not neglect business in the court here, but I would see the court and the lawyers on the opposite side, and, as they had always been very kind and indulgent to me, I had no doubt they would make an arrangement with me, as it was not likely I should be gone more than three days, and then, if a satisfactory fee was assured to me, I was willing to go and defend him. He said he thought I was the best person to go. So, after making this consultation with these gentlemen and the court, they agreed to indulge me, and I went up and announced to him that I was ready to go, provided the fee was satisfactory. Mr. Blair said he was not authorized—that is, he did not expect to bind himself for the fee in any way at all, but told me who this gentleman was; and I learned from others who had heard from him that he answered to the character he gave as a gentleman high in his profession and high, socially, in the Boston community; and I asked him if he felt himself authorized to make an engagement on behalf of this gentleman. He said he did; and I told him that was perfectly satisfactory. We agreed on the fee. I went up and rendered what services I could. Through Mr. Andrew, I was paid the fee that was contracted to be paid me.

Question. Did Mr. Blair show you the letter from Mr. Andrew?

Answer. He read the letter to me; I am not sure that I saw it; I had one letter from Mr. Andrew afterwards, I think.

Question. Have you that letter with you?

Answer. No, sir; I do not think I could find it.

Question. Can you state the contents of the letter from Andrew to you?

Answer. Well it was something in relation to carrying up his case to the court of appeals of Virginia—a mere inquiry. It was not a matter of any sort of importance, and I did not either take care of the letter or charge my memory with it. Mr. Andrew and myself corresponded rather through Judge Blair. I think I received two letters through him—one was on some matter of inquiry about taking the case up, the costs, &c., for employing additional counsel in Richmond, and Mr. Green was employed; the other was in relation to my fee.

Question. You say the fee was paid to you by Mr. Andrew?

Answer. Through Mr. Andrew. It was paid by two or three different drafts, I think, drawn by Mr. Blair upon Mr. Andrew—Mr. Blair showing me Mr. Andrew's letter authorizing him to draw from time to time.

Question. Have you any knowledge of whence the money was derived that made up that fee?

Answer. I have not. I understood through Mr. Blair, I think, and Mr. Andrew, that the money was raised by those whom he denominated the relatives and friends of Brown. He said there was a small number of them, and that he stood responsible for the fee that was contracted to be paid me and for the expenses of the suit.

Question. Did that engagement as counsel extend to any other person than Brown of those who were arrested there?

Answer. No, sir.

Question. The engagement was confined to appearing as counsel for Brown?

Answer. Entirely to Brown's case. I was written to while in Richmond attending to Brown's case by young Mr. Hoyt, and the records in two cases sent to me, with a request to Mr. Green and myself that if we were successful in getting a writ of error in Brown's case we would appear, saying that fees would be secured to us, but no definite amount was named, and there was nothing beyond that at all. As we failed in Brown's case, which we thought much stronger than the other cases, we made no application at all in the others. Judge Blair wrote to Mr. Andrew for me, and I recommended that assistant counsel in Richmond be employed, and recommended Mr. Raleigh T. Daniel. Then a telegraph came to know if Mr. Daniel would appear for a definite sum which was named, and I telegraphed to him and wrote also, and I got a letter from him stating that his grand jury court—he is the prosecuting attorney for the city of Richmond—was coming on about the time we made this application; he had a very large docket, and he did not think he could give his attention to it, but he recommended Mr. Green, whom he had consulted, and who said he would be content to take the fee offered and attend to the case. I went down and carried the record, and remained there about a week. We prepared the petition with a good deal of care, and the court considered it and overruled it. That was the end of it.

Question. In your intercourse with Brown during his trial, did he disclose to you, or state to you, the names of any persons out of the State of Virginia, or in it, who were connected with him in this assault upon Harper's Ferry, other than those who were present as his party at Harper's Ferry?

Answer. He did not. I did not ask him any questions. I listened to whatever he chose to communicate. I had a very long conversation with him on Sunday. I got there Saturday morning.

Mr. COLLAMER. I do not suppose we can develop anything material, especially after the answer the witness has made; but really, ought we to ask counsel as to communications from his client?

The CHAIRMAN. I do not think that question will arise on the interrogatory I put to him. Whether we could ask him to communicate anything that his client told him in reference to the matter pending before the court in which he was counsel is one question, but as to matters unconnected with his trial, is a different question.

Mr. COLLAMER. I do not wish to make a point of it; but what is communicated by the client—the client not knowing exactly what relates to his case and what does not—is always privileged.

The WITNESS. I was going to remark that I have a proper understanding of the relation; and if a question were asked me, the answer to which would disclose anything of that sort, I should very respectfully decline to answer it; but there was nothing in the world. I think it is much shorter to say that he never did make a communication to me of anybody. I did not ask him any questions about it. My questions were confined entirely to his defense there in court, and whilst he gave me a long narrative of his life pretty much, in the general, an interesting one too, he did not mention the name of a single individual. In fact, he did not mention the names of those in jail with him. I did not see but one of them, a man who was in the room with him, Stevens.

SAMUEL CHILTON.

FEBRUARY 1, 1860.

Hon. HENRY WILSON sworn and examined.

By the CHAIRMAN:

Question. Mr. Wilson, will you be good enough to inform the committee whether you had an interview with a certain Hugh Forbes in the city of Washington, when it took place, what led to it, and what resulted from it, and what communication he made, if any?

Answer. I was sitting in my seat in the Senate early in May, 1858, the first week I think of May, 1858, and Mr. Forbes came to my seat and introduced himself, the Senate not being in session. I think it was on Saturday. At any rate, the Senate was not sitting and I was sitting at my desk franking documents or writing. He came to my seat, introduced himself to me. I had never before heard of him, knew nothing whatever of him. He said that he had been employed the year before, I think he said about a year before that, by Brown to go to Kansas, or somewhere near there, to drill some men for the defense of Kansas. That was the idea he communicated to me. I understood him that he was employed by Brown. He spoke of Brown and said he had left him, I think the fall before, I cannot fix the time. He seemed to be in a towering passion, greatly excited; said he had been abused and treated badly. Brown had discharged him or they had parted; Brown had failed to pay him what he ought to be paid; that he thought those persons in the East, and he mentioned Dr. Howe among them, who had made contributions for Kansas were under obligations to pay him; that his family was suffering in France, I think he said Paris, but I am sure it was in Europe. He said his family were starving. He spoke very nervously and excitedly about it. Then he said that some of the arms that had been contributed by these people in the East for the defense of Kansas had got into Brown's hands and were somewhere in the West, in Ohio or in Iowa. I do not remember whether he said these arms were at Tabor or not, but I am confident he said they were at Iowa; and he said that Brown was not a fit man to have arms, and that they ought to be got out of his hands. He said very decidedly that they ought to be got out of his hands, that he was not a fit man to have those arms, or something to that effect. That was the idea he conveyed to my mind. He said that he was a revolutionist by opinion; and he had no objection to going into anything of that nature. He then left me and I have never seen him since. He remained in the city, as I understood, some days afterwards, but I do not know how long. I think that was the latter part of the week when the Senate was not in session. Afterwards I saw Dr. Bailey, I think I called at Dr. Bailey's on the Sunday evening following.

Question. Who was Dr. Bailey?

Answer. The editor of the "Era." It was the evening I wrote, and I know I wrote Sabbath evening. He asked me if I had seen a man by the name of Forbes here. I told him I had. Dr. Bailey said to me that Forbes had said to him that John Brown had got some arms in his possession that were contributed for Kansas. Dr. Bailey then said to me that those people ought to get those arms out of Brown's hands, and that I had better write to some of them to that effect. I told him I had the same impression. I sat down that evening and wrote a letter to Dr. Howe, of Boston, which is the letter referred to here. When I was here the other day, I told you I had sent to Dr. Howe for the letter. Dr. Howe has written me that he has searched everywhere but cannot find the letter. He states, however, that he recollects substantially the contents. I have stated the circumstances under which I wrote and the knowledge I had. I had no knowledge whatever of anything like an organized invasion, or anything of the kind. I had the impression that Brown belonged to the class of men who had been in Kansas who entertained the idea that when any attacks were made on Kansas in any way, they ought to be retaliated by going over the line into Missouri, and I supposed this had reference to imprudent acts that might be perpetrated on the frontiers between that State and the Territory of Kansas. Nothing else ever entered my mind; and believing that policy was wrong, and that the only proper policy was a defensive one, I wrote the letter under those circumstances. The letter, as near as I can recollect—I am very sorry it cannot be produced, because I should like to have the identical words—was very brief, and to this effect: that I wrote to him for the purpose of saying it was rumored that some of the arms that had been contributed by gentlemen in the East for the defense of Kansas had passed into the hands of John Brown, and were held somewhere in his hands, and that they ought to get them out of his hands and put them in the hands of some reliable men in Kansas, who would use them only for the purposes of defense, for which they were contributed; that if these arms should be used for any illegal purpose, they would involve the men who contributed for the other purpose in difficulties. That was the substance of the letter; that if they should be used for any illegal purpose whatever, they would be involved in difficulty, and they should get them out of his hands at once. I received a letter, three or four days after I wrote mine, from Dr. Howe, to this effect: that they had sent to Brown to deliver the arms into the hands of somebody in Kansas; at any rate, they had sent to him to take the arms into Kansas, or deliver them up in some way; and I supposed, at the time, the arms were those referred to as being in Iowa, which were sent out there and stationed on the way. I received this letter a day or two after I wrote. That was the substance of it. The whole matter, I supposed then, was a quarrel between Brown and Forbes, and I paid but little attention to it; and never, until the outbreak took place, dreamed or heard from any quarter whatever anything in regard to it. I heard nothing from Forbes or Brown or any other source. When, some months afterwards, I think it was in the autumn or the first of the winter following, Brown made a raid into Missouri, after the troubles in the south part of Kansas—the capture and murder of some free-State men—I thought that was probably what Forbes referred to in saying that the arms ought to be out of his hands. That is my whole knowledge of the matter.

Question. In Forbes's interview with you, did he tell you at all the cause of quarrel between himself and Brown?

Answer. No, sir; but I had the impression it was on account of want of pay; that Brown had no men to drill; that he went out to drill some men and they had none, and Brown did not pay him; that he had been employed in New York, I think, in teaching the use of arms in fencing, and that he had lost his place by going west. I saw him but a very few minutes, and he was very much in a passion with Brown and the men in the East.

Question. Did he tell you who those men were in the Eastern States whom he looked to to pay him, and who had declined doing it?

Answer. He said he thought that men like Howe and Sanborn, and I think he mentioned Mr. Lawrence, and that class of men who had made contributions for Kansas, ought to pay him, and I think he told me that he had written them to that effect; at any rate, he spoke of them and Brown with a great deal of bitterness. He was a very nervous man, and seemed to be in a great passion. I told him I knew nothing of it whatever—had never heard anything about it, and could do nothing about it.

Question. Will you state, if you please, why you wrote to Dr. Howe—what control he had over the subject?

Answer. Forbes had mentioned his name to me, as among the men in the East who had made contributions of money or arms for Kansas. I had not direct communication about it, but I knew by the newspapers and common rumor that Dr. Howe had been very active in contributions for Kansas. I knew he was an active man in the matter; I had never had any communication with him myself about it.

Question. Do I understand you correctly, that in your communication with Forbes, and afterwards with Dr. Bailey, you derived the impression that Brown intended to make some illegal or improper use of those arms?

Answer. I had an impression of this kind from what he said to me about getting the arms, and from the manner in which Dr. Bailey spoke to me, saying that the arms ought to be got out of his hands; that there might be border difficulties in Kansas, raids over the line; that he might strike back; that he might go over the line if anything should happen, and in retaliation, capture, and run off slaves. I had this impression from what Forbes and Dr. Bailey said, and from my past knowledge that there were a class of men in Kansas who had the idea that when there was any attack on Kansas, it ought to be retaliated by an attack over the line into Missouri. My own opinions were, that that was a fatal policy, and an illegal one, and ought not to be tolerated for a moment. I had the apprehension when this remark was made to me, that these arms were in Brown's hands, and ought to be got out; that he might, in retaliation, use them for that purpose; that was my idea. I had this feeling, that it was a thing which ought to be discouraged; that the arms which had been sent, as I supposed, for defense, ought not to be used for any illegal or aggressive purpose; that it was illegal and wrong so to use the arms, and so far as the men were concerned who contributed the arms, they ought to take them out of Brown's hands, and give him no encouragement, but keep clear of him. I had no idea of any general organization for the invasion of Missouri or any other State; but I supposed reference was made to mere border difficulty which a few men might get up between Kansas and Missouri.

Question. You received a letter afterwards from Dr. Howe, informing you that he had taken measures to perfect what you had suggested?

Answer. I received a letter within three or four days, I think as soon nearly as the mail could carry my letter and bring his back, in which he said substantially—I cannot give the exact words, but I remember distinctly about it, because I felt that the thing I had written for was accomplished—that he had sent an order to Brown either to carry the arms into the Territory or deliver them to somebody in it. The idea was, that an order had been sent by a gentleman who had control of them. I do not know that he had control of them, but that such an order had been sent. Dr. Howe further said in the letter that there was a man in Washington, a disappointed and malignant man, by the name of Forbes, who he supposed had communicated any information upon which I might have written the letter. I did not mention any source of information in my note to him, and therein is where I supposed Mr. Realf might have mixed the contents of Dr. Howe's letter, in which he sent mine to Brown, with the contents of my letter.

Question. Did you hear anything at any other time from Dr. Howe of whether these arms had been taken out of Brown's possession?

Answer. I never heard about it. I never made an inquiry afterwards. I supposed it was done, and never paid any attention to it, or thought any more of it; and, in fact, the whole subject then dropped out of mind. I saw nothing, and heard nothing from any other sources, in regard to it. I supposed it was a matter that was settled, as things were getting peaceable in the Territory, and everything was quieting down. The idea of an invasion at Harper's Ferry, or organization for an invasion of the South, had never been entertained by me any more than I entertain to-day the idea of an invasion of Boston from France or England. I never heard Dr. Howe say anything about it.

Question. Had you any acquaintance with John Brown?

Answer. I met John Brown in Boston, in the spring of 1859.

Question. Do you remember the month?

Answer. The last of May or the first of June; I met him at the Parker House, at Boston. There were a dozen persons there. Brown came in with somebody and was introduced to quite a number of gentlemen who were there. I was introduced to him and he, I think, did not recollect my name, and I stepped aside. In a moment, after speaking to somebody else, he came up again and, I think, he said to me that he did not understand my name when it was mentioned, and he then said, in a very calm but firm tone, to me: "I understand you do not approve of my course;" referring, as I supposed, to his going into Missouri and getting slaves and running them off. It was said with a great deal of firmness of manner, and it was the first salutation after speaking to me. I said, I did not. He said, in substance, I understand from some of my friends here you have spoken in condemnation of it. I said, I had; I believed it to be a very great injury to the anti-slavery cause; that I regarded every illegal act, and every imprudent act, as being against it. I said that if this action had been a year or two before it might have been followed by the invasion of Kansas by a large number of excited people on the border, and a great many lives might have been lost. He said he thought differently, and he believed he had acted right, and that it would have a good influence, or words to that effect. I saw him a night or two afterwards, on the stage of a large meeting in Tremont Temple, at which I was in the audience. Mr. Cheever, of New York, was delivering an address. That was all the conversation I ever had with Brown. In this conversation he spoke with great frankness, and I supposed that he referred to what I had said in regard to his going over the line, and taking away slaves from Missouri, which I had condemned, but he may have also referred to my letter to Dr. Howe.

Question. Did you learn from any credible source, or from anything that transpired, what was the object of his mission to Boston at that time?

Answer. I did not know anything of it. I never heard anything said about it in any way. I supposed that he was there as he was about the country generally, and I never heard that he was collecting funds for any special object; did not know anything about it.

Question. Do you remember the month?

Answer. I think it was the latter part of May or the first of June, because it was the week of the anniversaries in Boston, when the various religious and other societies hold meetings. It is called anniversary week with us. I think it is the last week in May in which a great number of societies—religious, and tract, and charitable, and benevolent societies—hold their anniversaries. The meeting at which I saw him on the stage, was the Church Anti-Slavery Society, an organization of ministers connected with the church. Dr. Cheever, of New York, was delivering an address before them, and I remember attending to hear him. Brown sat on the stand.

Question. Did Brown make any address?

Answer. He was called out, and said a word or two, but it was very brief indeed. I have little recollection of it, but it did not amount to much any way. It was just before Mr. Cheever got up, and there seemed to be a great desire to hear Cheever, and Brown sat down very abruptly. The meeting was called to hear Mr. Cheever deliver an address, and before they got ready, there being a very full audience, there was a call for Brown, and he got up, but he had hardly said a sentence or two before there was a call for Cheever, and he sat down saying he was more accustomed to action than to speaking.

HENRY WILSON.

FEBRUARY 2, 1860.

EDWARD K. SCHAFFER sworn and examined.

By the CHAIRMAN:

Question. Are you a member of the firm of Schaeffer & Loney?

Answer. I am.

Question. Will you state where your place of business is?

Answer. Our place of business is at Nos. 1 and 3 Hanover street, Baltimore.

Question. Will you please to say whether, at any time in October, and on what day, you sold a parcel of percussion caps to a man named Meriam; in what way, and under what circumstances, they were bought?

Answer. On the 13th of last October a young man came into our place of business and inquired for military percussion caps. The young man who waited on him asked him a certain price for them, and he stated that he wished to get a large quantity. He came back into the counting-room, where I was, and asked me whether I could sell them at any less price than he had offered. I then went out into the store to see who it was that wanted them. I found a young man there, who stated that he wanted quite a quantity of them. I tried to find out from him for what purpose he wanted them. He gave me no satisfaction in regard to the object.

Question. What did he say? Can you recollect?

Answer. He said he had an order for them. I was not favorably impressed with the young man's appearance. I thought it was rather an unusual quantity for any legitimate purpose; and I told him the price was so and so, as our young man had told him, and that I would not sell them for less. In fact, we had not as large a quantity as he wanted.

Question. What was the quantity he wanted?

Answer. I think he wanted forty or fifty thousand. We had on hand only about twenty and one fourth thousand at the time. He objected a little to the price we asked, but said he would see whether he could make up the quantity. I told him there were other houses in the city who kept them. As he went out of the store, he passed by where samples of spades and shovels were hanging up, and wished to know the price of spades and shovels. I suspected that he was furnishing supplies for some fillibuster expedition, though I knew that New Orleans and New York were generally the places where they got up those expeditions. Still, the appearance of the young man was unfavorable, and I refused to give him the price of the spades and shovels. I walked with him towards the door, and told him there were other houses in the city where he could probably procure a supply of them, and he went out. The next day he came into the store, went to the same young man, and told him he would take the percussion caps we had on hand of the kind he had looked at. He requested that a bill should be made out, and that the caps should be packed up and sent to Barnum's hotel, to his room, giving us the number, I think 204. He gave us his name—F. J. Meriam. The young man furnished him the caps and went out to make the bill. Meriam's manner appeared to be rather excited. He pulled out his money, which was in $20 gold pieces; and there were $14 or $15 of change coming to him, and in his hurry he went off without getting it. The young man called him back and gave him the change. The caps were packed up and sent to Barnum's, according to his request.

Question. What was the amount of the purchase?

Answer. Some $45. That was all we thought of it at the time, until a few days afterwards the outbreak occurred at Harper's Ferry. Then we suspected that our caps had probably gone there, and that the man who bought them was one of Brown's men. The name corresponded with one given in the accounts of the outbreak. Feeling anxious to know whether our suspicious were correct, we got a friend at the Ferry to inquire, and our caps were identified, among some that were found there, by the private mark that was on the papers, I think.

The CHAIRMAN exhibited to the witness a box-lid produced by A. M. Kitzmiller, marked:

20¼ M. caps. From
S. & L.,
B.

F. J. MERIAM,
204 *Barnum's Hotel*,
Baltimore.

and asked: Do you recognize this?

Answer. That is the identical lid of the box we sent.

Question. Did you know anything more of this man Meriam at any time afterwards; I mean anything that would lead you to a knowledge of who he was or what became of him?

Answer. I know nothing except what I saw in the papers.

Question. Did you make any inquiries in relation to who he was anywhere?

Answer. No, sir. The only inquiries we made in regard to it, I think, we stated in our letter to you. We found out from Barnum's that the young man had, the day before he made the purchase, received a sum of money by express from Boston.

The CHAIRMAN. It is unnecessary to go into that.

EDWARD K. SCHAEFFER.

FEBRUARY 3, 1860.

JOSHUA R. GIDDINGS affirmed and examined.

By the CHAIRMAN:

Question. Will you please to state, sir, your place of residence?

Answer. Jefferson, Ashtabula county, Ohio, is my residence.

Question. At one time you represented that district in the House of Representatives?

Answer. I represented that district in the House of Representatives for twenty-one years.

Question. Will you please to say whether you were acquainted with John Brown, who was recently executed under the laws of Virginia for offenses against that State?

Answer. I saw John Brown on Saturday afternoon—I cannot give the date—in the spring or summer last past. He appeared on the ground, where several gentlemen were engaged in playing ball, with a proposition to lecture in our village the next day.

Question. Where was that?

Answer. In the village in which I live. He appeared there for the purpose of making arrangements for his lecture. I was called aside to consult with our friends, for the purpose of making arrangements for the lecture, and introduced to Mr. Brown. This was the first time I saw him.

The CHAIRMAN. I did not hear the date distinctly of that.

Answer. I cannot give the date. It must have been in May or June last, I think. As I say, I was introduced to him for that purpose, and was consulted in regard to making arrangements for his lecture. I said at once, let him come and lecture. I did not like the idea of undertaking to say, in dollars and cents, what we would give Mr. Brown. My proposition was adopted. I did not leave playing ball, probably, more than three or four minutes. He left, appeared the next day, and lectured in the church where I worship. After the lecture, I made an appeal to the people present, stating Mr. Brown's past sufferings in Kansas; his trials and the persecutions to which he had been subjected there; that he was now without any regular employment on which to depend for a living; and for my own part, I was willing to contribute. Our friends generally contributed. The sum I cannot state, but I think it was satisfactory to all. It was less than twenty dollars; it was over ten, I should think. After this was done,

I invited him to my house to tea. He took tea with me and with my family, and, I think, one or two other gentlemen. We conversed from a half to three quarters of an hour after tea, in the common sitting-room of my residence, when suddenly his carriage drove up to the door and he left me. I never saw Mr. Brown at any other time or at any other place. That was the extent of my acquaintance with him.

Question. Mr. Giddings, will you look at this note? Probably it may refresh your recollection as to time. [Exhibiting the following letter:

JEFFERSON, OHIO, *May* 26, 1859.

MY DEAR SIR: I shall be absent during next week, and hope to be at home during the summer. Shall be happy to see you at my house.

Very truly, J. R. GIDDINGS.

JOHN BROWN, Esq.]

Answer. It corresponds very near to the date which I had stated.

Question. I wanted to know if that was the period you referred to?

Answer. It would not fix the time of his appearance at Jefferson or of his lecture. It fixes the time at which I solicited him to come there, but the date of his being there would probably be within three weeks from that time. This I state entirely without any date, with nothing but an impression as to the time.

Question. Do I understand that that was a note inviting him to come to the village where you lived?

Answer. Yes, sir. Mr. Brown was regarded as a man of some considerable distinction, or notoriety, if you please. He had lectured in the surrounding villages, except the seat of justice of our county. Perhaps I go too far in saying almost all the surrounding country. He had in a portion of it at all events. Our people were anxious to hear him, and his son, who was said at that time to reside in the town of Andover, had intelligence of this, and the first time I had any encouragement to invite Mr. Brown was on the receipt of a letter from him, saying that he would be in Andover at some time, but the time I cannot fix. In the letter he wrote, I should think he expressed a willingness to lecture for us. The note now presented me was written to say that I hoped to see him at that time, &c., as it now reads. In pursuance of this he called, but the date I cannot fix. The date of this note corresponds with my impression very well as to the time it was written, but not as to the period he was there.

Question. You say he had been lecturing in most of the adjacent villages?

Answer. No. He had lectured at Cleveland, at Painesville, and in those places which would correspond in point of population with ours.

Question. Was the purpose of those lectures, as far as you know, to get contributions of money?

Answer. I do not know anything of those lectures, further than what was stated by him in the lecture at our place. I do not know that I ever heard any analysis or description of those delivered in other places.

Question. Did he say anything in the lecture at your place which would show that the object of his lecturing was to collect money—to get contributions?

Answer. Not any further than stated. I did understand he was lecturing, and received compensation for it, for the purposes of his support. I got no idea that he lectured for any other purpose but to receive such sums as would sustain himself and family.

Question. What was the subject of his lecture in your village?

Answer. Slavery entirely. The duty of Christians in relation to the institution of slavery; the obligations which Christians were under to do to the slaves as they would have the slaves do to them under an exchange of circumstances. In the course of which he spoke of his professions of religion and the religious obligations which we were under to the slaves. He carried that to an extent that we were bound to aid the slaves in escaping, so far as we could, even in the slave States. That was the distinguishing feature in which he differed perhaps from our people.

Question. Did he develop any plan or propose, either directly or by intimation of his own purposes, to take measures in any way for the liberation of slaves in the Southern States?

Answer. From the time of his arrest to the present day I have not only thought and reflected on that, but I have inquired of other gentlemen who heard his lecture, and I am not only authorized to say that I have no recollection of any such thing, but I have the word of those who were present to say that men who were there had no impression of his expressing anything of the kind.

Question. Was there no suggestion as to the modes by which the great object to be attained could be effected?

Answer. Not the remotest.

Question. That, I understand, was the first time you ever saw him?

Answer. The only time I ever saw him. You will recollect he called on Saturday and lectured on Sunday. Those were the only two instances I ever saw him.

Question. Did you have any correspondence with him?

Answer. No further than this: his note to me and mine to him, which has been shown me. How he learned that we wanted him to lecture I did not know, but he intimated he would be in Andover, and would visit us if we wished.

Question. Are you acquainted with his son, John Brown, jr., of whom you spoke just now?

Answer. I have been since that, but not before. John Brown, jr., was educated in the town adjoining me, and my eldest daughter was educated at the same institution, and knew him when he first came to my house; but I had no idea of ever having seen or heard of him until that time. At this time, which was about the period of his father's lecture, he was at my house. I was introduced to him, and knew him for the first time. Since then, I have probably seen him two or three times, but you are aware I have not been there constantly. I have been absent a great deal.

Question. Did John Brown, jr., after that period, when you saw his father, at any time call upon you and advise or suggest to you anything about raising money for his father for any purpose?

Answer. He did; as stated in my lecture at Philadelphia. He called with a statement that his father was in want. I helped him with three dollars most cordially, the same as I contributed at the time of his lecture, for the same purpose and same object.

Question. Was that in a personal interview with you, that he told you that his father was in want?

Answer. These were the circumstances: I had my carriage ready to start for some place, I think, now, it was Cuyahoga, and I had started, actually got as far as my door, when he stopped me and told me, in a very few words, that his father was in want. I dare say I should not have given more than a dollar if I had the change, but three dollars was the lowest I had, and I gave it without any hesitation.

Question. Can you tell when that was?

Answer. I cannot say. It must have been, I should think, as late as August last, but I wish you to understand that this fixing of dates is one of the most difficult things in my own mind, unless I have something on which to predicate it. I have no data on which I can fix it in this case; but that is my impression, and I give it as my impression. It may have been September, and may have been July, but my impression is that it was August.

Question. Did he speak with you at that time, or any other time, about forming societies through that country for the purpose of making contributions?

Answer. I am not aware of anything of the kind. If he did, it has passed from my recollection. I should think that I said to him that I would ask friends where I was going to, and prevent his father from being in want. That is my impression; but as to forming any society, I do not know that I ever heard of the thing suggested until your question. I have seen some publication of a letter from John Brown, jr., giving some intimation that I would form associations. That was unquestionably an error. I was on my way to Portage county. At Portage, I had been told, there was an association for aiding in all those charitable and humane purposes connected with the escape of slaves fleeing from bondage. At least, I understood they were associated for that purpose—for the purpose of giving aid to the needy men and women who were escaping from bondage. This giving money for such purposes by individuals was very common. A former student of mine, who had originally read law with me, was said to be at the head of that association, and my reference was to that association.

Question. What is the name of the town?

Answer. The town of Ravenna, in which I was to lecture. I was unwell at the time, and when I had closed my lecture before the Eclectic Institute of Hiram, Portage county, and before I went to Ravenna, I returned from Hiram on account of ill health directly home as quick as I could go by railroad. I did not visit Ravenna at all, nor did I fill my appointment to lecture there.

Question. In the conversation to which you refer with John Brown, jr., in connection with associations of the sort you have mentioned, was the idea conveyed to Brown that money might be obtained from those associations for the purposes of his father?

Answer. It was not. Permit me also to say that in the conversation with John Brown, jr., no allusion was made to any other association upon earth than that referred to. I suppose Brown's impression must have arisen from my intimation to him that O. P. Brown, who was a lawyer there, had been instrumental in forming an association there. I suppose it must have been that he alluded to. I can account for his language in no other way. I merely give this as a supposition. I have not seen Brown since the affair at Harper's Ferry, and have had no opportunity to inquire his ideas.

Question. Were you a member of any of those associations?

Answer. I never was a member of any association of that kind. I wish to explain, that whatever I have given for the aiding of fugitive slaves, or for any such purpose, I have always done openly and undisguisedly, without any hesitation, and have taken pains, at all times, to proclaim it publicly.

Question. Will you state, as far as you know, what were the exact and definite objects of the associations of the character such as you have spoken of?

Answer. I would not be willing to undertake to say what were the exact objects of the association to which I refer. My inclination is that it was originally formed for the purpose of aiding fugitive slaves, who, having left the slave States, were continuing their flight through our State; such is my impression, and I only give that, without any specific authority.

Question. You have spoken of going through that country lecturing. Will you state the subject of the lectures?

Answer. My lectures were uniformly such as I deliver before lyceums. They are mostly upon the principles of our government; the legitimate powers and constitutional duties of human governments. That is one of my lectures. The higher law constitutes another.

Question. Will you explain the meaning of the higher law?

Answer. I will do so with great pleasure. What I mean by the higher law is that power which for the last two centuries has been proclaimed by the philosophers and jurists and statesmen of Germany, Europe, and the United States—called, in other words, the law of nature—by which we suppose that God, in giving man his existence, gave him the right to exist; the right to breathe vital air; the right to enjoy the light of the sun; to drink the waters of the earth; to unfold his moral nature; to learn the laws that control his moral and physical being; to bring himself into harmony with those laws, and enjoy that happiness which is consequent upon such obedience.

Question. In your lectures, was the theory of that law applied to the condition of African slavery in the United States?

Answer. Unquestionably, to all. Wherever a human soul exists, that law applies. I mean by the term "*soul*," that immortal principle in man that exists hereafter, which is called the human soul; and wherever such soul exists, there is the right to live; the right to attain knowledge; the right to sustain life, obey the laws of his Creator, and enjoy heaven or happiness.

Question. Was that theory or doctrine of a higher law in your lectures applied specially to the condition of African slavery in this country?

Answer. To all human beings, wherever they are.

The CHAIRMAN. I do not think you answer the question. You do not mean to evade it, I am sure; but you do not answer.

The WITNESS. I do not intend to evade it, certainly. Then I will say that the meanest slave who treads the footstool of God holds from his Creator the same right to live and attain knowledge and to liberty that you and I possess.

Question. In those lectures, were the doctrines you now speak of, applied directly to the condition of slaves in the United States?

Answer. Certainly; to all human beings.

Question. Will you state to what extent you exemplified the duty as you understand it of the people of the United States to carry that law into effect?

The WITNESS. You want my views.

The CHAIRMAN. I want the views given in your lectures.

Answer. The views given in my lectures go to this extent: that whenever, without going into any other State, we have the opportunity to sustain the right of a fellow being, it is our duty to do it. I have never felt myself called upon to advocate nor to encourage the entering into other States to speak thus to slaves; but wherever in my own State, where I can do it without violation of law, or enactments erroneously called *law*, I uniformly arm the slave; I uniformly tell him to defend his life and his liberty; I uniformly teach him his rights so far as I can.

The CHAIRMAN. The object of my question was not so much to get your individual opinions as those inculcated in your lectures.

Answer. I would be understood that these are the sentiments that I inculcated in my lectures.

Question. You were delivering lectures of that character during the summer and fall of 1859, in that region of country?

Answer. I am not aware of delivering that lecture during the summer past. I think in almost every instance during the summer I lectured on the trial of John Quincy Adams. The lecture which I delivered before the institute in Hiram, Portage county, was on the trial of Mr. Adams. My opinion is that I was advertized to lecture in Ravenna on the same subject.

Question. Will you be good enough to say whether, in propagating these doctrines which you call the doctrines of a higher law, there was inculcated in the lectures also the duty of the citizen to regard that law in preference to the laws of the country, if they came in conflict.

The WITNESS. You want the subject of my lectures.

The CHAIRMAN. As to that point.

Answer. I think, perhaps, I will publish some of them. My lectures inculcate this: what all writers upon natural law in Europe generally, and in the United States, for the last two centuries, have declared that any act, command, or enactment, violative of those eternal principles of right and liberty are void; that they have none of the essence or elements of law; that they are the mere mandates of despots; that it is not only the right of the people to disregard such mandates, but it is their duty and a high virtue to maintain the principles of enduring truth and justice, although the legislature, or the men acting in a legislative capacity, should overstep the bounds of their authority and command otherwise; that any enactment to disrobe a human being of the right which God has given him is just as wrong, as criminal, in him who perpetrates the crime, as it would be if no such enactment existed; that right and wrong are established by Heaven's law; that man may *obey* this law, but cannot modify or alter it; that in treating of these mandates of despotism, men are not bound to incur greater penalties than they believe it their duty to incur; but whenever they can, with safety, express their disapprobation or their opposition, or even resistance, with impunity, without incurring too great penalties on themselves, it is their duty to do it. I will go a little further in that respect: the man who would disrobe his fellow-man of any of the rights God has given him stands precisely in the character before the Christian world as other despots and criminals. There is no distinction between Nero and him who, at this day to the same extent, denies the equal right of his fellow-man to life and liberty.

The CHAIRMAN. That is rather beyond the scope of the question. The question was, which law was to be regarded, if they came in conflict: the laws of the country, or the higher law to which you have alluded?

Answer. Permit me again to explain. There can be no *law* which invades the right of any innocent human being to life, liberty, and happiness. The mandate or the enactment has none of the elements of law; it is a mere command to violate God's will or the laws of nature. I make this as an explanation.

Question. Will you be good enough to say whether at any period after you first became acquainted with John Brown, you were aware, from any source, of his purpose to attempt the liberation of slaves in the South?

Answer. In that respect, you will permit me to answer in this way, if you please: that I had an impression that he would do as he had done in Missouri; and I think that the general idea, the general impression, (for I have conversed with many of our leading men,) was that he would do the same again if the opportunity presented itself. That was the impression; but that I had any authority for it, except by way of inference, is not the fact.

Question. What do you refer to as having occurred in Missouri?

Answer. I suppose the history of the day has shown him to have taken off slaves from Missouri, as he himself stated in Virginia that he had done, and that that was his object in entering Virginia. He had taken away some twelve or fifteen slaves from Missouri.

The CHAIRMAN. I knew the instance; but I did not suppose but that you had a specific reference.

The WITNESS. I had reference to that.

Question. How did you derive the impression; you say you had the impression that that was his purpose?

Answer. That impression was an inference from what he had done, and from the fact that he was known as an outlaw; a price had been set upon his head, as he stated; that he had no fixed place of residence; that he was destitute of regular employment; that he was advocating the right of all men to liberty, and particularly that it was the duty of Christians to aid slaves in the slave States to escape. It was inferred from that. That is my impression.

Question. Did you know from John Brown, or any authentic sources, whether, in that descent upon Missouri, there was any violence used in getting the slaves?

Answer. I did not. I understood that there was no personal violence, no bloodshed, nor anything of that kind. There were threats. He was surrounded by force, and the marshal of Kansas, or the deputy marshal, or some officer with a *posse* attempted to surround him, and he, by display and address, came off without the shedding of blood. That was my impression. I did also get the impression, but whether it was from the publication or from any other source I do not know, that he was very much opposed to shedding blood.

Question. Have you been present at a meeting of any of the associations you refer to in your country, or anywhere else, that were organized for the purpose of facilitating the escape of slaves?

Answer. The Chairman evidently labors under a misapprehension of what I have stated. He uses the plural number, speaking of "associations." I have referred to but one. I know of but one, and that I only know by hearsay, as I have stated. It is the one to which I have alluded. I know of no other.

By Mr. DAVIS:

Question. Did you, in inculcating by popular lectures the doctrine of a law higher than that of the social compact, make your application exclusively to negro slaves, or did you also include minors, convicts, and lunatics, who might be restrained of their liberty by the laws of the land?

Answer. The interrogatory presupposes what has not been stated, that I inculcated a law higher than the Constitution of my country. My first answer is, distinctly, that the Constitution of my country is founded on that law, is not contradictory to it, and is essentially in all its bearings distinctly in favor of it. So far as the interrogatory professes to speak of insane persons and lunatics, it is the very safety of the people that they should be restrained from committing depredations on them; it carries out the objects of government to secure the whole people in the enjoyment of life and liberty, including the lunatic. It is not only consistent with the higher law and with the Constitution of the country, but with the common sense of the people.

Question. And if the law of the land should deem it equally necessary for the safety of the country to restrain other persons, does the higher law resist?

Answer. The proposition is of itself a contradiction to the common teachings of our reason.

Mr. DOOLITTLE. Right here, Colonel Davis, I must interpose. I think, although this colloquy between you and Mr. Giddings is very instructive to us all, perhaps, it is no evidence. You are simply asking for his opinion.

Mr. DAVIS. It is evidence only in this sense, that I wish to get at the sentiment which was inculcated by the lecturer and received with approbation, as connected with our present inquiry as to how far combinations exist to destroy the institutions of the country. That is the object.

The WITNESS. Propound your question directly, Colonel Davis. I will take great pleasure in answering it.

Mr. DAVIS. Did the doctrine inculcated teach that it was right to liberate any person who was restrained by the laws of the land from those liberties which you say belong to all as the endowment of nature?

The WITNESS. Permit me, with all due deference, to suggest, so that I may understand you, do you intend to inquire whether those lectures would indicate whether your slaves of the slave States had a right at all times to their liberty?

Mr. DAVIS. I will put the question in that form if you like it.

Answer. My lectures, in all instances, would indicate the right of every human soul in the enjoyment of reason, while he is charged with no crime or offense, to maintain his life, his liberty, the pursuit of his own happiness; that this has reference to the enslaved of all the States as much as it had reference to our own people while enslaved by the Algerines in Africa. At that time, as a nation, for the enslaving of our citizens by Algerines, we sent a navy there to butcher them. In all my lectures I inculcated the right of the Africans in the United States to their liberty, as standing upon precisely the same level that the claim of Americans enslaved by Africans stood at the time we sent our navy to Algiers.

Question. Then the next question is, whether the same right was asserted for minors and apprentices, being men in good reason, yet restrained of their liberty by the laws of the land?

Answer. I will answer at once that the proposition or comparison is conflicting with the dictates of truth. The minor is, from the law of nature, under the restraints of parental affection for the purposes of nurture, of education, of preparing him to secure and maintain the very rights to which I refer; and therefore, to say that the child compares with the slave of mature age is doing violence to the common sense of the land.

Question. How with apprentices?

Answer. The apprentice stands in the same condition as the child—so in law, and so in reason, and so in common sense. The apprentice is merely transferred to another parent, as it were, to teach him the mode of sustaining himself, to educate him and prepare him for usefulness; and when he shall attain to the age at which he is supposed to be capable of knowing and maintaining his rights, he becomes free, and holds the right to assert his natural prerogatives.

Question. This doctrine then is, that the laws of the land must conform to the higher law, and if they do not conform they are void?

Answer. By this indiscriminate application of the term "law" to all enactments that are of themselves despotic, we confuse the minds to whom we address ourselves. Such enactments have, I repeat, none of the qualities, the essential elements of law. For two centuries, all Christian writers have defined law to be a rule of action commanding that which is right, and prohibiting that which is wrong, and any enactment violating the law of nature, or the plainly revealed and well understood will of the Creator, is not only void but criminal.

Question. But who is to judge whether the laws of the land violate or conform to the laws of nature?

Answer. In our nation the people are made the judges. Our government was based upon the doctrine that it was constituted to secure the people in the enjoyment of life, liberty, and happiness; that whenever any form of government becomes destructive of life, of liberty, or of happiness, it is the right and the duty of the people to alter or abolish it, and reorganize its powers in such form as to them shall seem most likely to secure their interest and happiness. They do this in our government in the regularly constituted mode of turning out such officers as disregard the laws of nature, and placing those who hold to the doctrines asserted by the founders of our republic.

J. R. GIDDINGS.

SAMUEL G. HOWE sworn and examined.

By the CHAIRMAN:

Question. Will you please to state where you reside, and what is your profession?

Answer. I reside in Boston, Massachusetts; I am by education a physician; I do not practice my profession. I have charge of two charitable public institutions.

Question. Will you state whether you were acquainted with John Brown, who was recently put to death in Virginia, for offenses against the laws of Virginia?

Answer. I was.

Question. When did you form his acquaintance, and under what circumstances; what led to it, and where did you form it?

Answer. My acquaintance was first formed by correspondence in the year 1856 or 1857. I cannot state clearly which. I would premise that my memory is singularly deficient about dates. I became acquainted with him in 1857, as the agent of the Kansas Aid Committee.

Question. Who was the agent?

Answer. John Brown.

Question. You say you became acquainted with him as agent; was he the agent?

Answer. He was the agent, and in one sense I was the agent, inasmuch as I went to Kansas for the purposes of the committee.

Question. What was the style of the committee?

Answer. The Kansas Aid Committee of Massachusetts.

Question. Did you form his acquaintance in Kansas?

Answer. I think I did.

Question. Will you state what was the object and the occupation or employment of that committee? What was the object of raising the committee? What were their functions?

Answer. I was connected with two committees. One committee was raised for the purpose of getting clothing and money for aiding the suffering inhabitants of Kansas; that was the express object of the committee of which I was chairman. Another committee of which I was a member, was raised for the general purpose of aiding the inhabitants of Kansas in the defense of their freedom then invaded, and repelling invaders.

Question. For distinction sake, can you give us the distinctive names of those two committees?

Answer. One was the Boston committee, usually called the Faneuil Hall Committee, inasmuch as the original meeting was at Faneuil Hall; it had no official name; it was not an incorporated body; it was called just what people chose to call it. The other was the Massachusetts Kansas Aid Committee.

Question. Was that incorporated?

Answer. No, sir.

Question. When was this Massachusetts Kansas Aid Committee formed?

Answer. I think after the Boston committee, but I cannot say precisely.

Question. In what year, as near as you can recollect.

Answer. 1857.

Question. Did you say that Brown was the agent of that committee?

Answer. He was not then the agent of the committee. I said I had formed his acquaintance then. I was then the agent. He afterwards became the agent of the Massachusetts committee.

Question. What was his agency. What was he to do?

Answer. The first express business, as far as I can recollect, is that he called on the State committee, of which I was only a member, whereas of the other committee I was the chairman and the efficient working man. The first business that I can recollect was that, in consequence of the difficulty of getting arms and provisions and clothing up the Missouri river, it was found expedient to transport them across the country, and Mr. Brown, or Captain Brown as he was called, was appointed the agent to transport those articles of various kinds—arms, provisions, and clothing—into Kansas. That was the first, as far as I can recollect.

Question. Will you state what was the character and the quantity of the arms that were intrusted to him in that way?

Answer. I can state nothing with any precision. I have no knowledge about it that would enable me, under oath, to say. I can recollect distinctly that, to the best of my knowledge and belief, two hundred Sharp's rifles were committed to his care.

Question. Were those rifles at Chicago at the time they were committed to his care, do you recollect?

Answer. I cannot say.

Question. Were there any revolving pistols in the same parcel?

Answer. I cannot positively say, but my impression is that there were.

Question. Do I understand that those arms were the property of this Massachusetts Aid Committee?

Answer. They were the property of that committee.

Question. How were they procured?

Answer. They were purchased by subscriptions, contributions raised and sent in voluntarily by the people.

Question. For the purpose of purchasing arms?

Answer. No; for the aid of Kansas. The committee, in their own discretion, purchased the arms. I recollect that distinctly, because the objection was made that some of the contributors to the Boston committee would be unwilling to have any of their money appropriated for arms.

Question. And thus, according to your recollection, the arms were purchased as the act and at the discretion of the committee, with funds belonging to the committee?

Answer. Yes, sir, that is my impression.

Question. Can you tell us when those arms were purchased?

Answer. I think, in 1857; but I must fall back on my imperfect memory about dates.

Question. They were committed to John Brown, as agent of the society, for the purpose of transporting them across the country to Kansas, as the river was not considered accessible?

Answer. Yes, sir.

Question. Do you know at what point he deposited them after their transportation?

Answer. My impression is that he deposited them at Tabor, in Iowa.

Question. Can you tell us under whose control they were while they were at Tabor; who had charge of them there?

Answer. I cannot recall the person's name. I saw him, but cannot recall his name.

Question. By what authority were they put in the custody of the person whose name you cannot recall?

Answer. I have no knowledge of that.

Question. Were they recognized as under the control of that person by the Kansas committee?

Answer. I cannot answer that question with precision.

Question. Do you know in what way John Brown afterwards got possession of them, when he brought them to the region of Harper's Ferry?

Answer. I have no means of knowing that he did bring them there.

Question. Do you know then what became of them?

Answer. I do not know, of my personal knowledge, what became of them. I can state my impression.

Question. Tell us what knowledge you have of it, and how it was derived, as to what became of them?

Answer. In the year 1858, I received a communication from a Mr. Forbes, then in Washington, and information from other quarters, that Captain Brown had in his possession arms belonging to the committee, which he would probably use for purposes not intended by the committee. A meeting was called. The committee had then virtually dissolved; it had nothing more to do; but the members were called together. A vote was passed, instructing the chairman to write to Captain Brown and direct him, if he held any property, arms or otherwise, belonging to the committee, to take them into Kansas, there to be used only for the defense of freedom in Kansas. Such a vote was passed, such a letter was written, and, I have no doubt, received by him. I think that was the last record of the committee which was made.

Question. Has a copy of that letter been preserved?

Answer. A copy is with the records.

Question. Have you got it with you?

Answer. I have not. I was not chairman of that committee, nor secretary. The records are accessible.

Question. What committee are you speaking of now?

Answer. The Massachusetts Aid Committee.

Question. I thought you said you were the executive officer or the chairman of that committee.

Answer. Of the Boston committee—of the Faneuil Hall committee. I was only a member of the other committee.

Question. Are those records under your control?

Answer. I have no manner of doubt I can have access to them. I inquired of the person in whose charge they were, before I left the city of Boston, and his reply was that they were all straight, and in order, and in his safe. I presume they are accessible.

Question. Will you say in what form that communication was made to you by Captain Forbes? Was it by letter, or personally?

Answer. By letter.

Question. Written from where?

Answer. From Washington city.

Question. And your direction to Captain Brown was that the arms should be taken into Kansas and used only for purposes there?

Answer. Used only for the defense of freedom in Kansas.

Question. Do you know whether he did or did not comply with that direction?

Answer. I have no means of knowing; but, from my confidence in his character, I have no doubt that he did conform to it.

Question. You wrote that letter, as I understand, as the chairman of the Boston committee?

Answer. I did not write the letter. The chairman of the Massachusetts Aid Committee wrote the letter.

Question. Who was the chairman who wrote the letter you refer to?

Answer. I should prefer not to answer that question.

The CHAIRMAN. I see no reason why you should not answer the question.

The WITNESS. I am here to answer as to all I have done myself, freely and frankly, but I would respectfully ask to be excused from answering any question touching the actions of anybody else. I can only answer for my view as one of the committee.

Mr. DAVIS. The witness confounds his position. He is not here arraigned to answer for what he did, but to give information as to what everybody did.

The CHAIRMAN. The subject referred to the committee by the Senate is to make inquiry into all the facts attending the late incursion at Harper's Ferry, and connected with it in any way.

The WITNESS. Perhaps I am over sensitive about it, and inasmuch as the gentleman's name is perfectly well known as chairman of the committee, and is in print, I will give it, Mr. George L. Stearns.

Question. Did you see the letter?

Answer. I think I saw a copy of the letter afterwards.

Question. Did you receive a letter from Mr. Wilson, of the Senate, in reference to those arms?

Answer. About the same time.

Question. Will you state whether you preserved a copy of that letter?

Answer. I preserved a copy of the letter until recently, when, in the general destruction of my papers of no consequence, at the beginning of the year, I destroyed it, among others, but I have a distinct recollection of the contents.

Question. Will you state the contents?

Answer. It was that he had reason to believe that Captain Brown had in his possession arms belonging to the Massachusetts Aid Committee, which he would be likely to use for purposes not contemplated by the committee; that he, Wilson, considered the original movement for procuring anything of the kind to have been a very mistaken and unfortunate one, and he advised by all means that measures be taken to prevent Captain Brown using those arms for any purposes not contemplated in their original purchase. It was a short letter, and that was the amount of it; but I recollect he distinctly expressed his disapprobation of the fact of such arms being in existence, and his disapprobation of John Brown's general career.

Question. In what capacity, so far as you were concerned, was that letter addressed to you? Why was it addressed to you?

Answer. I suppose because I had been long acquainted with General Wilson, and he knew that I was interested in the matter.

Question. How did you act upon it, or did you act upon it?

Answer. The information came from these two quarters about the same time, and upon that information we acted as I have just described.

Question. The two quarters referred to, as I understand, were a communication made to you by Mr. Wilson, of which you have just spoken, and a communication made to you by Captain Forbes?

Answer. Yes. I did not heed much Captain Forbes's information, because it seemed to me to be ill-natured and spiteful.

Question. Did you preserve a copy of Forbes's letter?

Answer. I think I have a copy of it, but I have not got it with me. I think I have it; but I know I have a copy of my answer.

Question. Will you state the contents of Forbes's letter, if your memory will admit?

Answer. It was to this intent, that he had been engaged during the active war in Kansas by Captain Brown, or by, as he called them, the Northern Abolitionists, to go to Kansas and drill men; that he never got any money for it; that he was in great distress; that he must have money; that Captain Brown was not a reliable person; that his plans, if intrusted to a man of head and prudence, might come to something; and he seemed to intimate that he, Captain Forbes, was a man of head and prudence; that if Captain Brown was allowed to go on it would be disastrous; that he would denounce it; and other things to that effect. It was a letter rather threatening in its general character. I did not heed it so much. It was a very long letter, full of vituperation and abuse. I had never seen Captain Forbes, nor heard of him before.

Question. Did I understand you to say that a copy of that letter is preserved?

Answer. I think a copy of that letter is preserved. I am sure a copy of my answer is preserved. The letter was a long document; and I have the habit, when I am writing to persons with whom I am not very intimate, and on business that seems to be of any importance, to have my letters copied. General Wilson's was a short note, and I had several letters from him on various subjects.

Question. Did you communicate the contents or substance of Forbes's letter to Brown after you received it?

Answer. I have no distinct recollection of having done so. It is possible I may have done so. I have been told since that I did so; but I have no distinct recollection of it, and it seems to me improbable, inasmuch as it would have defeated the purpose which I had of preventing the contributions of persons who might have contributed, from being diverted. I may have answered his letter, but I have no recollection of having done it.

Question. How would your communication of the contents of that letter to Brown, in any way affect the use of the contributions you speak of?

The WITNESS. I do not understand your question.

The CHAIRMAN. I understand you to say that you think you did not let Brown know what Forbes had communicated to you, for the reason that it might be the means, in some way, of perverting from their legitimate use the contributions that had been made by the Massachusetts people.

The WITNESS. I remarked that such letter might prevent my purpose being executed, which purpose was to prevent such a perversion of the use of the arms.

Question. In what way would your communicating that to him prevent the execution of your purpose?

Answer. I do not know. I cannot say what passed through my mind. It looks to me now, thus: that if Captain Brown supposed the order which was given to him, which was peremptory, to take the arms into Kansas, was occasioned by such an act as this, he might suppose that by removing that objection the committee would be satisfied, or something of that kind. The order was peremptory, without any reason being assigned for it; and it appears to me that if we had assigned a reason and he could obviate that reason, he might say to himself "I am free *in foro conscientiæ.*"

Question. Had you any personal interview with Captain Forbes after his quarrel with Brown?

Answer. I never saw him at all.

Question. Did you know of any engagement he had made to go out to the West for any purpose connected with Kansas affairs?

Answer. I never heard of it until he stated it in his letter.

Question. Can you have the letters to which you have referred—Forbes's letter to you and your answer to it—together with the records of that committee, brought from Boston, without going there again in person?

Answer. It would be very difficult to get the copies of the letters. I have not now in my employment a person who wrote for me for many years, and who took care of my papers. Within the last three weeks I have been obliged to have a change, and I should have no certainty of finding those papers.

The CHAIRMAN. I wanted to know if you could get at them without going back?

Mr. DOOLITTLE. Perhaps the witness could procure copies of the letters there and attach to them his affidavit made in Boston that they were true copies, and send them here, and also a copy of that record.

The WITNESS. My impression is that all can be had, except possibly Forbes's letter, which may have been mislaid.

The CHAIRMAN. Can it be done without the necessity of your returning to Washington from Boston?

The WITNESS. I will make every possible effort to do it, and I think I shall be successful. I think everything can be done that would be done by my coming back.

Question. Will you state whether you saw John Brown, the man who has been referred to, in Boston, or elsewhere, at any period during the year 1858?

Answer. I saw him in Boston in the year 1858.

Question. Will you state the object of his mission in coming there?

Answer. He gave me no definite information.

Question. What information did he give you as to the purpose of his coming, the objects he had in view in visiting that part of the country?

Answer. The impression I got from him was that he wished persons to render him what assistance they chose to give him, as a man having suffered in the cause of Kansas, for the defense of freedom in Kansas, and as a man disposed to devote himself to the defense of the cause of freedom.

Question. In a former part of your testimony you said those arms were to be taken to Kansas for the purpose of aiding in the defense of the cause of freedom; will you state what you mean by "the defense of the cause of freedom?"

Answer. In Kansas, repelling invasions such as had been frequent there.

Question. What sort of invasions?

Answer. What are usually called border ruffian invasions.

Question. How was your "defense of the cause of freedom" to affect it?

Answer. To protect it by repelling those who would invade it.

The CHAIRMAN. I do not still distinctly understand you. How would the cause of freedom be affected by any invasion in Kansas?

Answer. The cause of freedom in Kansas would be. If the polls, for instance, at an election were surrounded by armed men from other States, and the freemen of Kansas were prevented by fear from voting,

I should call the man who attempted to repel those invaders a defender of freedom.

Question. A defender of the freedom of the white people in Kansas?

Answer. I know no distinction of color in freedom. I know no distinction of color in men.

Question. The negroes were not permitted to vote in Kansas, were they?

Answer. They ought to have been.

Question. Were they?

Answer. I supposed they would be.

Question. Were they at the time?

Answer. I have no means of knowing. I think, in the territorial condition, the people there were not responsible for the form of government imposed on them.

The CHAIRMAN. I want to get at this fact. In stating what you mean by defending the cause of freedom in Kansas, you instance it by referring to defending the freedom of voting at the polls. The negroes, we know, were not admitted at that time to vote at the polls, and I inferred, therefore, that when you spoke of the cause of freedom in Kansas, you meant the freedom of white people, for negroes were not interested in that subject.

Answer. I suppose, if Kansas was a free State, the consequence would be that her vote would be on the side of freedom in other States. It was a general term.

The CHAIRMAN. I must put the question in a different way——

Mr. DOOLITTLE. In relation to this question, it is a mere matter of argument between the chairman and the witness as to his opinions. I supposed our purpose was to get at the facts.

The CHAIRMAN. If you make an objection, we will entertain it.

Mr. DOOLITTLE. I do not propose to raise an objection.

Mr. COLLAMER. Other witnesses have been prevented from telling what took place in Kansas.

The CHAIRMAN. The object of the question is to see what use was intended to be made of these arms which we have shown were in Brown's possession and brought to Harper's Ferry. I understand the witness to say that Brown was instructed to take them into Kansas to aid in the cause of freedom. It is certainly pertinent, in my judgment, to ascertain what the witness means by the cause of freedom in ulterior connection with the use that was actually made of the arms.

The WITNESS. At that time I had no thought about anything but the freedom of Kansas as such, without any thought of any colored men at all, for as far as I had seen there were few or no colored men there.

Question. Then it was for the freedom of the white people in Kansas?

Answer. The freedom of Kansas, or the freedom of the white people of Kansas, because I knew of no others there. I was in Kansas and I do not recollect seeing any others.

The CHAIRMAN. Well, to come back again to Boston, I understand you to say that Brown's mission was in part, as you heard it, to obtain contributions of money in Boston?

Answer. I presumed so.

Question. Do you know for what purpose those contributions were wanted by Brown?

Answer. I have no definite knowledge.

Question. Have you any knowledge?

Answer. I have no positive knowledge that I can say on oath, any more than hearsay.

Question. Anything that you derived from him? What was Brown's statement of his object?

Answer. He never gave me a definite statement of any plan or purpose that he had definitely fixed upon.

Question. Did he tell you why he wanted to collect money there; for what purpose he wanted the money?

Answer. In no definite shape did he tell me any plan that he had.

Question. In any shape?

Answer. He appealed to me as an anti-slavery man to help an anti-slavery man.

By Mr. COLLAMER:

Question. Was that before this order about the arms?

Answer. I think it was before.

By the CHAIRMAN:

Question. Do you know whether or not the president of the Massachusetts State Kansas Committee gave Brown an order for those arms at Tabor, in 1857?

Answer. I think he did; I think they had been in the possession of another person whose reliability was called in question; that is my impression.

Question. Who was that other person?

Answer. I cannot recollect.

Question. You say you think he did. Will you state definitely whether you know that an order was given to Brown for those arms or not?

Answer. It is impossible for me say. I have a very strong impression; I know of the fact, that they were transferred from another person to him, and therefore I infer the order was given by which they were so transferred.

Question. Who was the secretary of the Massachusetts State Kansas Committee? Was it Sanborn?

Answer. No, sir.

Question. Who was it?

Answer. I can furnish the name, but I cannot remember it now.

Question. When you saw Brown in Boston, in 1858, can you tell what period of the year it was?

Answer. I can say that he had a fire in his room, and therefore I infer it was in the winter. I have no other recollection. My memory of dates is imperfect. It was probably in the winter months, February or March.

Question. Did you see him in Boston in the months of May or June, 1858?

Answer. I cannot say.

Question. Did you know that there had been a convention held by Brown and others, at Chatham, in Canada?

Answer. I think my first knowledge of it was derived from the newspapers.

Question. Then you had at no time any conversation with him on the subject of that convention?

Answer. Never; to my recollection.

Question. Dr. Howe, here is a paper that purports to be in your handwriting. Will you look at it and see whether it is or not? Exhibiting the following note:

"DEAR FRIEND: Our friend from Concord called with your note. I begin the investment with fifty dollars inclosed, and will try to do more through friends.

"DOCTOR."

Answer. I believe it is.

Question. Can you tell the date? There is no date to it. When was that written?

Answer. There is no evidence on the face of the paper to show when it was written.

Question. Will you say to whom this was addressed?

Answer. I presume it was addressed to Captain Brown.

Question. Do you remember the fact of writing it?

Answer. I did write it.

Question. Do you remember the fact of the time?

Answer. I do not remember the time.

Question. Will you please to say why your name was not signed to it?

Answer. Perhaps it was not signed because the investment to which it referred might have been of a character that he, perhaps, would not like to have known at the time, though he would have no objection afterwards. Captain Brown was considered to have suffered a good deal in Kansas, and a subscription was raised to purchase him a homestead, and a thousand dollars was raised for that purpose. He was a proud man, and perhaps would not like to have had it known.

Question. Was this fifty dollars part of that thousand?

Answer. If it was given in 1857 or 1858, it might have been.

Question. You say a thousand dollars was raised?

Answer. A thousand dollars or thereabouts was raised.

Question. In the years 1857 and 1858?

Answer. 1857 or 1858.

Question. To whom was that money paid, or what use was made of the money?

Answer. I only know that the homestead was bought; I presume by the chairman of the Massachusetts Aid Committee; I am not certain.

By Mr. DOOLITTLE:

Question. Is that homestead at North Elba, New York?

Answer. Yes, sir.

By the CHAIRMAN:

Question. This letter says: "I begin the investment with fifty dollars inclosed." If this fifty dollars was for the purpose of buying that homestead, the money was sent to Brown directly?

Answer. I think not. That is my impression.

The CHAIRMAN. Then that would be contradictory to the paper, "I begin the investment with fifty dollars inclosed, and will try to do more through friends," if this referred to that purchase.

Answer. That I cannot answer, unless I could know the dåte of the letter. I contributed to his aid at various times.

Question. His aid in what way?

Answer. In the same way that I contribute to the aid of other anti-slavery men; men who give up their occupations, their industry, to write papers or to deliver lectures, or otherwise to propagate anti-slavery sentiments. I give as much money every year as I possibly can afford. I am in the habit of contributing in that way.

Question. Was this money contributed to Brown because of his personal necessities, or not?

Answer. Because of his personal worth, and because he had no ostensible means of reimbursing himself for his losses in Kansas. I am not in the habit of questioning very closely, when I have confidence in the character of men whom I aid.

Question. Will you be good enough to say if that is in your hand-writing? It is not signed. [Exhibiting the following paper:]

"Horse-cars leave Tremont House every half hour.

"Get out at the Jamaica Plain, and inquire for house of *George R. Russell.*

"The steam-cars leave Providence depot.

"Get out at the Jamaica Plain station."

Answer. That is my hand-writing.

Question. Can you tell the date when that was written?

Answer. I cannot; but I remember the circumstances.

Question. What were the circumstances?

Answer. Mr. Sanborn wanted to take Mr. Brown to Mr. Russell's house. I was going with him; but I could not go, though I had promised to go. I remember now, seeing that, that I wrote this direction; but I cannot say that I gave it to Sanborn. He was not alone.

Question. That paper, then, was written for Mr. Sanborn?

Answer. It was written for Mr. Sanborn.

Question. Was it at the time when Brown was there?

Answer. It was.

Question. Do you know whether Sanborn wanted that for the use of Brown as well as himself?

Answer. I presume he did. Brown was with Sanborn at the time.

Question. Who is George R. Russell, to whom this refers? What business had Brown with him, do you know?

Answer. I do not know what business he had with him at all. Mr. Russell is one of our wealthy, liberal men, who is in the habit of contributing to the promotion of anti-slavery sentiments—a liberal contributor.

Question. Will you state what you mean by that phrase "contributing" for the promotion of anti-slavery sentiments? What is the meaning of that idea?

Answer. In the same way that I would promote the Gospel among the heathen; I could not precisely say what. The means are various—lectures, writing, talking, discussing the matter.

Question. What ends are to be attained by promoting that anti-slavery sentiment? What is the object in view?

Answer. The promotion of freedom among men; the same object as the fathers in the revolution.

Question. Was one of its objects the means of attaining the freedom of the African slaves held in this country?

Answer. That would be the natural and desired result.

Question. Was that one of the ends to be attained by propagating this anti-slavery sentiment by lecturing and otherwise?

Answer. It was. I answer these questions out of courtesy to the chairman, but I must think they are rather wide.

The CHAIRMAN. If you are disinclined to answer any question, you have only to state the fact. No question is asked you, of course, which is not deemed pertinent to the inquiry which is required of this committee, in the judgment of the gentleman who puts it.

Question. Will you be good enough to inform the committee whether you were acquainted with a man named J. H. Kagi?

Answer. I never saw him.

Question. Did you have any correspondence with him?

Answer. I never corresponded with him that I recollect.

Question. Will you please to say whether you have any recollection of this telegraphic dispatch? [Exhibiting the following dispatch:

[By telegraph from Boston, June 6, 1859.]

To J. H. KAGI.

CLEVELAND, *June* 6, 1859.

He got the needful, and left three (3) days ago, direction unknown.

S. G. H.]

Answer. I have not the slightest idea. My initials are "S. G. H.," but I have no recollection whatever of that.

Question. Have you no recollection of being in communication with Kagi at all?

Answer. I am certain, because Kagi struck me when reading the names in the public prints, and I asked myself the question who he was.

Question. Did you know John Brown, jr., the son of the Brown we have spoken of?

Answer. I saw him once.

Question. Where did you see him?

Answer. He called upon me at my house, I think, early in 1859. I cannot recollect the date.

Question. Here is a letter which has been proved to be in the handwriting of John Brown, jr., dated at "Syracuse, New York, Thursday,

August 17, 1859," addressed to "Friend Henrie," which has been shown to mean Kagi, in which he says:

"While in Boston, I improved the time in making the acquaintance of those staunch friends of our friend Isaac. First, called on Dr. H.——, who, though I had no letter of introduction, received me most cordially. He gave me a letter to the friend who does business on Milk street. Went with him to his house in Medford and took dinner. The last word he sent to me was, 'Tell friend (Isaac) that we have the *fullest confidence in his endeavor*, whatever may be the result.'"

Was it in that month of August, 1859, you saw John Brown, jr., in Boston?

Answer. It would be impossible for me to recollect any further than that it was not in cold weather.

Question. Did he introduce himself?

Answer. He introduced himself.

Question. Did you receive him as the son of old John Brown?

Answer. I did, and was very glad to see him as the son of John Brown.

Question. Did he tell you the object of his visit to Boston?

Answer. He did not.

Question. Did he tell you that he was there endeavoring to collect money?

Answer. He did not.

Question. Did he apply to you for money?

Answer. He did not.

Question. Do you remember having given him a letter to a gentleman who does business on Milk street?

Answer. Very likely I may have done so.

Question. Do you know to whom he refers?

Answer. Mr. George L. Stearns.

Question. Does Mr. Stearns live in Medford?

Answer. Yes, sir.

Question. You say that John Brown, jr., did not tell you the object of his visit to Boston?

Answer. He did not.

Question. Did you learn it from any other source?

Answer. I did not. I had a very strong impression that he had some business of his own, and otherwise wanted to see the friends of his father.

Question. Do you remember why you gave him the letter to Mr. Stearns?

Answer. As one of his father's friends.

Question. Was it at your own suggestion, or at his request?

Answer. I cannot say. Probably I might have heard his father speak of Mr. Stearns. He was a warm friend of the father and mother and the whole family.

Question. Did you inquire from him, or learn from him at that time, where his father was?

Answer. I did not learn.

Question. Do you remember to have inquired where he was, or what he was doing?

Answer. I do not remember whether I did or did not. I infer that I did not, because I had no knowledge of his whereabouts. Of course, if I had gained it from him or anybody else, I should probably have remembered it.

Question. Did you have any knowledge of where John Brown was, from the time you saw him in Boston, in 1859, until the outbreak at Harper's Ferry?

Answer. Not the slightest knowledge of his whereabouts. I was probably as much astonished when I heard of his turning up at Harper's Ferry, as the chairman of this committee was.

Question. You said you would probably recollect the name of the secretary of the Massachusetts State Kansas Committee. Can you give it to us now?

Answer. His name is Parnell.

Question. Where does he reside?

Answer. He does business in Boston, and probably resides in one of the small towns in the vicinity.

Question. Did you know a young man named Francis J. Meriam?

Answer. I did, very slightly.

Question. Have you seen him since the affair at Harper's Ferry?

Answer. I have.

Question. Where did you see him?

Answer. At the St. Lawrence Hotel, in Montreal.

Question. Did he tell you of what occurred at Harper's Ferry?

Answer. He did not.

Question. Did he make any reference to it in any form in conversation?

Answer. He began to talk about it, but I checked him. He applied to me. I do not know that I should volunteer any information further than to show that I am in a position to aid all purposes of proper investigation. He called upon me in the summer of 1859, and made my first acquaintance. I saw that he was in a state of mental excitement bordering on insanity. He wanted to know if I knew the whereabouts of John Brown. I told him that I did not; and I thought, moreover, that if I did, I would not tell him. When I next saw him, he was in the same wild state. I had my luggage all packed, and the porter was in the room to take it down to the cars in order that I might return to Boston, when he came in. That was at Montreal. I saw that he was in a state of painful excitement, and declined talking to him.

Question. Did he tell you, or undertake or commence to tell you, of any reason why that attack was made precipitately at an earlier day that had been contemplated?

Answer. He did not. He began a wild talk, and I stopped him. I said, "Mr. Meriam, you see I am going away;" and I knew too much of excitable men to wish to lead them on to a topic which was exciting to them.

Question. Do you know a negro man named Lewis Hayden?

Answer. I know him slightly.

Question. Do you know of any connection that he had with the affairs of Brown?

Answer. Nothing but what I have seen in the papers.

Question. Can you tell the committee where Meriam is now?

Answer. I cannot.

Question. You have spoken of having made various contributions to John Brown, and one is given here of $50; can you recollect any others, and at what time they were made?

Answer. I cannot recollect at what time, but I looked at my check book before I came away, and I found that I had given what I supposed was to John Brown $50. I set apart as much of my income every year for anti-slavery purposes as I can.

Question. Were you the recipient of contributions from others for John Brown?

Answer. During the time I was on the committee.

Question. Were those contributions for Brown, or for what you have spoken of—aiding in the defense of freedom in Kansas?

Answer. They were for John Brown, to be used at his discretion. During the Kansas troubles, the contributions were for aiding the cause of freedom in Kansas, and afterwards I received contributions for him.

Question. Up to what period did those contributions come in?

Answer. I think during all of 1858.

Question. Were they transmitted to John Brown?

Answer. I think not directly. I think that only part of them went to John Brown. The late transactions I cannot speak of with so much certainty; but the former ones I know were never directly transmitted to him.

Question. Was a memorandum kept of the mode in which the money was disposed of?

Answer. Certainly. Everything was clearly recorded; I mean everything in regard to Kansas matters.

Question. Amongst the records you can send to this committee?

Answer. Yes.

Question. Can you state the time when you last saw John Brown the elder?

Answer. It was in the spring or early summer of 1859.

Question. Did you ever hear at any time that he was passing elsewhere by the name of John Smith?

Answer. Never.

Question. You have spoken of having seen John Brown the younger in Boston; can you tell when it was that you last saw him anywhere?

Answer. I am sure I never saw John Brown, jr., but once, and that was when he called on me at Boston.

Question. You say that the money of which you have spoken was given to John Brown, to be used at his discretion after the Kansas troubles were over, for I presume they would be considered as over in 1858. What disposition was it expected he would make of it?

Answer. I do not know that I could say what disposition I thought he would make of it; I supposed that he was a practical anti-slavery man, and I was not inclined to scrutinize, having great confidence in him as a man.

Mr. COLLAMER. To prevent any misunderstanding about these contributions, I desire to ask a question. Were not the contributions received by the committees, which were made by the people in Boston and Massachusetts, for and during the Kansas troubles?

Answer. For that definite purpose.

Mr. COLLAMER. Was any money of those contributions ever sent to Brown after 1858?

Answer. Not that I know of.

The CHAIRMAN. But there were other contributions that were sent to him after the fall of 1858, and I understand you that they were to be used at his discretion.

The WITNESS. I had personal knowledge of several small sums.

Question. Was there any limit imposed upon his discretion, as far as you know, by your act or that of others, in the use of the money that was given to him?

Answer. No further than the confidence he inspired among his friends by two opinions entertained by him, one of which was that he was opposed to promoting insurrection among the slaves, and another was that he was opposed to shedding human blood except in self-defense.

Question. Where did he make those declarations?

Answer. More than once, in the presence of my friends and in my own presence, because I had arguments with him on the general matter of practical anti-slavery, and I knew his sentiments; his declarations were clear and explicit, and I had the utmost confidence in them.

Question. Do you know of any plan he had devised, or proposed to devise, to get the slaves off from the Southern States without promoting insurrection—abducting them, or seducing them away, or anything of that sort?

Answer. I know of no definite recent plans of his; he was secretive.

Question. What do you mean by his being secretive?

Answer. I mean that he was a man not accustomed to reveal his thoughts unnecessarily to any one, that he was not a communicative man. Secretiveness, I recognize as one of the human faculties; that word I use, though in no improper or disrespectful sense.

Question. You have spoken of a fund of a thousand dollars, or thereabouts, that was subscribed to purchase a homestead for Brown's family; was that investment actually made?

Answer. It was actually made, I think.

Question. Was it made by passing that money through Brown's hands?

Answer. My impression is that it did not pass through his hands; on that point I cannot answer definitely.

Question. Then the fifty dollars spoken of in your note could not have been for that purchase?

Answer. About that I cannot say; I have ransacked my memory to know whether it was, and I cannot say distinctly whether it was or not; I contribute all the money I can for what I consider good purposes, and I am not accustomed to make any distinct memoranda about it.

By Mr. Doolittle:

Question. In all your conversation or communication with Brown, had you ever, at any time, from him any intimation of an organized attempt or effort, on his part to be made, to produce an insurrection among the slaves in the slave States of the South?

Answer. Never.

S. G. HOWE.

I would add that, on careful consideration, I think I was mistaken about the application of the fifty dollars referred to in the torn paper. It could not have been part of the money for the purchase of the homestead. I am convinced, moreover, that the purchase of the homestead was a *bona fide* transaction, and the money was given mainly, I think, by persons who would not have countenanced any direct interference with slavery in the slave States.

S. G. H.

APPENDIX TO DR. HOWE'S TESTIMONY.

Boston, *February* 20, 1860.

Gentlemen: Agreeably to the request of your chairman, the Hon. J. M. Mason, in his note of the 3d instant, I have endeavored to obtain copies of the correspondence referred to in my testimony; also of the votes of the Kansas committee, touching the matter of the arms intrusted to John Brown.

Inclosed you will find copies of all the documents and letters bearing on the subject which I have been able to obtain.

No. 1 is a letter from the chairman of the Massachusetts State Kansas Committee, dated January 8, 1857, inclosing an order on the agent of the committee, to deliver to John Brown two hundred Sharp's rifles, with ammunition, &c. These he was directed to hold as agent of the committee, and subject to their order.

No. 2 is a letter, dated April 15, 1857, inclosing No. 3, being a vote of the committee, by which Brown was authorized to dispose of one hundred of the committee's rifles to free-State men in Kansas, at not less than fifteen dollars apiece; and to appropriate the proceeds to relieve the suffering inhabitants. Also, authorizing him to draw upon the treasurer for five hundred dollars, for the same purpose. These show the original destination of the arms and the money.

Next should be a letter from Hon. Henry Wilson, alluded to in my testimony, but the original cannot be found.

The correctness of the statement I made before the committee is confirmed by conversation with persons who saw the letter. It was dated May 9, 1858, and was very brief. It stated that he (Wilson) had been informed that Brown contemplated some unlawful expedition, and would use arms belonging to the Kansas Aid Committee. That he (Wilson) had considered the original purchase of those arms to be an unwise measure; but, at any rate, he advised that they be taken from Brown, and that the committee keep clear of him.

Nos. 4 and 5 are my replies to Mr. Wilson, in which are these words: "prompt measures have been taken, and will be followed up, to prevent any such monstrous perversion of a trust as would be the application of means, raised for the defense of Kansas, to a purpose which the subscribers to the fund would disapprove and vehemently condemn."

No. 6 is a letter of the chairman of the Kansas committee to John Brown, dated May 14, 1858, inclosing a copy of Mr. Wilson's letter, and containing these words: "You have the custody of the arms alluded to, to be used in the defense of Kansas, and it becomes my duty to warn you not to use them for any other purpose."

No. 7 is a letter, dated May 15, 1858, informing Captain Brown of arrangements made for taking possession of the arms.

Next should be a letter from Captain Forbes, but the copy sent to me was not preserved. That person's language was so intemperate and vituperative that I would not write to him a second time, or read his letters further than to see their abusive character. The New York Herald, October 27, 1859, contains what purports to be correct copies of them. The one dated May 6, 1858, is probably the one which I answered, though I think some sentences have been omitted in the printed copy.

No. 8 is my answer to Captain Forbes's letter, in which are these words. "I said that *I had confidence* in the *integrity* and ability of Captain Brown," but it is utterly absurd to infer from that any responsibility for his acts. I have confidence in the integrity and ability of scores and hundreds of men for whose words and acts I am in no wise responsible. Neither as a member of the Kansas committee, nor as an individual am I responsible, either legally or morally for any contract between Captain Brown and you. I never heard your name connected with Kansas until quite recently. I was an active member of the committee from its foundation, (until it ceased active operations which was long ago,) and never heard of any contract with you; and I know that the committee never delegated power over any one to bind it by any legal or even moral obligation with you. So! the brains are out of *that* charge; and I will not heed any ghosts of it which you may parade before me or the public.

"Your mistaken notion about my being in any way responsible for Captain Brown's actions is the key, I suppose, to certain enigmatical allusions in your last letter to some projected expedition of his; as though I was to be responsible through all time for him!

"I infer from your language that you have obtained [in confidence] some information respecting an expedition which you think to be commendable provided *you* could manage it; but which you will *betray* and denounce if he does not give it up!!

"You, sir, are the guardian of your own honor, but I trust that for your children's sake at least you will not let your passion lead you to an action which might make them blush."

It is my belief that Forbes or some one else, did inform the President, or the Secretary of War, of Brown's plans; and that the knowledge of this fact led Brown to abandon the plan whatever it was.

I think that satisfactory proofs of the first part of this statement are in existence, and can be obtained by the investigating committee.

The examination I have given to these matters enables me to correct some parts of my testimony before the committee, and I wish to do so. The memorandum shown to me by the chairman, which was in my handwriting, being a direction to Mr. Russell's house, and which I said might have been given to Mr. Sanborn, was not given to him, but to some person who accompanied Captain Brown, and whose name I cannot recollect.

The letter in my handwriting, mentioning fifty dollars sent to Brown, and promising more, I testified might refer to my subscription for purchasing a farm for Brown's family; but I could not tell certainly, because the date had been torn off.

I have examined the list of subscriptions for that purpose, and do not find my name there. The letter, therefore, must have referred to some other transaction.

As doubts have been expressed whether the purchase of the farm was a *bona fide* transaction, and as it is pertinent to your inquiry about funds raised for John Brown, I inclose an extract from the original correspondence:

"BOSTON, *November* 7, 1857.

"MY DEAR FRIEND: Your most welcome letter came to hand on Saturday. I am very glad to learn that, after your hard pilgrimage, you are in more comfortable quarters, with the means to meet present expenses. In my opinion, the free-State party should wait for the border ruffian moves, and check-mate them as they are developed. Don't attack them; but if they attack you "give them Jessie," and Frémont, too! You know how to do it.

"I inclose a copy of the subscription list that you may know who has been so kind to you, with their address, that you may write to them if you wish to do so. The original I will keep until I see you.

"*Subscription list.*

"The family of Captain John Brown, of Ossawatomie, have no means of support, owing to the oppression which he has been subjected to in Kansas.

"It is proposed to put them (his wife and five children) in possession of the means of supporting themselves, as far as is possible for persons in their situation. The undersigned, therefore, will pay the following sums, provided one thousand dollars shall be raised. With this sum a small farm can now be purchased in the neighborhood of their late residence, in Essex county, New York.

"Captain J. BROWN, *Kansas.*"

Then follow the names of sixteen subscribers, whose joint contributions amount to one thousand dollars, which sum was applied to the purchase of a small farm in New York State.

As neither the names of the writer nor of the subscribers are important for the purposes of the committee, they are not given.

S. G. HOWE.

The foregoing statement and extracts of letters are, to the best of my knowledge and belief, correct. I can make affidavit to this in a formal manner, if required.

S. G. HOWE.

The SELECT COMMITTEE
Of the Senate of the United States.

No. 1.

MASSACHUSETTS STATE KANSAS COMMITTEE ROOM,
Boston, January 8, 1857.

DEAR SIR: Inclosed we hand you our order on Edward Clark, Esq., of Lawrence, K. T., for two hundred Sharp's rifled carbines, with four thousand ball cartridges, thirty-one military caps, and six iron ladles—all, as we suppose, now stored at Tabor, in the State of Iowa.

We wish you to take possession of this property, either at Tabor or wherever it may be found, as our agent, and hold it subject to our order.

For this purpose you are authorized to draw on our treasurer, Patrick T. Jackson, Esq., in Boston, for such sums as may be necessary to pay the expenses as they accrue, to an amount not exceeding five hundred dollars.

Truly yours,

GEORGE L. STEARNS,
Chairman of Massachusetts State Kansas Committee.

Mr. JOHN BROWN, *of Kansas Territory.*

No. 2.

BOSTON, *April* 15, 1857.

DEAR SIR: By the inclosed vote of the 11th instant, we place in your hands one hundred Sharp's rifles, to be sold in conformity therewith, and wish you to use the proceeds for the benefit of the free-State men in Kansas, keeping an account of your doings as far as practicable.

Also, a vote placing a further sum of five hundred dollars ($500) at your disposal, for which you can, in need, pass your draft on our treasurer, P. T. Jackson, Esq.

Truly, yours,

GEORGE L. STEARNS,
Chairman Massachusetts State Kansas Committee.

Mr. JOHN BROWN,
Massassoit House, Springfield, Mass.

No. 3.

BOSTON, *April* 15, 1857.

At a meeting of the executive committee of the State Kansas Aid Committee of Massachusetts, held in Boston, April 11, 1857, it was

Voted, That Captain John Brown be authorized to dispose of one hundred rifles, belonging to this committee, to such free-State inhabitants of Kansas as he thinks to be reliable, at a price not less than fifteen dollars, ($15,) and that he account for the same agreeably to his instructions for the relief of Kansas.

GEORGE L. STEARNS,
Chairman Massachusetts State Kansas Committee.

At the same meeting, it was

Voted, That Captain John Brown be authorized to draw on P. T. Jackson, treasurer, for five hundred dollars, ($500,) if, on his arrival in Kansas, he is satisfied that such sum is necessary for the relief of persons in Kansas.

GEORGE L. STEARNS,
Chairman Massachusetts State Kansas Committee.

No. 4.

Copy of a letter to Hon. Henry Wilson, dated May 12, 1858.

DEAR SIR: I have just received your letter of the 9th. I understand perfectly your meaning. No countenance has been given to Brown for any operations outside of Kansas by the Kansas committee. I had occasion, a few days ago, to send him an earnest message from some of his friends here, urging him to go at once to Kansas and take part in the coming election, and throw the weight of his influence on the side of the right.

There is in Washington a disappointed and malicious man, working with all the activity which hate and revenge can inspire, to harm Brown, and to cast odium upon the friends of Kansas in Massachusetts. You probably know him. He has been to Mr. Seward. Mr. Hale, also, can tell you something about him. God speed the right!

Faithfully, yours,

S. G. HOWE.

No. 5.

Copy of a letter to Hon. Henry Wilson, dated May 15, 1858.

DEAR SIR: When I last wrote to you, I was not aware fully of the true state of the case with regard to certain arms belonging to the late Kansas committee.

Prompt measures have been taken, and will be resolutely followed up, to prevent any such monstrous perversion of a trust as would be the application of means, raised for the defense of Kansas, to a purpose which the subscribers of the fund would disapprove and vehemently condemn.

Faithfully, yours,

S. G. HOWE.

No. 6.

BOSTON, *May* 14, 1858.

DEAR SIR: Inclosed please find a copy of a letter to Doctor Howe, from Hon. Henry Wilson. You will recollect that you have the custody of the arms alluded to, to be used for the defense of Kansas, as agent of the Massachusetts State Kansas Committee. In consequence of the information thus communicated to me, it becomes my duty to warn you not to use them for any other purpose, and to hold them subject to my order as chairman of said committee. A member of our committee will be at Chatham early in the coming week, to confer with you as to the best mode of disposing of them.

Truly your friend,

GEORGE L. STEARNS,
Chairman Massachusetts State Kansas Committee.

Mr. JOHN BROWN,
Chatham, Canada West.

No. 7.

BOSTON, *May* 15, 1858.

DEAR SIR: I wrote to you yesterday, informing you that a member of the Massachusetts State Kansas Committee would visit you at Chatham, to confer about the delivery of the arms you hold.

As I can find no one who can spare the time, I have to request that you will meet me in New York city, some time next week. A letter to me, directed to care of John Hopper, 110 Broadway, New York, will be in season. Come as early as you can. Our committee will pay your expenses.

Truly yours,

GEORGE L. STEARNS,
Chairman Massachusetts State Kansas Committee.

Mr. JOHN BROWN,
Chatham, Canada West.

Dr. Howe will go on as soon as he knows you are in New York.

No. 8.

Copy of a letter to Hugh Forbes, dated Boston, May 10, 1858.

SIR: Some two or three weeks ago I received a long communication from you. I had scarcely read a dozen lines, when I saw that it was written in the passionate and vituperative style which had characterized some communications of yours to friends of mine. I therefore threw it aside without reading it. I did not wish to be drawn into correspondence with one who could rudely violate the courtesies becoming gentlemen. It was the same with your second letter. Your third arrived three days ago. I have reconsidered my determination, and read your letter. I have concluded to write to you, in the hope, though perhaps the vain one, of disabusing your mind of certain errors which seem to be growing into insane belief. Your railings at "New England humanitarians" do not affect *me*, for I make no profession to "humanitarianism" *par excellence*. Your vituperative epithets, intended to be insulting to me personally, pass me as the idle wind, which I respect not; for I long ago settled in the belief that what are called insults affect him only who utters them. "That which cometh *out* of the mouth defileth the man," and him only. I confess that I have been pained by the thought that one who must sometimes behave like a gentleman, in order to enjoy the respect of a dear friend of mine, should so lower himself and injure his children. Indeed, I cannot think about your children without being moved at the thought of your standing like a madman between them and aid. All the allegations and assertions and claims; all the superstructure, indeed, of your long letter to me, falls to the ground, because built on an entire fallacy. I said to Senator Sumner that I *had confidence* in the *integrity* and ability of Captain Brown; but it is utterly absurd to infer from that any responsibility for his acts. I have confidence in the integrity and ability of scores and hundreds of men for whose words and acts I am in no wise responsible. I never made myself responsible, as a member of the Kansas committee, or as an individual, neither legally nor morally, for any contract between Captain Brown and you. I never heard your name connected with Kansas until quite recently. I was an active member of the committee from its formation until it ceased active operations, (which was long, long ago,) and never heard of any contract with you; and I know that the committee never delegated power to any one to bind it by any legal or even moral obligation, with you. So the brains are out of that allegation, and I will not heed any ghosts of it which you may parade before me or the public. Your mistaken notion about my being in any way responsible for Captain Brown's actions is the key, I suppose, to certain enigmatical allusions in your last letter to some projected expedition of his; as though I was to be responsible through all time for him! I infer from your language that you have obtained (in confidence) some information respecting an expedition which you think to be commendable, provided *you* could manage it, but which you will *betray* and *denounce* if he does not give it up! You are, sir, the guardian of your own honor! but I trust that, for your children's sake, at least, you will never let your passion lead you to a course that might make them blush. In order, however, to disabuse you of any lingering notion that I, or any of the members of the late Kansas committee (whom I know intimately) have any responsibility for Captain Brown's actions, I wish to say that the very last communication I sent to him was in order to signify the earnest wish of certain gentlemen, whom you name as his supporters, (in your letter and in the anonymous one,) that he should go at once to Kansas and give his aid in the coming elections. Whether he will do so or not, we do not know. I may, perhaps, save you trouble by declaring that, though I am willing to do my uttermost to aid your family, or any distressed family, and though I am willing to listen to any supposed claim of yours upon me, or any of my friends, I will not read letters couched in such vituperative and abusive language as you have hitherto used to Mr. Sanborn and me. I will read only far enough to see the spirit of the communication, and if it is similar to that of your former letters, I shall put it in the fire, with a real feeling of regret at seeing a man of ability and acquirements willfully injuring himself and family by his own passions.

Yours,

S. G. HOWE.

FEBRUARY 7, 1860.

RALPH PLUMB sworn, and examined.

By the CHAIRMAN:

Question. Will you please to state where you reside and what is your occupation?

Answer. I reside in Oberlin, Ohio. I practice law.

Question. Please examine this letter, and say if it is your handwriting. [Exhibiting the following letter:

OBERLIN, *August* 23, 1859.

DEAR SIR: Yours of August 9 came to hand this morning, and I hasten to reply, and should have replied to your first letter before, but it was so long reaching me that I was afraid you would have left Chambersburg. My pecuniary condition is such, (having made a loss in consequence of being in jail, of about $1,200 on property shipped west,) that I regret to say I cannot advance the money to save your father's land. It would give me great pleasure to do this, and I am very sorry I cannot.

Next, with regard to the last proposition. Our people have been drained of the last copper to pay expenses for the Oberlin trials, and are now sued by Lowe for $20,000 damages for false imprisonment. We have, in all probability, got to have another clinch with the scoundrels, and money, money, money, will be needed at every step. If I could possibly do so, I would send you the required amount; but, in my opinion, it will not be possible to raise it. By visiting other places and interesting other parties, it might be done, but not here. I have to go Missouri in a few days to look after my business there, which has been left in a disastrous condition by imprisonment.

Yours, truly,

R. PLUMB.

Answer. That is my handwriting?

Question. It is addressed to "J. Henrie, Esq." Will you state to the committee who he was?

Answer. He was J. H. Kagi.

Question. Why was it addressed to him by the name of Henrie?

Answer. It was a request of his in a letter I received. I received two letters from J. H. Kagi, and in the last one, I think, he requested me to reply to J. Henrie.

Question. Where were those letters addressed to him by you?

Answer. I replied but once. I will state the whole circumstances in connection.

The CHAIRMAN. Certainly.

The WITNESS. The first letter I received from him was sometime in reaching me, for some reason or other. I was absent from home a portion of the time, and for some other reason, I know not what, it was a good while before it came to me; but the object of both letters was the same. He wrote me that his father had made a purchase of 160 acres of land in some place near to Omaha or Nebraska City, I have now forgotten which, in this way: he had loaned of a Cincinnati banker a land warrant, for which he had agreed to give $200, and had given his note due in a year, and had entered the land warrant upon the land, and had executed a mortgage to the Cincinnati banker for the payment of the $200 at the end of the year. He said in his first letter that his father was not able to raise the money, and he wished me to enable him to take up the mortgage, and he proposed to have the mortgage assigned to me for security for so doing. I did not answer the first letter for the reason that I have stated. The second letter was in the forepart of August, I should think.

The CHAIRMAN. This is dated the 23d of August.

The WITNESS. It seems to be in response to a letter of the 9th of August. When that letter came to hand I replied to it.

By Mr. COLLAMER:

Question. Was that substantially the same thing?

Answer. Yes, the same thing, with this exception, in the second letter he expatiated more fully upon the wants of his father. The last letter stated that he himself could get along well enough, but his father was an old man, and he was very anxious to raise the money, and he wanted, if I could not raise it myself, that I should interest somebody else to raise the money for him. I then replied in the language of that letter.

By the CHAIRMAN:

Question. Where were his letters written from?

Answer. From Chambersburg, Pennsylvania, both of them.

Question. Did he give you any reasons for being there?

Answer. He did not.

Question. Did you know why he was there?

Answer. I did not. I did not know that he was there until I received the letter from him.

Question. In this letter of the 23d of August, you reply to his request for money to aid his father: "Next, with regard to the last proposition. Our people have been drained of the last copper to pay expenses for the Oberlin trials, and are now sued by Lowe for $20,000 damages for false imprisonment. We have, in all probability, got to have another clinch with the scoundrels, and money, money, money, will be needed at every step. If I could possibly do so, I would send you the required amount, but, in my opinion, it will not be possible to raise it. By visiting other places and interesting other parties, it might be done, but not here.'' It would appear from the letter that, after disposing of the application to aid his father, you proceed to his request for money for other purposes: "Next, with regard to the last proposition.'' What was that other proposition?

Answer. It was to interest other parties to raise the $200. He wrote to me for $200 for that purpose, and that alone.

Question. Did your correspondence, as shown in this letter, refer to raising money for no other purpose than that of aiding his father?

Answer. For no other purpose. It was the $200 he wished for that purpose. He stated to me particularly in his first letter the amount of money.

Question. Did you know a man named L. S. Leary?

Answer. I did.

Question. Was he a white man or a negro?

Answer. He was a negro.

Question. Where did he live?

Answer. He lived in Oberlin.

Question. Did Leary apply to you at any time in the fall of 1859 for money?

Answer. He did.

Question. What did he want with the money?

Answer. I understood that he wanted to engage in assisting slaves to escape.

Question. Did he say that?

Answer. No, sir; he did not.

Question. Did he tell you what plans he had?

Answer. No, sir; he did not tell me any particular plans he had in view. I can state fully if it is desired.

The CHAIRMAN. Anything that is pertinent and relevant to the scope of the question you can state.

The WITNESS. He called on me first, and borrowed a small amount of money, without informing me anything about what he wanted it for, except that he wanted to use it. He called again afterwards, and told me he would like to keep the amount I had given him, and would like a certain amount more for a certain purpose, and was very chary in his communications to me as to how he was to use it, except that he did inform me that he wished to use it in aiding slaves to escape. Circumstances just then transpired which had interested me contrary to any thought I ever had in my own mind before. I had had exhibited to me a daguerrotype of a young lady, a beautiful appearing girl, who I was informed was about eighteen years of age——

The CHAIRMAN. Is that in connection with your lending money to Leary?

Answer. In connection with it as to time, and it had an influence on my mind in giving him the money. He did not inform me that his object was in connection with that.

Question. Then, what connection has this daguerrotype of a beautiful young girl to do with your lending money to Leary for any purpose whatever, unless it was connected with that young girl?

Answer. I was going to explain.

The CHAIRMAN. But tell us, before you give the evidence, what connection it has with it?

The WITNESS. I was desirous to say that in the community where I live we are in the habit of giving money for aiding slaves to escape, and whenever application is made to us for that purpose, there is but little said. If we are satisfied it is wanted for that purpose, and can give it, we give it.

Question. Did you know a man named Copeland?

Answer. I did.

Question. Was he white or black?

Answer. He was a black man.

Question. Where did he reside?

Answer. He resided in Oberlin also.

Question. What was his occupation there?

Answer. A carpenter and joiner.

Question. What was Leary's occupation?

Answer. He was a harness maker. I must say, however, that my acquaintance with them was very slight. I merely knew them. I do not know that I ever had any conversation with either of them in my life, excepting in this case of Leary when he applied to me. I knew him on the street.

The CHAIRMAN. On the second application, as I understand, he told you he wanted to retain what he had, and wanted you to let him have more, and then said the purpose was to aid in the escape of slaves?

Answer. Yes, sir.

Mr. DOOLITTLE. If the witness desires to state any fact by way of explanation of his having given money to aid slaves to escape, would it not be right for him to state if he is particularly interested in anything about some young woman?

The CHAIRMAN. I asked him if there was any connection between his giving money to Leary to aid slaves to escape and the daguerrotype of this particular girl he spoke of. Was it to aid her to escape?

The WITNESS. I wish to state the whole facts.

The CHAIRMAN. Answer that question. Did he ask for money to enable him to aid this girl to escape?

The WITNESS. He did not. My impression was——

The CHAIRMAN. Then what connection has it with the subject?

The WITNESS. My impression was it was in connection with that, because, as I said, we are not in the habit of talking about these matters. When money is wanted, it is our custom to give the money and say nothing.

Question. Is it your custom to give the money without inquiring in what way it is to be used, except the object to which it is to be applied?

Answer. Certainly. We do not know the ways in which it is to be used. The object is the only thing. We are not in favor of promoting insurrections, and had no such thought or intention, and should never give any money for that purpose in our community.

Question. Will you state the amount of money that you lent to Leary altogether?

Answer. I let him have $17 50 altogether.

Question. When did you make him the last advance?

Answer. I cannot state now. I do not know. I remember that it was not long before he was said to have gone away. I did not know when he went away.

Question. Was it in the month of August or September of last year?

Answer. It must have been later than that. It must have been the last of September; if not later.

Question. Did he and Copeland go off together?

Answer. I do not know.

Question. Do you know when they left there?

Answer. I do not.

Question. Was there any person present when you advanced this money?

Answer. Yes, when the last money was given.

Question. State who it was?

Answer. My brother.

Question. State his name?

Answer. Samuel Plumb.

Question. Does he reside at Oberlin?

Answer. He does.

Question. Did he interest himself to get the money from you for these men?

Answer. No, sir.

Question. Were you aware at the time that these two men were going away together?

Answer. I was not.

Question. Did Copeland apply for money at all?

Answer. He did not.

Question. Did you know that Leary was to share with Copeland what you advanced to him.

Answer. No, sir.

Question. Will you state where the money came from that you advanced; whether it was your own funds or others?

Answer. Part of it was my own funds; but I will state again that, according to our custom in the place, I went to certain individuals and said I wanted to raise certain money. I requested one man to give me three dollars, another a dollar; generally a dollar apiece. I asked them if they had faith in me, and they said they had, and gave the money.

Question. Did you state what you wanted with it?

Answer. I did not. That is according to our custom in that place when we wish to raise anything for such a purpose.

Question. Is it the custom to solicit money from people of that place without telling them what use is to be made of it?

Answer. Yes, sir; especially when it is to be applied for the purposes of enabling slaves to escape.

Question. How are they to know it is to be so applied if you do not tell them so?

Answer. We know each other well, and have confidence that whatever is wanted in this way is properly applied.

Question. Why is not the use that is to be made of it disclosed to each other?

Answer. Well, the opposition to the practice of aiding slaves to escape, on the part of some of our citizens, is such that it is not thought advisable to say anything about it.

Question. Did you know John Brown, who was recently put to death in Virginia by the laws of that State?

Answer. I saw him twenty-four years ago. I have not seen him since.

Question. Have you been in correspondence with him?

Answer. I never had a letter from him, or wrote him a letter in my life.

Question. Did you know where he was at any time during the last fall?

Answer. I did not.

Question. Did you know this man Kagi?

Answer. I did. I will state the circumstances under which I knew him. I was arrested under a charge of violating the fugitive slave law and thrust into jail, and I laid there eighty-four days. That was at Cleveland, Ohio, on the 5th of April, 1859. During my stay in jail, in company with a number of other men, Kagi came there to the jail, and was there several times. I never saw him at any other time or place except then and there.

Question. Did you know of any associations or societies in your neighborhood, or in that region of country, for the purpose of raising funds to enable slaves to escape?

Answer. No, sir.

Question. Are there no such societies existing there?

Answer. None, that I know of, except this general understanding I speak of among the neighbors.

Question. Did you know John Brown, jr.?

Answer. Yes, sir. I once lived within three miles of him, when I lived in Trumbull county, before he moved to Kansas. Since that time I have moved to Lorain county, and he has returned from Kansas to Ashtabula county.

Question. What is the distance from your residence to his now?

Answer. It is about ninety or one hundred miles.

Question. Has John Brown, jr., been at Oberlin since he came back from Kansas?

Answer. I saw him there once.

Question. Do you remember when that was?

Answer. It seems to me as if it was about the 1st of August last. I am not positive as to the time, but it was somewhere in that neighborhood.

Question. Was he at Oberlin then?

Answer. I met him in the streets of Oberlin. I had been away attending a law suit. I returned at a late dinner hour for us, about one o'clock, and met him about two blocks from my residence walking in a hurry. I invited him to go back to my house and take dinner, and he said he could not; he would like to see me, but he could not go with me then. The cars were due at perhaps quarter past one, and he was then walking rapidly towards the station. I merely shook hands with him and bade him good bye, and he passed on to the cars.

Question. Was there no further conversation?

Answer. No further conversation.

Question. Did you see him at any subsequent time?

Answer. I saw him immediately after his return from Kansas; the next day after his return from Kansas.

Question. In what year was that?

Answer. It was in November, 1856; I recollect it because it was just before the Presidential election.

Question. Do I understand that you did not see him again until the interview in Oberlin?

Answer. I did not.

Question. Have you seen him since?

Answer. No.

Question. Have you had any correspondence with him?

Answer. Not on this subject; it was not a recent correspondence.

Question. Were you aware he was soliciting money anywhere through your region of country last fall?

Answer. He has not done so to my knowledge; he has not from me or any person that I know of.

Question. Did you hear from any sources, so that you can state, what brought him to Oberlin at the time you last saw him there?

Answer. I cannot; I do not know what brought him to Oberlin; I was anxious to see him; I formerly had an acquaintance with him, and I wanted to see him.

Question. Did you hear any person speak of his being there, and what his business was there?

Answer. I did not; my daughter said that such a man as he came to my door and rang the bell just before I came, but she did not know him.

Question. Did you ever hear, or were you aware, that John Brown, jr., had formed, or was seeking to form, associations through that part of the State for the purpose of raising money?

Answer. I was not aware of it; I was in jail at the time.

The Chairman. I mean in August or September, the time you last saw him.

Answer. No, sir; I never knew him to raise associations for the purpose of raising money for any object whatever; there was a kind of association called the Sons of Liberty that was formed in various parts, but, so far as I know, their object had nothing to do with that.

Question. What was the object of the Sons of Liberty?

Answer. It was expressed in these words, I think I saw in one of their constitutions: "That no person shall be deprived of life, liberty, or property without due process of law, if we have power to prevent it."

Question. Where was that society formed? Was it at Oberlin?

Answer. I was informed that there was an association at Oberlin, though I was in jail at the time it was formed, and do not know myself anything about it.

Question. Had you any knowledge of any intention on the part of Brown or any others in the fall of last year to make a descent on any of the slave States for the purpose of running off slaves?

Answer. No, sir.

Question. Did you hear the fact spoken of that those two negroes, Leary and Copeland, had left Oberlin on a mission of that kind at the time they went away?

Answer. I do not think I did until the news came of the attack upon Harper's Ferry by telegraph, which astonished us very much.

By Mr. COLLAMER:

Question. I wish to ask this man what he means by aiding the escape of slaves?

Answer. The meaning of that term is assisting those who are fleeing towards Canada.

Mr. COLLAMER. Assisting slaves who have escaped. That is what you mean by it?

The WITNESS. Yes, sir; I have never known anything further than that.

By the CHAIRMAN:

Question. Would there be any objection on the part of yourself, or those with whom you act, to contributing money to assist in the escape of slaves from a State where they are held in slavery?

Answer. Well, we are generally opposed to that; my answer to that is that I have extensively traveled, while I was in business as a merchant, at the South, and I have steadily refused ever to do anything in the direction of suggesting or aiding a slave in a slave State to escape, while I have always assisted them where they have been fleeing in our own State.

RALPH PLUMB.

FEBRUARY 9, 1860.

JOHN A. ANDREW sworn and examined.

By the CHAIRMAN:

Question. Will you please to state where you reside, and what your occupation is?

Answer. My home is Boston, Massachusetts, and I am a practitioner of law in Boston.

Question. Will you state whether you engaged counsel to defend John Brown, who was recently executed in Virginia for offenses against the laws of that State, on his indictment and trial?

Answer. I engaged the Hon. Samuel Chilton, of Washington, who assisted in the defense of Captain John Brown, at Charlestown, and also the Hon. William Green, of Richmond, Virginia, who assisted Mr. Chilton in relation to the prosecution of a writ of error. The fact of the action of these gentlemen is not personally known to me of my own knowledge; I only know it by correspondence and public report. I never had the pleasure of being in Virginia.

Question. Will you please to state under what circumstances you engaged them as counsel; what led you to do it; what was the reason why you engaged them?

The WITNESS. The operative motive on my mind?

The CHAIRMAN. Any reasons connected with it—who employed or engaged you, or why you did it.

Answer. If my motives are deemed——

The CHAIRMAN. Not your motives at all. What I want to know is, at whose instance were counsel employed in Virginia, and who furnished the compensation to the counsel?

The WITNESS. As I was about to remark, if it is desired by the committee to know what operated on my own mind, and led to the employment of these gentlemen through my intervention, I will state with entire freedom, and I hope the gentlemen of the committee will not regard anything I may say as intended to be at all disrespectful to them or to Virginia. When the intelligence reached Boston by telegraph that the local court in Jefferson county, Virginia, was proceeding to the trial of John Brown and one of his associates, with such speed and hurried action on its part as to render it probable that there was to be no sufficient opportunity to make a full and complete defense, and under such circumstances as that the physical condition of the men themselves seemed to render it entirely improbable that they could prepare a defense with propriety, it struck my mind, and the minds of various other gentlemen whom I met with in the ordinary avocations of my business, in the street, the office, the court rooms, and otherwise, as being a judicial outrage. I certainly felt it to be such. It was wholly unlike anything I had ever known or heard of in my practice as a lawyer. When some persons had been indicted for kidnapping, in Massachusetts, last September, the court gave General Cushing, their counsel, two or three months after their arraignment before he was required even to file a plea. Various gentlemen said to me, without respect of party or person, "You, Mr. Andrew, are known to be a lawyer of anti-slavery sentiments, or of Republican sentiments, and of considerable readiness to act on any occasion which seems to you to be proper; why do not you go to Virginia and volunteer to defend Captain Brown?" Without remembering the names of the persons who spoke to me, I should not think it strange if twenty men, of all shades of opinion, might have made that remark; and many persons thought that the circumstances under which this proceeding was going on in Virginia were such as to tend to increase rather than to diminish the ill feeling that the unfortunate foray of Captain Brown had already excited. I said to others, and said to myself, "If I should go to Virginia, I, a Republican lawyer and a Massachusetts man, should be before a court and jury so little in sympathy with myself that I should be quite as much on trial as my client would be. Besides that, I am a stranger to the local jurisprudence and practice of Virginia," (although I was somewhat familiar with the reports, and not unfamiliar with some books, particularly I remember Mr. Robinson's practice, which I read with a great deal of pleasure.) Knowing nobody sufficiently well to take that liberty with him, save Judge Montgomery Blair, of Washington, I at once wrote to him a letter, of which I think I kept no copy—I feel very sure I did not—stating to him how I felt about it and how other gentlemen felt, and I think I also suggested that I thought if Captain Brown was in Massachusetts, charged with any crime, he would not only have a long time given to him to enable his friends to examine into the state of his mind, with a view to testing its sanity, but that it did seem to me an investigation would result in finding testimony, all the way from Boston to Kansas, which would tend to prove him insane. That suggestion I made in the letter, and I made it merely as the result of an inference, not as the result of any facts of which I had personal knowledge. I also said that if Judge Blair would himself go to Virginia, undertake the cause, and see that Captain Brown had a complete and appropriate defense, according to the laws of the jurisdiction where he was indicted, raising whatever questions of law ought to be raised, and having them heard before the tribunal of ultimate resort, I would guaranty to him a proper and honorable compensation; or if he was not of opinion that he ought to go, or if he could not go in person, I would adopt his selection of any other gentleman of the bar, and would guaranty his compensation; that I desired a gentleman familiar with the institutions, practice, and jurisprudence of Virginia, and whose personal presence would not prejudice his client. The result was the employment of Mr. Chilton. After Mr. Chilton had retired from Charlestown, either in consequence of a letter written by himself to me, or a letter written on his behalf by somebody else, I was led to offer, in the same feeling and with the same general view and purpose, a fee of $300 (fixing it in my own mind, because there was but little time to make any bargain about it) to any gentleman from Richmond whom Mr. Chilton should himself select as an associate. Mr. Daniel, of Richmond, and Mr. Green were both spoken of. Mr. Daniel declined, on account of his other engagements, and his letter was sent to me. He recommended Mr. Green. Mr. Green was retained, and I honored the drafts to the amount of Mr. Chilton's fee of $1,000, and Mr. Green's fee of $300. In undertaking to retain and pay these gentlemen, I acted self-moved, except in so far as my own opinion and judgment was influenced by the general remarks of which I have spoken, made to me by friends and neighbors and fellow-citizens of Boston, of various descriptions and opinions. In my letter to Judge Blair, I said I make this application to you in behalf of the friends or of friends of Captain Brown. I felt justified in using that expression; because I could safely call all of us who desired a fair trial of a man—of whom we had, for a long time, entertained a good opinion as an honest man—his *friends*. I felt, also, that I could fairly say, if it were needful, that the application was made in behalf of his family, because I was sure that I must be serving the welfare of a man's family, in seeking to secure for him a good defense. I wished, also, so to express myself as not to place Mr. Blair, nor any other counsel whom he might employ in his stead, in any relation of delicacy towards myself, of the same profession. If I had offered the money as out of my own pocket, or upon my own risk, my friend, Mr. Blair, or any other lawyer, would have doubtless felt a certain delicacy in accepting the retainer, coming from a brother lawyer, influenced only by public or benevolent considerations. I adopted phraseology, therefore, which would steer clear of that delicacy of relation which a direct statement of my precise position would have involved.

Question. Will you state how this money was furnished and by whom furnished? If you can, give the names.

Answer. Without regard to my being in full possession of the funds or not, I accepted the drafts, as they were drawn on me, and the money was furnished by A., B., & C., whom I might happen to meet in business, or in pleasure, or at church.

Question. Was the money furnished at your request, or was it voluntarily proffered?

Answer. I stated to various gentlemen—gentlemen whom I might meet at dinner; gentlemen whom I met at church, in the court house; and others whom I might perhaps take pains to fall in with—what I had done, making the remark, "If you approve of my conduct and think it is right, please to give anything towards the fund which you feel free to give." Various gentlemen, friends of mine, I remember, came in and offered me money which they had collected on the street, as they told me, on State street, on 'Change, anywhere, having said to people: "Mr. Andrew has assumed responsibility for the defense of John Brown, stating the circumstances, do you desire to give anything towards relieving him from the pecuniary responsibility he has undertaken?" In that way, the money came in. Some gentlemen, perhaps, would give five dollars, and some fifty dollars. I knew some of the donors; others I did not know. For example, I remember that I asked a gentleman to state the fact of what I had done, as he might have opportunity, among the members of the legislature, the general court of Massachusetts, then in session, and almost everybody in the legislature knew me personally or knew something about me. The result was that some of the money came from them. It came from lawyers, and merchants, and legislators, and perhaps ladies, although I do not know that any ladies gave anything towards it of my own knowledge.

Question. Will you state, sir, whether your reason for volunteering your aid in this matter and the representations that you made to others, or what induced you to act as you state you did act, was founded on the impression that Brown was not going to have a fair or just trial, or was it founded on a disposition to aid in his defense, because of his career against the institution of slavery?

Answer. Well, sir, I know——

Question. In other words, if you had no impressions that the trial was not one fairly and properly conducted, would you have acted as you did, in getting money for his defense, only from a desire to serve him because of the career in which he was embarked?

Answer. I am quite clear on that point, putting the question in that way. As you, sir, first proposed the question, it was a little complex and intricate. Had I felt that Captain Brown and his associates were in the way to a full and complete opportunity for a fair judicial investigation into all their rights according to the laws of the jurisdiction within which they were, I have no reason to suppose that I should have interfered. I should have felt that I had no occasion to interfere. I had known about old Mr. Brown for several years, and I approved a great deal which I had heard of touching his career in Kansas; I thought he had been an honest, and conscientious, and useful assistant of the free-State cause. My impression of him was derived from many sources. I had never seen him but once in my life, and then only for a few moments. I say in frankness that I felt a certain sympathy for a man who had, as I thought, been useful in behalf of a great cause in which I was interested. I had no sympathy with his peculiar conduct touching which he was then indicted. I felt injured by that, personally, as a Republican.

Question. Suppose the only difficulty connected with his trial as you heard, had been the want of means, would you and your friends then have volunteered to furnish the means to employ counsel?

Answer. It is not easy, Mr. Chairman, for one man to speak as to another's motives. I can only speak as to my own; and you have now put a question which embarrasses me to this extent: It is always unpleasant for a man to blow the trumpet of his own virtue, and I am sorry to be asked to state to what extent I may be a benevolent man, or otherwise. I can only give you one little circumstance, as an illustration of what I might do under such circumstances. Last year a man was convicted in Boston for piracy, and sentenced to be hanged. I had never seen him, to speak to him, in my life, nor did I know by sight any person related to him in any way. After other efforts had been made, I devoted some week, at least, to preparation, and came to Washington, at my own expense, without fee or reward, or the hope of any, in order to press upon the Attorney General and the President those considerations which I deemed proper to be considered in support of the application for executive clemency. The man's life was saved. I never spoke to him until I accompanied Mr. Marshal Freemen to his cell, and assisted in the reading of the President's warrant of commutation. I have sometimes done just such things as that on other occasions. I do not profess to be a particularly benevolent man, but I mention that as an illustration of what I *might* do, even for a stranger.

Question. You have spoken of your opinion that evidence might have been obtained from Boston to Kansas to show that Brown was insane. Will you say whether, as far as you know, it was his general reputation in Massachusetts, that he was insane?

Answer. I cannot answer to that. I took that position in my letter to Judge Blair, in consequence of an inference drawn by myself from circumstances attending the outbreak at Harper's Ferry—the outbreak itself and the circumstances attending it. It was my own inference. I am not aware that I had ever heard it suggested by any man that Captain Brown was insane. I have since been informed that some twenty-five or thirty affidavits were taken in different parts of the country and submitted to the executive of Virginia, in support of some theory of insanity, in behalf of Captain Brown.

Question. Were you aware that a young gentleman named Hoyt had been sent to Virginia as counsel for Brown and his associates?

Answer. I knew that Mr. Hoyt went to Virginia. I personally know Mr. Hoyt. He is a very young man, a very excellent young man, a gentleman of talent, but inexperienced as a lawyer, and he would not regard himself, nor would he be regarded by others, as a gentleman of that degree of professional experience to be placed in a position of such responsibility as the defense of a capital cause, in a strange State, under foreign laws.

Question. Were you aware or cognizant of who sent him, who employed him to go, at whose instance he went?

Answer. To the extent of my knowledge, I can speak, and I have no doubt that I, substantially, know the facts. I think Mr. Hoyt went without any compensation, and I think his expenses, which of course would be small, were paid by gentlemen whom he knew. It is customary with us, as I suppose it is everwhere, for gentlemen of the bar, particularly younger members of the bar, to act as volunteer counsel in capital causes, and even in other important criminal causes, where the parties are not able to procure counsel by compensation. Mr. Hoyt went to Virginia before Mr. Chilton, and when he left Boston I think he had no means of knowing, or suspecting, probably, what I intended to do. He went suddenly, probably upon an impulse. There might have been a little professional aspiration, for aught I know, mingling with his motives.

Question. You have spoken of a a custom prevailing at Boston, and probably at the bar generally, for junior members of the bar to volunteer in criminal causes where the party is not able to pay counsel; is it customary for them to volunteer their services to go out of their own State and to a remote State for that purpose?

Answer. I do not remember any other instance save one, and that occurred in this very case of Brown and his associates, in the person of Mr. Sennott, who is a Democrat, and a supporter of the Democratic federal administration.

Question. What did he do?

Answer. He went in the same way. I think Mr. Sennott had no compensation at all when he went to Virginia—that is, no promise of any, and I do not know that he has ever been paid anything. I do not know whether, in his recent visit to Virginia within a few days past to defend Stevens, Mr. Sennott went as a mere volunteer or upon the promise of compensation; but I am very sure that Mr. Sennott and Mr. Hoyt both went to Virginia originally, without any expectation of pecuniary compensation.

Question. How did you derive that information?

Answer. I am very sure that both Mr. Hoyt and Mr. Sennott told me so. It was a case of a great deal of public impression, as you perceive, and it is not very strange that young men might perceive, or think they perceived in it, an opportunity for some exercise of professional prowess, and that, added to a sentiment of humanity or pity for a man deemed to be in circumstances of hardship and misfortune, would be a sufficient motive to operate on many minds.

Question. Will you inform the committee whether, at any time during the years 1858 or 1859, you contributed money in any form to be paid over to John Brown for any purpose? I mean before the Harper's Ferry affair.

Answer. I never saw Mr. Brown until some time in the spring of 1859. I never contributed any money in aid of any purposes of Mr. Brown's whatsoever, unless contributions which I may have made to the Emigrant Aid Society or to the Kansas committee may have indirectly reached him, of which last fact I am, however, wholly without any means of information. But after having met Captain Brown one Sunday evening at a lady's house, where I made a social call with my wife, I sent to him $25 as a present.

Question. Was that in the spring of 1859?

Answer. Yes, sir. I do not know the date, but it was sometime in the spring of 1859. I do not know whether anybody else gave him any money or not. I sent him $25. I did it because I felt ashamed, after I had seen the old man and talked with him and come within the reach of the personal impression, (which I find he very generally made on people,) that I had never contributed anything directly towards his assistance, as one whom I thought had sacrificed and suffered so much for the cause of freedom and of good order and good government in the Territory of Kansas. He was, if I may be allowed to use that expression, a very magnetic person, and I felt very much impressed by him. I confess I did not know how to understand the old gentleman fully, because when I hear a man talk upon great themes, touching which I think he must have deep feeling, in a tone perfectly level, without emphasis and without any exhibition of feeling, I am always ready to suspect that there is something wrong in the man's brain. I noticed that the old gentleman in conversation scarcely regarded other people, was entirely self-poised, self-possessed, sufficient to himself, and appeared to have no emotion of any sort, but to be entirely absorbed in an idea, which preoccupied him and seemed to put him in a position transcending an ordinary emotion and ordinary reason. I did not regard him as a dangerous man, however. I thought that his sufferings and hardships and bereavements had produced some effect upon him. I sent him $25, and in parting with him, as I heard he was a poor man, I expressed my gratitude to him for having fought for a great cause with earnestness, fidelity, and conscientiousness, while I had been quietly at home earning my money and supporting my family in Boston under my own vine and fig tree, with nobody to molest or make me afraid.

By Mr. Doolittle:

Question. Was the whole amount of money you paid refunded to you, or how much were you left out of pocket?

Answer. I have not carefully examined, for I came to Washington without having any information as to the point towards which the examination of the committee would tend. I have not examined my accounts. Perhaps I am out of pocket $100. If I do not lose more than $50 or $100, besides conducting the correspondence, I am satisfied.

By Mr. DAVIS:

Question. You state that your sympathy with Brown arose from the useful service rendered by him in Kansas for the preservation of good order and government. Will you state what the character of the service was which you so denominate?

Answer. At a time when, according to the best and all the information which I possessed, there was no law, nor official of the law, to protect, or who did protect, the free-State settlers from Massachusetts and from the South, too, I am led to believe that Mr. Brown was efficient, with other men, in the attempt to guard and protect and secure them against unlawful violence from marauders, resident or pretending to be resident in Kansas, and invaders from adjoining slaveholding States.

Question. Did you include in those services what is known as the Pottawatomie murders?

Answer. No, sir; for I have always understood that Captain Brown was not present at the Pottawatomie transaction. I, however, have heard that Captain Brown said that he approved the transaction at Pottawatomie as an action of necessary self-defense, though he was not himself personally present. I was never in Kansas in my life, and am dependent wholly for my opinions on those who have visited Kansas, and who have given me information.

Question. There was another feat of his, that of kidnapping negroes in Missouri, and running them off to Iowa. Was that a part of his services which commanded your sympathy?

Answer. The transaction to which you refer is one which I do not, from my point of view, regard as justifiable. I suppose Captain Brown did, and I presume I should not judge him severely at all for that transaction, because I should suppose that he might have regarded that, if not defensive, at least offensive warfare in the nature of defense—an aggression to prevent or repel aggressions. And I think that his foray into Virginia was a fruit of the Kansas tree. I think that he and his associates had been educated up to the point of making an unlawful, and even unjustifiable, attack upon the people of a neighboring State—had been taught to do so, and educated to do so by the attacks which the free-State men in Kansas suffered from people of the slaveholding States. And, since the gentleman has called my attention again to that subject, I think the attack which was made against representative government in the assault upon Senator Sumner, in Washington, which, so far as I could learn from the public press, was, if not justified, at least winked at throughout the South, was an act of very much greater danger to our liberties and to civil society than the attack of a few men upon neighbors over the borders of a State. I suppose that the State of Virginia is wealthy and strong, and brave enough to defend itself against the assaults of any unorganized unlawful force.

Mr. DAVIS. My purpose is to learn whether the witness and those who aided him in their contributions had their sympathy for Brown excited by deeds of murder and robbery, or whether those acts did not diminish their sympathy.

The WITNESS. I think I ought to say in reply, that I was not aware that I ever heard of the Pottawatomie transaction until since Captain Brown's trial. Therefore, the Pottawatomie transaction could not have affected my mind at all either way. I have not been accustomed to discriminate much between one and another of the Kansas conflicts. They were general; and there were many of them. I had heard of the Ossawatomie affair, but I do not remember to have heard about that transaction at Pottawatomie. I undoubtedly had read of it because I read the report of the investigating committee in 1856. It, however, had passed out of my mind, and I remember that in the affidavits taken by Mr. Oliver on that committee there was but one man who professed to identify Captain Brown as connected with that transaction, and I am not sure that he expressed himself with certainty.

Question. Had you heard of his stealing horses, to be taken into Ohio and sold?

Answer. I had heard it frequently said that, sometime during the controversy between the free-State men and the pro-slavery men, they were accustomed, when they prevailed against each other, to treat their horses as fairly the spoils of war. I am quite confident that I had heard this statement made in connection with Captain Brown, but I did not regard him singular in that respect, and I always believed and do now believe that the free-State men were acting defensively in substantially all that was done by them in Kansas.

Mr. DAVIS. Then it was sympathy for a soldier engaged in such a war as you have described?

The WITNESS. Your question is incomplete, Mr. Davis.

Mr. DAVIS. I will give it any form which will enable you to answer it more satisfactorily to yourself.

The WITNESS. You said it was sympathy——

Mr. DAVIS. The sympathy which you say you expressed or felt towards John Brown, is that which you felt for a soldier engaged in such a civil war as that which you describe in Kansas.

The WITNESS. That would hardly be a fair statement of my feeling.

Mr. DAVIS. I wish merely to get at what your feeling is. It is not a statement, but an inquiry.

The WITNESS. I am constitutionally peaceable, and by opinion very much of a peace man, and I have very little faith in deeds of violence, and very little sympathy with them except as the extremest and direst necessity. My sympathy, so far as I sympathized with Captain Brown was on account of what I believed to be heroic and disinterested services in defense of a good and just cause, and in support of the rights of persons who were treated with unjust aggression.

By Mr. FITCH:

Question. There is a question which, perhaps, would be germane. Without saying to the witness what has, or what has not been in proof heretofore before the committee, we could put this supposition to him: suppose that it had been known that Brown had had in contemplation precisely such a thing as he was guilty of in Virginia, for fifteen or twenty years; that he sought this Kansas service for the very purpose of educating himself and those who acted with him for this ulterior object, would the witness and those who sympathized with him, have sympathized with his Kansas operations, with that knowledge?

Answer. I have no reason to suspect that of myself, nor do I believe of any other gentleman with whom I agree or act, that the transactions of Captain Brown at Harper's Ferry would be deemed justifiable, nor would any such attempt made or contemplated, receive our sympathy.

Mr. FITCH. The answer does not go to the full extent desired. I intended to ascertain from the witness, whether, if he and those who acted with him, had supposed that Brown had contemplated this Harper's Ferry foray, using the means and men they were placing at his disposal in Kansas for that purpose, they would have given him those means, or encouraged him in his Kansas operations?

The WITNESS. Of course not. So far as a man can answer hypothetically, I say, of course not.

By Mr. DAVIS:

Question. You stated when you first saw Brown; will you state when you last saw him?

Answer. I never saw him but once, and I thought it singular that I should not have seen him, for I heard he was frequently in Boston. I was not a member of the Kansas committee or any Kansas association.

Question. Do you know when he was last in Boston?

Answer. I have never heard that Mr. Brown was in Boston since the time when I saw him, last spring. He may have been there, though.

By Mr. COLLAMER:

Question. In the Pottawatomie transaction which has been spoken of, as you understood the thing, did you understand that Mr. Brown was participating in it?

Answer. I will say that I never did believe, and from the best information I have ever received, I do not now believe, that Captain Brown was present, and a participator in the transaction. It would be fair for me to say, I think, with regard to other gentlemen who may have contributed towards this money, that I ought not, perhaps, to be taken as a representative of them all, because I may be a very much more ultra man in my opinions than they. I think there were Democrats who contributed towards that money, though I have not a personal knowledge of the fact. The money was handed towards my fund merely for the purpose of securing a fair trial. I am confident that some people gave under the impression that it would be better for the peace of the country to have it more apparent that Captain Brown was well defended.

JOHN A. ANDREW.

FEBRUARY 10, 1860.

CHARLES ROBINSON sworn and examined.

By the CHAIRMAN:

Question. Will you please to state where you reside?

Answer. I reside at Lawrence, Kansas.

Question. How long have you resided in Kansas?

Answer. Since September, 1854.

Question. Were you acquainted with John Brown, who was recently put to death in Virginia?

Answer. Yes, sir.

Question. State when your acquaintance with him commenced?

Answer. My first acquaintance with him was in November or December, 1855.

Question. Where did you meet him then?

Answer. At Lawrence, in Kansas.

Question. When did you last see him?

Answer. I think it was in September, 1856. It was about that time.

Question. Where then did you see him?

Answer. That was also at Lawrence.

Question. Did you have any conversation with him which would tend to show what his views and purposes were in reference to interference with slavery in the slave States?

Answer. I had a conversation with him at both times, more particularly the last time.

Mr. COLLAMER. You never saw him at any other time but on those two occasions.

Answer. I believe I never had a conversation with him at any other time. I am not positive that I ever saw him at any other time.

The CHAIRMAN. Give the substance of the conversation, as near as you can recollect it, in regard to the subject I inquired about.

Answer. I had a conversation with him about the general state of affairs in Kansas.

By Mr. COLLAMER:

Question. At what period was that?

Answer. That was in 1856. I do not remember about the first interview with him. I found that his purpose was different from mine in regard to Kansas matters. From that conversation I learned that his object was to rather create difficulties and disturbances than to establish a free-State government in Kansas, while mine was for the latter object.

The CHAIRMAN. Give the substance of the conversation—what he said, and what you replied, so as to get out the conversation.

Answer. He said that he did not come to Kansas for the purpose of settling at all. He would never have come there had it not been for the difficulties, and had he not expected those difficulties would result in a general disturbance in the country, and that was what he desired. He desired to see slavery abolished, and he hoped that the two sections would get into a conflict which would result in abolishing slavery.

Question. What sort of a disturbance or conflict did he have reference to?

Mr. COLLAMER. Did he explain it?

The CHAIRMAN. What did he say?

Mr. COLLAMER. Confine yourself, Mr. Robinson, to what he said, as near as you can.

The CHAIRMAN. Not to the language, but to the substance.

Answer. I cannot recall his language again; but I understood him that he expected the difficulties there would result in a collision between the North and the South, and I understood him to be in favor of encouraging or fanning the disturbances there until that would result. I understood that he thought that was an opportunity to get at slavery in the country and abolish it; and he came there for that purpose, and not simply to operate in Kansas, and for Kansas alone. That is where he and I differed, and we could not agree.

Question. Did he develop his views in reference to the mode in which he desired or hoped to carry out that policy?

Answer. No, sir, not definitely; no further than that by encouraging the difficulties there, they would gradually engage the States, and they would become engaged by sympathy with the different parties.

Question. Can you state whether he took any steps, or adopted any measures, in any form, to mold that sort of policy in Kansas, amongst any of the people of Kansas—to get up a party, or diffuse it in any way?

Answer. Not at that time. Governor Geary came at that time, and hostilities ceased for some time. I think they did not commence again until after Governor Denver came, or about that time.

Mr. COLLAMER. Fix your dates as nearly as you can. When was that?

Answer. I do not remember any general disturbances again until they occurred in the southern part of the Territory.

Mr. COLLAMER. At what time was that?

Answer. Governor Geary, I think, left in 1857. It was while Governor Denver was there that the disturbance occurred there; and I do not exactly remember the date that he was there. I know that I went south with him to see if we could not stop them.

By the CHAIRMAN:

Question. What did you find, so far as Brown's views were concerned?

Answer. Brown was not there when I went down. They said he was about there, but I did not see him. Mr. Montgomery was the principal actor that I saw at that time. I heard that Brown was or had been in the vicinity, but I did not come in contact with him.

Question. Were there any United States troops down there? Did any of them go with you or Governor Denver?

Answer. I think there was a company of troops at Fort Scott or its vicinity. They had been sent there, and I believe were there at the time.

Question. Have you any reason to know that any persons or any party in Kansas sympathized or united with Brown in that policy of keeping up hostilities there, with a view to their extension?

Answer. Yes, sir; I know that there were some others who were sympathizers with that policy.

By Mr. COLLAMER:

Question. And for that purpose—to bring on a general conflict?

Answer. Yes, sir. They avowed themselves in favor of it. That is, one man did particularly; and he said others were acting with him. There was but one man, I believe, that ever avowed himself to be in favor of it at any length.

By the CHAIRMAN:

Question. Will you state who that was?

Answer. That was James Redpath.

Question. Was he in the southern part of the Territory at the time you speak of, when you went down there with Governor Denver?

Answer. No, sir; he had left then. He had given up the contest before that. As he told me, he had despaired of accomplishing his object.

Question. What object?

Answer. Of getting up a general disturbance in the country, and abolishing slavery by means of it.

Question. Had you seen him in Kansas?

Answer. Yes, sir; frequently.

Question. Can you remember now during what years he was there?

Answer. The conversation I had with him was in the spring of 1858, I think, or early in summer.

Question. Do you know what his business or occupation in Kansas was?

Answer. He was a reporter, or a correspondent for newspapers.

Question. Of what newspapers?

Answer. When he first came out there, it was 1855, I think. He was then correspondent, principally, for the "Missouri Democrat." He was at times correspondent of the Chicago "Press and Tribune," and of the "New York Tribune;" and he might have corresponded for other papers; I am not certain.

Question. How long did he remain, according to your recollection, in Kansas?

Answer. He was there, off and on, until 1858. I think that he left there and has not been back since 1858. He published a paper awhile, called the "Crusader of Freedom," at Doniphan.

Question. Can you inform the committee whether the purpose and policy of Brown in his scheme of inciting insurrection in the States was known in Kansas, so far as you were informed, during the troubles there or at any time?

The WITNESS. Do you have reference to this matter in Virginia?

The CHAIRMAN. No, I do not mean in Virginia, but any of the

Southern States—this object to make insurrection for the purpose of making a descent upon them?

Answer. I do not believe it was generally believed or known what the purposes of these men were. I think the few I speak of did know, and were in sympathy with him, and were coöperating with him, but I do not think the people knew it. I do not think the people would sanction it at all. As soon as they suspected anything of the kind, they would fall away. There was a movement got up there at one time to massacre all the pro-slavery men of the Territory. It was by these same men, but as soon as it was known the people dropped off from them in an instant. They attempted it and started it, but they had to abandon it.

Question. Do you know where that movement took its origin—who originated it?

Answer. No, sir; it was developed at Lawrence; that is, the plan was announced there in my presence at one time, but it fell still-born the moment it was known.

Question. A plan to massacre all the pro-slavery men?

Answer. Yes, sir; and then to extend it, as I understood, into Missouri.

By Mr. COLLAMER:

Question. Into the border towns?

Answer. Yes, sir.

Mr. COLLAMER. That was after the Missouri people had invaded them and hunted them out?

The CHAIRMAN. You may understand it as it presents itself to you, I understand the witness to say that there was a plan origated at Lawrence, in Kansas, which had for its object the massacre of the pro-slavery men in that Territory.

The WITNESS. I say such an announcement was made; perhaps it was a plan.

The CHAIRMAN. And then I understand him the design was to extend the same policy into the State of Missouri as a slaveholding State.

Mr. COLLAMER. I can ask questions about it when the chairman gets through.

By the CHAIRMAN:

Question. Did you know any of the men who were with Brown at Harper's Ferry?

Answer. Yes; I knew Mr. Kagi, and I knew of Mr. Stevens, and I have seen him, but I do not consider myself much acquainted with him.

Question. Can you state how long Kagi had been in Kansas?

Answer. He was there more or less up to 1858. I do not know but that he was there in 1858; perhaps he was. Yes, he was there when Brown went into Missouri after slaves.

Question. Do you know whether or not he was an asscociate or confederate of Brown's in his plan, such as you have spoken of?

Answer. He was understood to be. He was with him more or less.

Question. What was his occupation in Kansas?
Answer. He was a reporter or letter writer.
Question. For the same class of papers that you have indicated?
Answer. Yes, sir.
Question. Do you know where he came from?
Answer. He was a foreigner, I think. I think he was from England. Most of these letter-writers that I speak of were foreigners.
Question. Is Redpath a foreigner?
Answer. Yes, sir.
Question. Do you know what countryman he is?
Answer. I think he is from England.
Question. Have you any other knowledge that will give to the committee information upon the subject connected with any plans of insurrection in any of the Southern States, except what you have mentioned?
Answer. No, sir. I know nothing about anything outside of Kansas. I know, by report, of their going into Missouri for some slaves. I was never a confidant of them, except Mr. Redpath. After he had had a falling out with General Lane, and gave up the contest, he stopped at my house a few days. He had been a bitter denunciator of me, and all associated with me, up to that time, and then came and unbosomed himself and apologized for the course he had taken in regard to me. He said, as a statesman, I could not have acted differently; but they had another object in view, and he told me what it was. He left the Territory soon after, and said he had given up the contest; he had no hopes.
Question. When did he leave the Territory, according to your recollection?
Answer. The last time I saw him was, I think, in 1858. I think I have never seen him there since.

By Mr. Collamer:

Question. These men of whom you speak, that is, these letter-writers, &c., you say were mainly foreigners?
Answer. Most of them were, that I understood to be in this arrangement.
Question. That is, you understood that from Redpath?
Answer. I do not say that he said they were foreigners; but I know them, and I know they are called foreigners.
Question. But that they sympathized with Redpath and his purposes, you understood from him? Those were the men that Redpath said did that?
Answer. Yes, sir. There were some who were not foreigners whom, he said, they relied upon.
Question. But the persons he was talking about were a limited number of people?
Answer. Yes, sir; quite a limited number, who were in the secret of his movements.
Question. Do you know that there was any connection between these men and Brown?

Answer. No, sir; no more than that Redpath counted on Brown as one of their allies.
Question. That is what he told you?
Answer. Yes, sir.

C. ROBINSON.

[By direction of the Committee, portions of the testimony of this witness, being hearsay only, and deemed irrelevant to the inquiries before them, are omitted.]

FEBRUARY 13, 1860.

MARTIN F. CONWAY sworn and examined.

By the CHAIRMAN:

Question. Will you please to state where you reside, and what your occupation is?
Answer. I reside in Lawrence, Kansas Territory. My occupation is that of a lawyer. I am the general agent of the New England Emigrant Aid Company in the Territory of Kansas.
Question. Will you state how long you have resided in Kansas?
Answer. About five years.
Question. You were the general agent of the Massachusetts Emigrant Aid Society—is that the title?
Answer. No, sir; the title is the New England Emigrant Aid Company.
Question. What were the duties of that office of general agent?
Answer. The duties were to have charge of the business of the company in the Territory, and that business consisted of a general supervision of the property of the company. It was an incorporated company. The property consisted of land and houses, and saw-mills. My duty was to dispose of the lands and houses, collect rents, pay taxes, &c.
Question. Were there no other duties than those of taking care of the property of the company devolving upon the agent?
Answer. No, sir.
Question. Did you receive remittances?
Answer. Yes, sir.
Question. You were of course in correspondence with the authorities of the company in New England, whoever they were?
Answer. Yes, sir.
The CHAIRMAN. The summons required you to bring any papers or correspondence you had, connected with that company.
The WITNESS. I did not so understand it.
The CHAIRMAN. That is my recollection.

Answer. I might as well have brought my private correspondence, I thought, because all my correspondence with that company is of a business character.

The CHAIRMAN. The summons required you to "bring any documents, correspondence, or papers in your possession," &c., "belonging to any associations or committees with which you may have been connected, and which may contain any information or reference, in any form, to the plans or purposes of John Brown, late of Kansas, and which plans or purposes had for their design operations in, or against African slavery in, any of the States, by inciting insurrection amongst the slaves, or aiding them to escape, or otherwise."

Answer. There was not a particle of correspondence in my possession, belonging to the New England Emigrant Aid Company, having the slighest reference to any of the plans of John Brown, or any other plans with reference to African slavery, either in Kansas or in the States.

Question. Was there no correspondence from any other source, connected with producing the effect of insecurity, or otherwise, upon African slavery in the States?

Answer. No, sir; it was of the most purely business character possible, all my correspondence with the New England Emigrant Aid Company. If the committee desire it, my letters are in Boston, in the hands of the secretary there.

Question. Will you state who the secretary is?

Answer. Dr. Thomas H. Webb, No. 3, Winter street. Copies of his letters to me are preserved there. My letters to him are in his possession. I have copies of them preserved in my office. The originals are with him. They can be had any day. I supposed, as they had no connection with this matter, it was of no importance to bring them.

Question. You have said that you received remittances, from time to time, from that company. Will you state what use was to be made of them, by their direction?

Answer. I received remittances, now and then. I remitted money oftener than I drew money. I drew small sums sometimes. When I first became agent of the company, I drew some $250. Afterwards, I drew small sums; I think $200 at one time, and recently, within the last three months, I drew a couple of hundred dollars. I occasionally remitted some. The object of these drafts was to meet demands on the company, for which I had no money in my possession, derived from collections, to satisfy them.

Question. Demands of what kind?

Answer. Demands for taxes and for compensation for agents.

Question. You mean demands relating to the property of the company?

Answer. Yes, sir; exclusively.

Question. Were you connected with any other society or committee, of any of the other States, for operations in Kansas?

Answer. Not a society, or committee, sir.

Question. Were you acquainted with John Brown?

Answer. Yes, sir.

Question. State, if you please, when you became acquainted with him, and where?

Answer. I met him for the first time in Boston, in the winter of 1856; I think, about December, 1856.

Question. What brought you to his acquaintance?

Answer. I was invited to Boston by Dr. Thomas H. Webb, in the fall of 1856. I was then sojourning at my old home, in Baltimore city. I was invited to Boston for the purpose of telling to the people there the story of Kansas. He had been there, presenting the cause of the people of Kansas to those in that region of country who were interested in that subject.

Question. Had you been previously in Kansas?

Answer. Yes, sir; I went to Kansas in 1854.

Question. And remained there until 1856?

Answer. Yes, sir. I remained there until the summer of 1855, when I was compelled to leave the country in order to save my life. It was then under a reign of terror. I had been away before, though, occasionally.

Question. Have you said by whom you were invited to Boston?

Answer. Yes, sir. I said, a few minutes ago, that I had no connection with any company, or corporation, or committee, in the least, in Kansas. That is true; but I was connected with a committee in Massachusetts there, the Massachusetts State Kansas Committee. My connection with it was of this description: when I went to Boston, under invitation of Doctor Webb, I was desired to act in connection with that committee, in going through the State of Massachusetts and attending meetings called through the agency of this committee, in different places.

Question. What is the style of that committee?

Answer. The Massachusetts State Kansas Committee, and telling the people at the different meetings what had transpired.

By Mr. COLLAMER:

Question. Was not that the Kansas aid committee?

Answer. It was for the purpose of aiding Kansas, but that was not the style of it. My action in connection with that committee was limited exclusively to the mere matter of telling the people who were assembled at the different places, from time to time, what had happened in Kansas, and what the merits of the cause of the people of Kansas were in the contest that was then going on out there. That was all the connection I had with that committee; I had nothing to do with any collections which were made, or with any disposal of the money obtained.

By the CHAIRMAN:

Question. Were collections obtained at these various meetings?

Answer. The object of that committee was to have Kansas presented to the people there, and then their *modus operandi* was to have committees appointed by the meetings; these committees to make collections, and these collections to be remitted to Boston to the Massachusetts State Kansas Committee, to its treasurer—Mr. Jackson, I think his name was.

Question. Were collections taken up at the various meetings which you addressed?

Answer. No, sir; but committees were appointed at these meetings, and sometimes the committees were already appointed. I remained in Massachusetts until April, 1857, when I returned to Kansas. During this time I met with John Brown; he was in Boston; he seemed to be engaged in a similar work, though I believe he was not acting in connection with this committee. I heard him tell his exploits at several meetings; I talked with him frequently there.

Question. Do you remember by whom you were introduced to Brown?

Answer. I cannot call that to mind.

Question. Did he go through the State, and was he present at these meetings you have spoken of?

Answer. No, sir; I do not think he was at any of the meetings at which I spoke; he had no association with me in the work at all; in fact, nobody had.

Question. Did you go alone to these meetings, or were you attended by anybody from Boston?

Answer. Generally, I was alone; sometimes I would go in company with others; once or twice I was attended by other gentlemen.

Question. You were deputed to that duty by this committee in Boston you have spoken of?

Answer. I was in one sense—that is, they fixed the times and places of the meetings, and I attended the meetings and spoke.

Question. Do you know what amount of money was collected by that committee in the whole?

Answer. I do not know, but I think there was quite a large sum; I do not know what was done with the money; I had never anything to do with the receipt or disbursement of the money; it never passed through my hands or came in contact with me at all.

Question. Did you meet John Brown afterwards in Kansas?

Answer. Yes, sir; I met him twice afterwards in Kansas.

Question. Do you remember in what years?

Answer. I met him in the fall of 1858; I met him once before then, incidentally; but I do not recollect anything about it any more than he was present, or came into a room where I was at one time, in Lawrence, with several others, and was there for a few minutes. I do not remember anything distinctively in connection with him at that time; it amounted to nothing, whatever it was.

Question. Did you ever talk with him, or he with you, as to what his views were in reference to slavery in the States?

Answer. Oh yes, sir; quite at large on one occasion.

Question. Will you state what his views were as to his own purposes in life in connection with the system of African slavery?

Mr. Doolittle. Give the time and place.

The Witness. Yes, that had better be fixed. I met him, I think it was, late in July, or early in August, of 1858. A friend of his, a person who seemed very closely connected with him, named Kagi, the same I believe who was killed at Harper's Ferry afterwards, called at my house and told me that Brown was in the town, and that he wished to see me; he told me he was staying at the house of Mrs. Killan, a public house in the town; I told him I would call; I did so, and Brown told me that he had been sick; he looked very feeble at the time; he said he had the fever in southern Kansas; had been lying there for several weeks, and had hardly recovered; that he was very much in need of money; he had received, he said, an order from the national executive committee for some large sum of money, which he had never been able to get; that the——

The CHAIRMAN. Was that the Boston committee?

Answer. No, sir; the national committee.

Question. What committee was that?

Answer. That, I presume, was the committee of which Thaddeus Hyatt was president.

By Mr. DOOLITTLE:

Question. The committee formed at Buffalo?

Answer. Yes, sir. He said he had received this order, but had never been able to get the money; that Mr. Whitman, who was an agent of that committee in that place, had not acted well toward him in the matter. Mr. Whitman, it seemed, was authorized to make collections of debts due the committee in the Territory, and was required, I believe, from Brown's representations, to satisfy that claim of Brown's out of the proceeds of his collections. He had not done it and Brown seemed to be aggrieved by it. He was complaining to me of it. Whitman, I think, was not in the town at the time, but about that I do not recollect distinctly. From that he went on to talk of his troubles in the Territory, how he had suffered there, and of the spirit that was manifested by the pro-slavery party, and of the manner in which they seemed disposed to dominate over the country. He thought that they would have to be met in a very decided way. He thought that there was only one thing could check them, and that was force.

By the CHAIRMAN:

Question. How did he propose to check them by force? Did he develop his plans at all?

Answer. No, sir; he seemed to talk all around something, but did not make any revelation whatever to me. I manifested no disposition to receive any suggestion from him of anything that he might purpose doing. His connection with me was that which brought us together in defense of the rights of the people who inhabited Kansas, and I had no disposition to go any further. I always felt satisfied that Brown was not content to limit himself to that work.

Question. That work in Kansas, you mean?

Answer. Yes, sir.

Question. What did you suppose his views were as to extending some of his operations beyond the defense of the people of Kansas?

Answer. As to facts I know nothing; he never told me anything about it.

Mr. DOOLITTLE. Would it be advisable to take the witness's supposition?

The CHAIRMAN. Not unless he knows, from conversations with Brown or facts brought to his knowledge, what his plans were.

The WITNESS. I know nothing whatever from any facts or conversations with him as to any plans on the subject. In the other conversations we had he confined himself entirely to matters of business and seemed not disposed to force any views on me of that kind. He seemed also disposed not to connect me with himself in any public way. Afterwards he called at my house, and took occasion to do so at a time when he might not be observed. I noticed it, because it seemed to me to suggest that he understood I was not in his line if he had any such purpose as I supposed he might have of general hostility in a forcible way to slavery in the States.

Question. Did you give him any encouragement to develop his plans by putting questions or anything of that sort?

Answer. No, sir; it was just the other way with me.

Question. You showed a disinclination?

Answer. Yes, sir.

Question. Was it not in the fall of 1858 that he made that last incursion into Missouri when he captured slaves?

Answer. I think that was after his interview with me.

Question. Did he refer to that purpose in any way?

Answer. No, sir.

Question. Did you see him afterwards?

Answer. Yes, sir.

Question. When and where did you see him?

Answer. He came again, I think it was about a month afterwards, perhaps not so long as that. He stopped at the same place, I believe. He called at my house with a bundle of papers, told me that the National Kansas Committee had passed a resolution sometime before upon which he based a right to act himself as agent for that committee in the Territory in the collection of debts due it, and as Mr. Whitman did not seem to satisfy him in that business, he had taken it upon himself to make collections.

Question. Had he received a commission to do it?

Answer. He claimed to have received a commission, and as a result of his labors he produced a package of papers, which he said were promissory notes from parties in the Territory, who had received provisions and clothing from this committee during the troubles in 1856. They had engaged to pay for them and they had given these notes, and he had got them, and he came to me to ask me a favor that I would take these documents and put them in my safe and keep them subject to his order. I agreed to do so.

Question. What became of them afterwards?

Answer. His friend Kagi sat down at my desk and wrote a receipt for them, enumerating and describing each note. I glanced over it and saw what the general character of the document was, and I signed it, and he took it.

Question. Do you remember the amount?

Answer. No, sir, I do not. I did not examine the papers at all; I glanced them over and put them in my safe. Mr. Whitman afterwards called on me several times and claimed them, saying that Brown had acted improperly in the matter; that he had never heard of Brown having a commission from that committee, and did not believe he had any, and desired me to give them to him. I refused to do so; told him it was none of my quarrel, that I wished Brown and him to settle the matter between themselves; that I had engaged to take the custody of these papers, and that they were subject to the orders of Brown, and I did not feel at liberty to give them up to him. So they remained with me. Finally, Brown was executed, and I concluded that there would have to be some disposal made of them, and I consulted with several persons about it. I had occasion to visit the East not long since, and I inquired of several gentlemen what I ought to do with them. A gentleman in Chicago, who was connected with some of the committees in some way, Mr. White, at the office of the "Press and Tribune," and Mr. Thaddeus Hyatt I spoke to. I spoke to Mr. Sanborn, who was connected in some way with a committee in Boston, and upon his advice, I forwarded these papers to the Massachusetts State Kansas Committee.

Question. You do not remember their amount?

Answer. No, sir; I never knew it.

Question. They remained in your possession until after Brown was executed?

Answer. Yes, sir.

Question. Can you state, from your recollection, whether from the appearance of those notes and the persons whose notes they were, they were good notes, notes which could be collected.

Answer. I think many of them were.

Question. Were any of them for a large amount?

Answer. Not over \$100 or \$200, and ranging, I presume, from \$25 to \$200. I should suppose so from what I saw of them.

Question. Can you tell whether there were any notes given for contributions, or otherwise than for supplies?

Answer. I am not certain for what they were given, but it is my impression that they had been given for supplies.

Question. You have no personal knowledge of them having been given for any other purpose?

Answer. I have no personal knowledge of them at all. They were brought to me and left in my hands, and they remained there, and I did not look at them at all until I came to forward them to Boston, and then I glanced over them.

Question. That was about a month after your interview with Brown, in July or August, 1858, you say?

Answer. Yes, sir; I shall not be precise about these dates, but it was somewhere about that time. It was late in the summer or early in the fall of 1858, when Brown first called on me in this matter.

Question. Did you see Brown after that?

Answer. I never saw him after that.

Question. Do you know what became of Kagi?

Answer. No, sir; except I saw his name in the papers connected with the Harper's Ferry difficulty.

Question. Did you never see him after he left you at that time?

Answer. He may have been at the hotel when I called to see Brown; I do not remember; but I had no conversation with him.

Question. Do you know anything of what his subsequent movements were in Kansas?

Answer. No, sir.

Question. Did you hear nothing of him or his movements in that section of the country?

Answer. I think I heard he was in southern Kansas a while after that, in some trouble they had down there. There was a party marched into Fort Scott, I think, and killed a man in retaliation from some provocation that had been given by people there, and I think I heard of his connection with that, though I do not know what foundation there was for that report.

Question. Did you ever see or hear of a man named Hugh Forbes out there?

Answer. I never saw or heard of him until his name appeared in connection with the Harper's Ferry affair. We supposed, in Kansas, that he was a myth.

Question. During the winter of 1857–58, or about that time, did you hear, or were you aware in any way, of the collection of a parcel of young men, Kagi amongst them, for purposes of military training?

Answer. No. I heard that Kagi, Cook, and Realf, and some others who had been in Kansas, had gone away with Old Brown somewhere, for some purpose. I think military drill, or something of that sort, was connected with it. There was such a rumor through the community, when it would be asked where this man was and that man was, Realf or Cook, or any of these young men, it was said they have gone off somewhere to be drilled, or something like that.

Question. Was there any rumor connected with the purpose for which they were to be so drilled—what was the object of their training?

Answer. No, sir.

Question. Any inquiry made into it, that you know of?

Answer. No, sir. There were general views in that matter. It was thought that other Territories were to be opened, and that there would be trouble hereafter in connection with the slavery question in the Territories; and my idea, when I heard of this, was that perhaps Old Brown, with these young men, was preparing for the opening of the Southern Territory below Kansas, or Arizona, or something of that sort, that was mentioned.

Question. You were aware, I suppose, of course, of a law passed by the territorial legislature creating a military board?

Answer. There was such a law.

Question. Were you aware that those officers were placed under Lane?

Answer. Yes, sir.

Question. Did you know what Lane's views were in connection with the services of that board; what objects he had planned or contemplated in connection with the use of that board?

Answer. No, sir, I did not. I had very little to do with that enterprise. I cannot say that I remember anything about that. The fact is that Lane was engaged in so many things of that kind, with such general and indefinite objects, that I could not tell. I do not know that he had any definite object himself.

Question. Were you a member of the legislature?

Answer. No, sir. I think it likely that Lane was looking out for the organization of the State government, and fixing things up for it.

Question. By means of that military board in any way?

Answer. Oh, well, probably. I think that these military boards were very frequently thought of as good instruments to control elections with, and were, therefore, designed as much for civil as for military operations. I do not suppose that Lane had any idea of invading the State of Missouri, or the State of Virginia, or any other State on the globe, by means of his military board.

Question. Have you any knowledge of the orders he gave to the officers of that board, or the service on which he sent them?

Answer. No, sir.

Question. Do you know anything else that would give to the committee information as to the purposes or plans of John Brown when he was last in Kansas, or after he left there?

Answer. No, sir; I cannot think of anything at all. I never saw him afterwards.

Question. Do you know of any contributions that he received, in money, in Kansas, in 1857 or 1858?

Answer. No, sir. I remember this, a while after he was at my house, he seemed to be suffering and in need, and said he was so, and that he would have to have money. He appeared to be very much distressed, and a while after that I received a letter from a gentleman in Boston, whose business agent I am in the Territory, saying to me in substance that, if Old Brown called on me or sent to me for money, to let him have $50, drawing on him for it. I supposed, as a matter of course, as this was a benevolent gentleman, he had heard of Brown's necessities, sickness, and need, and had sought to relieve him in this way. I mentioned that fact to several in Lawrence, that I had a little money for Brown, and, after the difficulty at Harper's Ferry, a great point was made of that, and some people seemed to think it made me one of Brown's insurgents. He sent a person to me for that money, and I gave it to that person.

Question. How long was it after you received the order to pay it, that he received it?

Answer. It must have been in January of 1859.

Question. Who was the person to whom you paid the money for Brown?

Answer. A man named Joel Grover, living in the neighborhood of Lawrence.

Question. Was it paid upon Brown's order?

Answer. Yes, sir.

Question. Will you state who the Boston man was that instructed you to pay it?

Answer. It was Dr. Howe.

Question. Then you received the order from Dr. Howe sometime in the winter of 1858?

Answer. Yes, sir; I must have received it in December, 1858, and I paid the money in January, 1859.

Question. Do you know of any other money that Brown received there?

Answer. Never another cent.

Question. What you stated, as to your supposition that it was from the benevolence and humanity of that gentleman, is supposition only?

Answer. Supposition only.

Question. He did not tell you for what purpose it was to be used?

Answer. No, sir.

Question. Dr. Howe accounted with you for the $50 afterwards?

Answer. Yes, sir. I had business relations with Dr. Howe, and charged that to his account.

By Mr. FITCH:

Question. Do you know what committee had charge of or exercised control or ownership over the arms which are said to have been used by Brown?

Answer. No, sir, I do not.

By the CHAIRMAN:

Question. Were there any arms under the control of or forwarded by any committee with which you were connected?

Answer. No, sir. There were arms forwarded to me at Lawrence, about a year and a half ago, from St. Louis—arms belonging to a gentleman in Boston, which were taken away from certain parties on the Missouri river in 1856, at Lexington, and remained there for a long while, and were afterwards recovered by action at law, and were sent up to me at Lawrence, and put in my charge.

Question. What amount of arms, and what description?

Answer. I think 100 Sharp's rifles.

By Mr. FITCH:

Question. What time did they reach you?

Answer. It must have been in 1858.

By the CHAIRMAN:

Question. What became of them?

Answer. They are in my charge now.

By Mr. FITCH:

Question. In whom is the ownership?

Answer. Dr. Samuel Cabot, of Boston.

By the CHAIRMAN:

Question. Are they still his property?

Answer. I presume so. I know nothing to the contrary. They are in my charge, subject to his order.

The CHAIRMAN. Perhaps it would be as well for the witness to state, as far as he knows, the names of the different committees that were organized outside of Kansas for the purpose of operations in that Territory?

Answer. The Massachusetts State Kansas Committee was one, the National Kansas Executive Committee was another.

Question. Where was that formed?

Answer. I think in Buffalo; a convention was held at Buffalo, and that committee appointed; the seat of its operations was in New York chiefly, yet the Massachusetts committee was mixed up with it in some way, by reason of the members of the Massachusetts committee being members of the national executive committee.

Question. Do you recollect any more?

Answer. No, sir.

The CHAIRMAN. There was the Emigrant Aid Company.

Answer. The New England Emigrant Aid Company was purely a business concern; that was an incorporated company.

Mr. FITCH. A real estate speculation company?

Answer. Yes, sir; nothing else in the world.

The CHAIRMAN. But I want to know what they were.

By Mr. COLLAMER:

Question. Had the Emigrant Aid Society, which was an incorporated company, ever any other agent in Kansas but you?

Answer. Yes, sir; I did not become its agent until the spring of 1858; I was appointed agent in March, 1858.

Question. Was that in operation in 1854, when you went there?

Answer. Yes, sir; I had nothing to do with it then, though.

Question. Had that society ever anything to do with any arms or ammunition, or war supplies of any kind?

Answer. From what I had understood of it, it had not; I remember when I first thought of going to Kansas I heard of this Emigrant Aid Company, but I did not know the nature of it, and I thought it a good idea, and I wrote to Mr. Thayer about it to see what advantages I could get in my emigration to Kansas by connecting myself with it in some way, but his answer was very brief and unsatisfactory; I learned then that it was simply a company for the purpose of making money by speculating in land, putting up saw-mills, and building hotels, and taking land as a consideration, and holding the land for the profit they could make on it in the end.

Question. You found, by being agent for it, that that had been its business?

Answer. Yes, of course.

By Mr. DAVIS:

Question. Did it not send out emigrants?

Answer. It sent out no emigrants; I was told that I could get my ticket to go to Kansas at less price, a few dollars less in price by means of it; that is, it would make me acquainted with this man, that man, and the other, and we would all go out together, and, by going out together, we would get our tickets at a lower price. I did not know they could do me any particular favor by making me acquainted with this man, that man, and the other; that did not strike me as particularly advantageous in my case; and as for the small reduction in the price of fare, that was nothing.

Question. Did they send persons to occupy the land which they were buying in Kansas?

Answer. The parties which they would get together in this way, by making people acquainted with each other, would be disposed to go to such place as they would direct, as people would wish to get together for mutual protection. The idea that the Emigrant Aid Company was interested in a particular place, and had put in a saw-mill there, and was going to build a hotel, a large moneyed corporation would leave the impression that that was going to be a great place, and every body would wish to go there, and so these companies that would be formed by them in this way would naturally go to those places. In that way they first went to Lawrence, then to Topeka, then to Ossawatomie, and then to Manhattan.

Question. Did the company never advance the transportation of those persons that went out?

Answer. Not to my knowledge.

Question. Did they have any lien on the land occupied by these persons after they went out?

Answer. No, sir; not to my knowledge.

By Mr. COLLAMER:

Question. I want to know when the Massachusetts State Kansas Committee was formed; what was the time and what was the occasion when it was formed?

Answer. As far as I know, it was formed at Faneuil Hall, in the latter part of May, 1856.

Question. What was the occasion of its being formed?

Answer. News was received of the sacking of Lawrence, and there was a great excitement in Boston, and a great meeting held at Faneuil Hall to consider what they should do, and the result was the appointment of this committee, to raise funds for the support of the people of Kansas in their contest with those who had invaded the Territory.

Question. And at the request of that association you went about the State of Massachusetts to state the condition of things in Kansas?

Answer. Yes, sir. I was not there at that time. I only speak of the manner in which the committee was formed by report. I went there afterwards. I had been driven out of Kansas.

Question. How were you driven out of Kansas?

Mr. DAVIS objected to the question as irrelevant, and, after consultation, the committee sustained the objection.

Mr. COLLAMER. Did you say you knew anything about any committees raising arms?

Answer. No arms.

Question. You did not know of any?

Answer. No, sir.

Question. You knew nothing about any arms sent, or attempted to be sent into the Territory, except those you have spoken of, belonging to a gentleman in Boston?

Answer. No, sir.

Question. When were those sent?

The WITNESS. Attempted to be sent?

Mr. COLLAMER. Yes, sir.

Answer. They were attempted to be sent in the summer of 1856, and were seized on the Missouri river, at Lexington. A large crowd assembled there, stopped the boat, and took the guns and everything else away from the men who were on board the boat, and sent them back.

Question. Do you know of Lane having any connection, or transacting any affairs in any way with Brown?

Answer. No, sir. I do not think Lane had anything to do with Brown, or that Brown had anything to do with Lane.

M. F. CONWAY.

FEBRUARY 17, 1860.

AUGUSTUS WATTLES sworn and examined.

By the CHAIRMAN:

Question. Will you please to state where you reside?

Answer. I reside in Moneka, Linn county, in the southern part of Kansas Territory.

Question. How long have you resided in Kansas?

Answer. About five years.

Question. Were you acquainted with John Brown, who was recently put to death in Virginia, under the laws there?

Answer. Yes, sir.

Question. When did your acquaintance with him commence?

Answer. I had a knowledge of him in Ohio, several years ago, but I was not intimate with him until 1855, in Kansas. I saw him first in Kansas in the fall of 1855.

Question. Did your acquaintance continue with him in Kansas from that time until he left Kansas?

Answer. Yes, sir.

Question. When did you last see him in Kansas?

Answer. I do not recollect the month, but it was in the winter of 1858–59.

Question. Where did you see him then?

Answer. I saw him at my house.

Question. Who was with him?

Answer. There was nobody with him at the time I allude to. I was confined to my bed by sickness, and he came in to bid me good-bye as he was leaving. He had been into Missouri and taken those slaves, and was going out of the Territory.

Question. Had he the slaves with him at the time?

Answer. I suppose they were under his charge at the time. I do not know particularly whether he had them with him or not. He had not them in my house with him.

Question. That was in the fall of 1858?

Answer. The fall or winter, I do not recollect the month.

Question. How long did he remain at your house at that time?

Answer. I suppose half an hour.

Question. Had he been a previous visitor at your house?

Answer. Yes, sir.

Question. Did he ever remain with you for any long time?

Answer. He frequently remained a considerable length of time.

Question. Were there other men with him at those times that he was your guest?

Answer. In the fall of 1856, when he was driven from the Territory by the United States troops, he came to my house to stop, and his sons and sons' wives. They were at my house some time; I cannot tell how long; I should suppose more than a week.

Question. How do you mean driven from the Territory; for that was in the Territory?

Answer. The troops were attempting to arrest him, and he came to stop with me on his way from Ossawatomie to Nebraska. He was attempting to collect his cattle together, and what property he had in Ossawatomie, and sell it; and he had the women there to take them down the river; and he had his sons there, and their wagons, to go back to Ohio.

Question. That was in 1856?

Answer. Yes, sir; that was in 1856.

Question. Did you see him there in 1857?

Answer. No, sir; I do not think I did. I have no recollection that he was in the Territory in 1857.

Question. Did you know a man named Kagi?

Answer. Very well, sir.

Question. Did you know Realf?

Answer. I knew him when I saw him. I was not intimate with him. I was not very intimate with Kagi; but Kagi was connected with the press, and used to be in the office very frequently.

Question. What office?

Answer. The printing office where I was.

Question. What press was that?

Answer. The "Herald of Freedom," published in Lawrence.

Question. Were you one of the editors?

Answer. Yes, sir.

Question. Was your residence in the town of Lawrence?

Answer. I lived about seven miles from Lawrence. I used to stay in Lawrence a good deal. My family lived out of the town at that time.

Question. What did you say was Kagi's connection with the press?

Answer. He was a correspondent of some paper here in Washington, and he used to be passing in and out to get newspapers to read.

Question. Was Kagi, as far as you know, connected with Brown in any of their fights or battles in Kansas?

Answer. Yes, sir; I think he was with Brown all the time, or at least connected with him; not personally with him all the time, but he always knew where he was, so that he could go to him. I understood it so generally.

Question. Did Brown ever, in conversation with you or otherwise, develop to you what his plans were in reference to the abolition of slavery either in Kansas or outside of Kansas?

Answer. No, sir; I never heard him speak on the general question in the manner in which you are presenting the subject. I have only heard him speak in reference to our plans there, as far as his action was concerned, which was simply defensive; but in discussing the principles of abolition, I have heard him give sentiments like these; in conversing with him on the subject once, he said: "I have been at your abolition meetings, mentioning in Massachusetts and Ohio, and your scheme is perfectly futile; you would not release five slaves in a century; peaceful emancipation is impossible; the thing has gone beyond that point." I recollect this distinctly from the ridicule which he attached to a remark I made. I said that a forcible emancipation was worse than slavery. He said that his plan was to put arms in the hands of the slaves; give them their choice, stand behind them so as to protect them in a free choice; give them a free choice, and if they chose to go into slavery, let them stay in it; but if they chose to go out, sustain them in it. I said it was an impossibility to give them arms, referring to the expense and difficulty of furnishing them. He said he had a plan for an arm for them better than a musket—a long pike. What he said as to emancipation in that way, I supposed was a mere matter of opinion, which I had no idea had anything practical connected with it.

Question. Did he tell you whether, and how, he proposed to carry out plans of that kind, of putting arms in the hands of slaves?

Answer. No, sir; I had no idea that he had any plan of the kind; never heard him allude to anything further than conversation which men frequently have.

Question. Were you at Fort Scott, or in the neighborhood of Fort Scott, at the time of the troubles or difficulties there?

Answer. No, sir; I was at home; that was at the time I speak of when I was sick. He went near to Fort Scott, and was remaining there while Kagi and others went to Fort Scott and released a prisoner who was a member of their company.

Question. Who was he?

Answer. They call him Ben Rice. I do not know whether his name is Benjamin Rice or Ben Rice. It is the familiar name he goes by.

Question. Do you know that that is his true name?

Answer. I presume it is; I have no means of knowing that it was not.

Question. He remained where, when Kagi went to Fort Scott?

Answer. I understood—this is conversation I had with them, I was not there—he remained on the Little Osage, at a fort that he had built.

Question. What is the distance of your residence from Fort Scott?

Answer. Twenty-five miles.

Question. Do you know whether General Lane was down there at the same time?

Answer. No, sir, not at this time. The time that I am speaking of is the last week that Brown remained in Kansas before he left; General Lane was there before, in the winter of 1857.

The CHAIRMAN. I do not remember what year it was, but a party with which I understood Brown was associated, made an attack on the village at Fort Scott, and took possession of it. When was that, or when did Lane do it?

Answer. Neither of them.

Question. Who were the party that did that?

Answer. I am not sure whether Captain Montgomery or Kagi commanded. I think they were both considered rather leaders of small parties that did it.

Question. Was neither Lane nor Brown present?

Answer. Neither of them. Lane was never at Fort Scott at any of those difficulties.

Question. Was there no further hostility or violence there except in liberating this prisoner.

The WITNESS. At Fort Scott?

The CHAIRMAN. Yes, sir.

The WITNESS. There was nothing only what grew out of that. Mr. Little was killed, and his store was robbed.

By Mr. COLLAMER:

Question. When was that?

Answer. That was in the winter of 1858–59, I do not recollect the month.

Question. It was just before Brown left the territory with those slaves?

Answer. Yes, sir. While they were doing that, Brown remained sick at his fort on the Little Osage, twelve miles north of Fort Scott. While he was remaining there, a negro man came up from Missouri, wandering along by chance, and fell in with Brown. He told Brown that he was looking for some man to help him to run away from Missouri; that his master was dead; that he owed no service to anybody in particular except heirs; that he did not know when he was to be sold with his family, and he wanted help to bring them away into Kansas, and Brown made arrangements with him to go down after him the next day. He went down. The story which is reported in the newspapers about that is correct, I suppose, substantially. At the time, I saw it in the newspapers, and I heard it conversed about also.

The CHAIRMAN. I do not know that that is important. I only wanted to get at Brown's connection, if there was any, with that attack on Fort Scott.

The WITNESS. I suppose Brown was advisory to it. I knew nothing about it until it was over.

Question. Did you get that information from Brown himself that he counseled it?

Answer. No sir; I did not. I wish to say here while on this subject that when Brown came to the Territory in the spring or summer of 1858, he came to my house to know if he could be of any use in defending the frontier. Hamilton and others had come in and had killed——

Mr. COLLAMER. From Missouri?

Answer. They were not Missourians; they were people that had been driven out of Kansas. There were three brothers Hamilton. They were Georgians. Most of these people, I suppose, had been driven out of Kansas in the fights on the border. They assembled in a company and came in from Missouri and took a number of prisoners, who were men at work on their farms and travelers on the road, and shot them; and it was a great shock to the community. They gave out word that they were going to take all the settlers in Linn county and shoot them in the same way. We all assembled; some 200 men, more or less, more I think, assembled on the line and detailed a company to stand guard all the time; to ride up and down the line and keep watch of this body of men and see that they did not break in. It happened in May, just at the time people should be plowing and planting, and it took citizens away from their work. Brown came in at this time and wanted to know if he could be of any service in guarding the line. I told him that he could, and we should be very glad to have him. We had sent to Governor Denver for arms, and to come down there; and the governor had promised to assist us. At my suggestion, a paper was drawn up, which Brown signed, and all the men who went into his company to guard the line signed, stipulating that he should not go into Missouri on any provocation whatever, and that no man in Kansas should be disturbed for his political opinions. I signed that also, and all the citizens to whom it was presented, who lived along the border, signed it. I do not know how many. I did not see it after there were eight or ten names to it. Brown went on to the claim where these murders had been committed—the Marais des Cygnes murders—bought the claim, and fortified it, and gave out word that he was there either to fight or be peaceable, as they might choose; that he was Old Brown; and that they could make as good a neighbor of him as they wanted, or as bad a one. He remained there a month or two, more or less, and these men passed out of the State of Missouri. Mr. Hamilton and Titus and others passed away, and the troubles ceased—that is, all danger from them disappeared. Brown went then away about his business. He was taken sick and came to my house and stayed, perhaps, two weeks.

By the CHAIRMAN:

Question. Was that in the month of May?

Answer. I think the murders were committed in May. Brown went there in June, and perhaps in August he came to my house to be nursed, while he was too sick to lay there.

Question. What year was that?

Answer. 1858.

Question. Did you hear him speak of any plan that he had of putting a number of young men to a military school or military training during that winter?

Answer. No, sir; I never heard him speak of it I heard some young man in Lawrence—I do not know who it was; whether it was Realf or some other one—say that they were going to take military lessons of an English officer. I am not positive who told me. It was a matter that I heard in conversation.

Question. Was the object of the training expressed?

Answer. No; I had no idea that it was anything more than is common all over the free States, where the young men drill and learn military tactics.

Question. Will you look at this paper and see if it is your handwriting, and, if it is, say to whom it was addressed? [Exhibiting to the witness the following letter:

"MONEKA, K. T., *March* 29, 1859.

"DEAR FRIEND: Your favor of the 10th instant was received last evening. We were gratified to hear from you and of your success. We had followed you with anxious hearts from point to point on your perilous journey. Be pleased to let us hear from you from time to time, as you have opportunity. We are all well, and have been neither frightened nor hurt, though in constant peril of assassination or arrest. The pro-slavery party has defeated itself, more by their own stupidity than by our smartness. We vote on the county-seat in June. Send all the abolitionists here you can.

"Please continue that writing which you begun at my house. I am a member of the historical society of Kansas, and am appointed on the department of *biography*. Please make a note of this, and act accordingly.

"Yours truly.

"Dr. Weaver killed himself, I presume you have heard, while bringing in guns from Missouri to murder his neighbors with. It was a providential interference for our protection, I have no doubt."]

Answer. That is addressed to John Brown, in answer to a letter that he wrote to me. It is my handwriting.

The CHAIRMAN. It is not signed.

Answer. It was published with my name to it. How did they know that?

The CHAIRMAN. That is a different matter. Why did not you sign your name to it?

Answer. I do not know now. I supposed it was signed. When I saw it in the newspapers, I supposed it was signed. I cannot give any reason why it was not signed, unless it was that I supposed my letters might be opened. It was a common thing for letters passing in and out, to and from Kansas, to be opened and read, to the disadvantage of the writer.

Question. Opened where, and by whom?

Answer. We supposed they were government officials, and that they were placed somewhere as spies, either at the distributing office or some other post office. That was the supposition.

The CHAIRMAN. The way it was known to be your letter, I suppose, was the fact that it is indorsed here, in Brown's handwriting, "A. Wattles's letter; answered May 18th."

The WITNESS. I have Brown's letters in my pocket, that he has written to me. According to your orders, I brought them all along that I could find.

Question. What "success" was that you referred to in this note?

Answer. In getting away from the enemies that were following him up through Kansas. I understood the United States troops were up after him, through the northern part of the Territory, and companies of men went from Atchison and elsewhere to arrest him.

By Mr. COLLAMER:

Question. Was that when he carried those slaves away?

Answer. Yes, sir.

By the CHAIRMAN:

Question. Why did you fear assassination or arrest?

Answer. That has no connection particularly with this. There was a *posse* got up there——

The CHAIRMAN. I do not put the question in reference to anything connected with your troubles and difficulties in Kansas, but only to ascertain if it was in anything connected with Brown and his fortunes.

Answer. No, sir; not at all. There was a *posse* got up there to drive the free-State settlers out of Linn county, certain obnoxious ones, Republicans or abolitionists, or whatever they were called, headed by Marshal Russell, a man who was sent for that purpose, I supposed, from the south part of the country. We supposed he came from Arkansas, and I was told by a man in the secrets of the lodge—it was a secret oath-bound society—a man who was in there, and told me, he said, at the peril of his life——

The CHAIRMAN. Unless that refers to your connection with Brown, it is not necessary to state it.

The WITNESS. I tell that in explanation of what follows there in that letter.

The CHAIRMAN. Well, go on.

The WITNESS. He said this man had come in there to take charge of the *posse* which was raised to arrest certain men in the county, or to kill them or drive them out, and that my name was on the list, perhaps the first on the list; and other prominent men, old settlers there, were on the same list; and this man who told me advised me to leave the county, as they were every day threatening to come over and kill me. I thought a good deal as John Brown thought about one thing, that I was worth as much to be shot there as any place, and I would let them act out their own plans; I took no measures against them but to go up and see Governor Medary. He said this *posse* was got without his consent, that Marshal Russell was acting without his orders, and he would put a stop to it.

The CHAIRMAN. That is in explanation of the part of your letter in which you speak of the threats of assassination?

Answer. Yes, sir; that is the whole of it. The *posse* at Paris were threatening to kill me every day, I was told.

By Mr. COLLAMER:

Question. And you say the governor stopped it?

Answer. Governor Medary stopped it.

By the CHAIRMAN:

Question. Have you the papers which the summons required you to bring?

Answer. I have brought all my papers connected with my operations in Kansas.

The CHAIRMAN. We only want those papers that would throw light on John Brown's ulterior plans after he left Kansas—anything of that kind.

Answer. I was speaking about John Brown being at my house when he was sick. When he left my house, after he got well, he went out, took some claims for his sons that he expected would move back to Kansas, and went to work on them. He mowed some hay on the government land and put it up, and afterwards sold it. He bought a cow of neighbor of mine.

Question. What has all that to do with the question? We only want to know if you have any papers or documents of any kind that will throw any light on his ulterior purposes?

The witness exhibited certain letters, among them the following in the handwriting of John Brown:

BOSTON, MASSACHUSETTS, *April* 8, 1857.

MY DEAR SIR: Your favor of the 15th March, and that of friend H. of the 16th, I have just received. I cannot express my gratitude *for them both.* They give me just that kind of news I was *most of all things* anxious to hear. *I bless God* that he has not left the free-State men of Kansas to *pollute themselves* by the *foul and loathesome* embrace of the *old rotten whore.* I have been trembling *all along* lest they might *back down* from the *high and holy ground* they *had* taken. I say, in view of the *wisdom, firmness, and patience* of my friends and *fellow-sufferers,* (in the cause of humanity,) *let God's name be eternally praised!* I would most gladly give my hand to all whose "garments are not defiled;" and I humbly trust that I shall *soon again* have opportunity to rejoice (or suffer *further* if need be) *with you,* in the strife between Heaven and Hell. I wish to send my most cordial and earnest salutation to *every one of the chosen.* My efforts this way have not been altogether fruitless. I wish you and friend H. both to accept this for the moment; may write soon again, and hope to hear from you both at Tabor, Frémont county, Iowa—care of Jonas Jones, Esq.

Your sincere friend,

NELSON HAWKINS.

AUGUSTUS WATTLES, Esq.,
Lawrence, Kansas Territory.

Question. What did you understand by his saying that he had been trembling all along in reference to his friends, lest they might back down from the high and holy ground they had taken?

Answer. Yes, sir; he was afraid that they would yield to the bogus laws—pay obedience to the bogus laws—which the government and Missouri were trying to enforce in Kansas.

Question. He says, also, "my efforts this way have not been altogether fruitless." What does he refer to there as his efforts in Boston and New England?

Answer. I do not know, positively; but my suspicion is that he meant that he was raising funds for the purpose of coming out to Kansas.

Question. The fighting was over in Kansas, in 1857, I suppose. What object could he have had in raising funds?

Answer. I suppose that he was collecting funds to sustain the free-State party in their political position in resistance to the bogus laws.

Question. Why did you suppose he had reference to that?

Answer. That letter, I presume, follows the other; and it was all we were engaged in, politically.

Question. Did you know of his intended visit to New England?

The WITNESS. When he went?

The CHAIRMAN. Yes, sir.

Answer. No, sir.

Question. Had you any information that he went to Boston before that letter?

Answer. No, sir; I only replied to his letters.

[The following letter was next exhibited in John Brown's handwriting:

HUDSON, OHIO, *June* 3, 1857.

MY DEAR SIR: I write to say that I started for Kansas some three weeks or more since, but have been obliged to stop for the fever and ague. I am now righting up, and expect to be on my way again soon. Free-State men need have no fear of my *desertion*. There are some half dozen men I want a visit from at Tabor, Iowa, to come off in the most QUIET WAY, viz: *Daniel Foster*, late of Boston, Massachusetts; *Holmes*, *Frazee*, a Mr. *Hill* and *William David*, on Little Ottawa creek; a Mr. Cochran, on Pottawatomie creek; or I would like *equally* well to see *Dr. Updegraff* and *S. H. Wright*, of Ossawatomie; or *William Phillips*, or CONWAY, or *your honor*. I have some very important matters to confer with some of you about. Let there be *no words* about it. Should any of you come out to see me *wait* at Tabor if you get there *first*. Mr. Adair, at Ossawatomie, may supply ($50,) fifty dollars, (if need be,) for expenses on my account *on presentation of this*. Write me at Tabor, Iowa, *Frémont county*.

Very respectfully, yours,

JAS. SMITH.

A. WATTLES, Esq.,
Lawrence, Kansas Territory.]

Question. Do you know the object Brown had in collecting these men at Tabor?

Answer. I did not know at the time. I took the letter to Mr. Phillips as he requested, and asked Phillips if he knew what the old man wanted. He said, no; he suspected it was some scheme or other he had, but he had not time to attend to it. I know more now about it, but I am speaking of what I knew at the time.

Question. Did you attend?

Answer. No, sir.

Question. Did you inform him you could not attend?

Answer. No, sir.

Question. How did you derive the information you now have?

Answer. From common fame.

The CHAIRMAN. That will not do. Did you derive it from Brown?

Answer. No, sir.

Question. Did any meeting take place there?

Answer. I do not know that it took place there. According to the testimony which I have seen in the papers, he collected a number of men there. That is all I know about it. That was the first information I had about this meeting. Cook's confession, I think, says these young men met there for the purpose of drilling or making arrangements to go into Canada to form some combination. My knowledge about the meaning of that has come to me since this thing has broken out at Harper's Ferry.

Question. You had no knowledge of it at the time?

Answer. No; I took the letter to Colonel Phillips and Captain Holmes, and they both said they did not know anything about it.

Question. Who is Daniel Foster, late of Boston, that he speaks of?

Answer. Daniel Foster was a man who lived in Bourbon county, in Kansas, and I think he was a preacher.

Question. Who is Holmes?

Answer. Holmes was the son of a New York broker, a young man about seventeen or eighteen years old, his lieutenant at one time, and afterwards captain of his company.

Question. Whose company?

Answer. Captain Brown's in the southern part of Kansas.

Question. Who is Frazee?

Answer. I never saw Frazee but once. He was old Brown's teamster. When he went out of Kansas, in 1856, he drove his four-horse team.

Question. The letter also mentions a Mr. Hill and Mr. David?

Answer. I did not see them.

Question. Did you know them?

Answer. Not particularly. I would know them if I saw them at that time, but I never spoke with them.

Question. Cochran, on Pottawatomie creek?

Answer. I never saw him. I think I heard his name.

Question. Dr. Updegraff?

Answer. Dr. Updegraff was one of Brown's company at the battle of Ossawatomie, and I think he is now president of the council in Kansas.

Question. Who was Mr. Adair, of Ossawatomie, referred to here?

Answer. He was a Presbyterian preacher. Brown was a half brother of his wife.

Question. Do you know from what funds that fifty dollars was to be supplied?

Answer. No, sir; only from inference. Jason Brown sold property there when he left, and left the notes with Adair to collect. I infer that was it.

Question. This letter concludes with a request that you would write to him at Tabor? Did you do so?

Answer. I do not think I did; I do not recollect; but if I did write, I presume the letter is with his papers. I do not recollect writing to him.

Question. This letter was written the 3d of June, 1857. You saw him afterwards in 1858?

Answer. The next time I saw him was in June, 1858.

Question. Had he any reference, in his conversation then, to the meeting which he had called through this letter.

Answer. No, sir.

Question. Here is a letter dated "Boston, May 18, 1859." What "inclosures" does he refer to in it as having been received from you?

Answer. He sent to me to send all the letters which had come to the office for him. I sent them, and put in at the same time the little note you showed me first.

Question. He speaks of kindness to him and his men. Did he mean his sons?

Answer. No. He refers there to Tidd and Gill, I think. They had the fever and ague when they were on the line, and they were brought to my house to be taken care of. That was in July or August, 1858.

By Mr. Collamer:

Question. You mean belonging to his company on the border?

Answer. Yes, sir. They were sick, and I took care of them as long as they chose to stay. The only allusion that Brown ever made to going to Harper's Ferry, in my presence, was the last conversation he had with me, and at the time I put no construction upon it, did not think anything about it until after I saw that announcement in the newspapers. He called in to see me, as I was telling you, in going out of the Territory, and I censured him for going into Missouri, contrary to our agreement, and getting those slaves. He said, "I considered the matter well; you will have no more attacks from Missouri; I shall now leave Kansas; probably you will never see me again; I consider it my duty to draw the scene of the excitement to some other part of the country." Said he: "Farewell, God bless you." He took hold of my hand, gave me a shake of the hand, and left me. I was lying on the bed, sick. I did not know particularly what he meant. I did not attach any definiteness to it. As soon as I heard of the Harper's Ferry attack, I remembered what he had said to me, and supposed he had had allusion to it. In one of his letters to me, he made an allusion which I did not understand at the time; but in reading it over now, I have supposed he might have alluded to that. He acted as much like a settler as any man in Kansas. He worked claims and lived on them, built houses, and nobody would suppose he had any other idea than settling his sons in Kansas, with their cattle and other property gathered around them.

By the CHAIRMAN:

Question. Why were these letters of his signed with fictitious names?

Answer. Because, as I have already said, we were afraid of our letters being opened.

Question. He speaks in this last letter about a writing which he had commenced at your house, and you in your letter speak of a writing. What was that?

Answer. The writing he mentions there was an autobiography which I requested him to write. We were conversing on the subject of the stories we heard about things in Kansas, as his murders, and things that I knew not to be true. There was another letter which I had, which I was anxious to bring here, but I could not find it. I put no particular value on the letters. They were not laid away for the purpose of saving them.

The CHAIRMAN. Does that refer to any of his ulterior plans?

Answer. No, sir; not at all. It just refers to clearing me of the suspicion that I knew of his invasion of Missouri from that letter. The letter was concerning some imported stock that I had bought of him, and others had bought, and I wanted a pedigree. He had two Morgan horses that he brought from Vermont. He agreed to send me the pedigree of his stock, and I wrote about it.

Question. Were you the agent or in any way connected with any of the societies that were got up in New England or elsewhere for contributing money to the people in Kansas?

Answer. Yes, sir.

Question. Which of them were you connected with?

Answer. I was agent for the Female Kansas Aid Society of Wisconsin, of which Mrs. Hinton, of Waukesha, was secretary.

Question. What was the object of that society?

Answer. To feed and clothe destitute people in Kansas, people who had been robbed by the invaders there in 1856.

Question. Were any arms furnished by that society?

Answer. No, sir, not the first one.

Question. Were you connected with or agent for any other society?

Answer. Professor Daniels, of Wisconsin, the State geologist, made me his agent to distribute clothing, and I distributed some for Mr. Arny.

Question. Were you agent for or connected in any way with any society that did furnish arms?

Answer. No, sir; not any that I know of. I never handled any, never kept any, and I do not think that I ever carried any in Kansas for more than two days.

The CHAIRMAN. I speak of arms being furnished abroad for other persons, not for you. Had you any knowledge of the arms, Sharp's rifles and pistols, that Brown brought with him to Harper's Ferry?

Answer. No, sir; I had not.

Question. Do you know how he got possession of them?

Answer. I do not. I tried to get a Sharp's rifle in Kansas, out of some that I heard had been sent there, but I could not; and I did not know who had care of them. I suspect now, from what has been developed, that he had them somewhere where they could not be got at. I heard of their being sent, but never knew of their arrival.

By Mr. COLLAMER:

Question. At any time, from anything Brown said or wrote, did you learn from him, in any way, or have any knowledge of Brown's designs to make an attack or create any disturbance, in relation to slavery, anywhere else than in Kansas?

Answer. I never had. I never saw so secretive a man as Brown. I never heard of his telling his plans to anybody. I have not the least idea that he disclosed them to any person, unless it was Kagi——

The CHAIRMAN. That is your inference?

The WITNESS. Yes, sir.

Mr. COLLAMER. I merely wished to know whether you knew of his plans, or of its having been known in Kansas that he intended any such project as creating an insurrection or disturbance in the slave States.

Answer. I never heard of it. I think he alludes to it in one letter to me, and in a remark that he made in a blind way, that his duty called him somewhere else.

Mr. COLLAMER. You say your impression of what he meant by that has only been entertained since the invasion of Harper's Ferry.

Answer. Yes, sir; I had no idea before of what he meant.

AUGUSTUS WATTLES.

NOTE. I desire to add that Captain Brown and his two oldest sons came to Kansas in the fall of 1854, selected claims, and commenced improvements. They spent the winter in Missouri, and in the following spring they returned to Kansas, with their wives and cattle and horses, with the intention of making it a permanent home. Jason Brown had a nursery and horses, and John Brown, jr., some blooded cattle and other stock. Captain Brown told me that he had no idea of fighting until he heard the Missourians, during the winter he was there, make arrangements to come over into the Territory to vote. He said to me that he had not come to Kansas to settle himself, having left his family at North Elba, but he had come to assist his sons in their settlement, and to defend them, if necessary, in a peaceable exercise of their political rights.

A. WATTLES.

FEBRUARY 24, 1860.

GEORGE L. STEARNS affirmed and examined.

By the CHAIRMAN:

Question. Will you state where you reside?

Answer. I reside in Medford, Massachusetts, about five miles from Boston.

Question. Will you state whether you were acquainted with John Brown, who was recently put to death in Virginia for offenses against that State?

Answer. Yes, sir.

Question. Where did you make his acquaintance, and when?

Answer. I made his acquaintance early in January, 1857, in Boston. It might possibly have been the last of December, 1856; but I think it was after the 1st of January, 1857.

Question. Will you state in what way you made his acquaintance; what led you to his acquaintance; what was his object in forming your acquaintance, or yours in forming his?

Answer. I was introduced to him by one of our Kansas men, meeting him accidentally.

Question. Who was the man who introduced you?

Answer. I do not recollect now. It was entirely accidental.

Question. Did Brown tell you what was the object of his visit to Boston at that time?

Answer. No, sir.

Question. Were you president of the Massachusetts State Kansas Committee?

Answer. Yes, sir.

Question. What was the object of that committee?

Answer. The object was to relieve the wants and sufferings of the men in Kansas.

Question. In what way was that done? By contributions of money?

Answer. Contributions of money and other things.

Question. What other things?

Answer. Everything which was needed. I cannot specify.

Question. Do you recollect that in January, 1857, you gave to John Brown an order for certain Sharp's rifled carbines, as the property of the Massachusetts State Kansas Committee?

Answer. Yes, sir.

Question. How was it that that committee were in possession of arms, if their object was only to relieve the sufferings of the people?

Answer. I have made a statement on paper, which, as I am unaccustomed to speak in public, or even to give evidence—for it is very seldom that I have been in courts as a witness—I would ask the permission of the committee to allow me to read as evidence, because it would be a clearer and more condensed statement than I could make in any other way.

[After consultation, the committee allowed the witness to read that part of his manuscript which he considers an answer to the question.]

Question. You say there, I think, that you made him your agent to receive those arms, and they consisted of two hundred rifled carbines, with a proportion of ammunition?

Answer. Yes, sir.

Question. Were there any revolving pistols?

Answer. No, sir.

Question. Did your committee possess any revolvers out in that country?

Answer. No, sir.

Question. Were those the only arms held by your committee?

Answer. Yes, sir.

The CHAIRMAN. It has appeared elsewhere, in evidence taken before the committee, that, together with two hundred Sharp's rifled carbines, Brown was in possession of two hundred revolving pistols. Can you tell the committee where he got them?

Answer. I think that the better way would be for me to read my statement throughout. That would open the whole question, and give you a better understanding. These questions involve my connection with Kansas affairs, and hence it will be as well to give the whole statement.

After consultation, the committee agreed to allow the witness to read his statement, which is as follows:

In the spring of 1856, I went to the Boston committee for the relief of sufferers in Kansas, and offered my services. I worked for them until June of that year, and then, being willing to devote all my time to the cause, was made chairman of the State Kansas Committee of Massachusetts, which took the place of the first-named committee, and continued the work throughout the State. In five months, including August and December of that year, (1856,) I raised, through my agents, about $48,000 in money, and in the same time my wife commenced the formation of societies for contributions of clothing, which resulted in sending from $20,000 to $30,000 more, in supplies of various kinds. In January, 1857, our work was stopped, by advices from Kansas that no more contributions were needed except for defense. If we had not been thus stopped, our arrangements then made would have enabled us to have collected $100,000 in the next six months. Soon after our State committee had commenced work—I think in August, 1856—a messenger from Kansas—who came through Iowa (for the Missouri river was then closed by the Missourians to all free-State travelers)—came to us asking earnestly for arms and ammunition for defense of the free-State party. Our committee met the next day, and immediately voted to send two hundred Sharp's rifles, and the necessary quantity of ammunition, which was procured and sent to the National Kansas Committee at Chicago, to be by them forwarded through Iowa to Kansas. From some cause, which I have never heard explained, these arms were delayed in Iowa; and in November or December of that year we directed an agent to proceed to Iowa at our charge, and take possession of them as our property. Early in January, 1857, John Brown, of whom I had heard, but had not seen, came to Boston and was introduced to me by one of our Kansas agents, and after repeated conferences with him, being strongly impressed with his sagacity, courage, and stern integrity, I, through a vote of our committee, made him our agent to receive and hold these arms and the ammunition, for the defense of Kansas, appropriating $500 to pay his expenses. Subsequently, in April of that year, we authorized him to sell 100 rifles, if expedient, and voted $500 more to enable him to proceed to Kansas with his armament.

About this time, on his representing that the force to be organized in Kansas ought to be provided with revolvers, I authorized him to purchase 200 from the Massachusetts Arms Company, and when they were delivered to him in Iowa, paid for them from my own funds; the amount was $1,300. At the same time I gave him, by a letter of credit, authority to draw on me at sight for $7,000 in sums as it might be wanted, for the subsistence of 100 men, provided that it should be necessary at any time to call that number into the field for active service in the defense of Kansas, in 1857. As the exigency contemplated did not occur, no money was drawn under it, and the letter was subsequently returned to me. In the summer of 1857, I contributed with others, $1,000 to purchase an addition to the farm then and now occupied by his family at North Elba. The money was paid by my agent for that purpose, and satisfactory evidence given me on his return that a proper conveyance of the land had been made to the family o John Brown. My subscription to that fund was $260, as appears by the subscription paper. Besides these transactions, which were for specific purposes, I have given him money from time to time, how much I do not know, as I never keep any account of my personal expenses, or of money I give to others; it is all charged to my private account as paid me. I should think it might amount to, say, from $1,500 to $2,000. About May, 1858, I saw a letter from Henry Wilson to Dr. Howe, and also one or two from a Mr. Forbes. I had never heard of Forbes until I saw his letters, which were so coarse and insulting in their language, and incorrect, in ascribing to others what I had done, that I concluded he was an adventurer whose only aim was to extort money; but at Dr. Howe's request, I wrote the letter to John Brown, dated May 14, 1858, of which he has forwarded to you a copy. In addition to what I have before stated, I raised money and sent an agent to Kansas to aid the free-State party in the Lecompton election, and again for the election in 1858.

Question. Was it at Brown's request that you put him in possession of those arms in January, 1857?

Answer. No, sir; but because we needed an agent to secure them. They were left in Iowa, and under circumstances that made it doubtful whether they would not be lost entirely, and we put them into his hands because it was necessary to have some agent to proceed there and reclaim them from the hands they were in, and take proper care of them.

Question. It is stated in the writing, "I, through a vote of our committee, made him our agent to receive and hold these arms and ammunition for the defense of Kansas?"

Answer. Yes, sir; of course they were intended for the defense of Kansas, and that was the object for which they were to be held.

Question. Do you know that the pistols were delivered to Brown?

Answer. The exact statement of the case is, that upon the delivery of the railroad receipt to me, promising to deliver them to him in Iowa, I paid for them.

Question. Do you know, from the admission of Brown or otherwise, that he afterwards got those pistols?

Answer. No, sir.

Question. There are copies of two letters here, among those forwarded by Dr. Howe, did you read them?

Answer. I have not read the whole, but I have read my own letters.

Question. There is a copy of a letter purporting to have been written by you, as chairman of the Massachusetts State Kansas Committee, to John Brown, dated at Boston, January 8, 1857, and another to John Brown from you, dated at Boston, April 15, 1857, and the third dated Boston, April 15, 1857. I will read them to you. I only want to know if they are correct copies of your letters?

The following letters were then read to the witness:

MASSACHUSETTS STATE KANSAS COMMITTEE ROOM,
Boston, January 8, 1857.

DEAR SIR: Inclosed we hand you our order on Edward Clark, Esq., of Lawrence, Kansas Territory, for two hundred Sharp's rifled carbines, with four thousand ball cartridges, thirty one military caps, and six iron ladles; all, as we suppose, now stored at Tabor, in the State of Iowa.

We wish you to take possession of this property, either at Tabor or wherever it may be found, as our agent, and to hold it subject to our order.

For this purpose you are authorized to draw on our treasurer, Patrick T. Jackson, Esq., in Boston, for such sums as may be necessary to pay the expenses as they accrue, to an amount not exceeding five hundred dollars.

Truly yours,
GEORGE L. STEARNS,
Chairman Massachusetts State Kansas Committee.

Mr. JOHN BROWN,
Of Kansas Territory.

BOSTON, *April* 15, 1857.

DEAR SIR: By the inclosed vote of the 11th instant we place in your hands one hundred Sharp's rifles to be sold in conformity therewith, and wish you to use the proceeds for the benefit of the free-State men in Kansas; keeping an account of your doings as far as practicable.

Also a vote placing a further sum of five hundred dollars at your disposal, for which you can, in need, pass your draft on our treasurer, P. T. Jackson, Esq.

Truly yours,
GEORGE L. STEARNS,
Chairman Massachusetts State Kansas Committee.

Mr. JOHN BROWN,
Massassoit House, Springfield, Massachusetts.

BOSTON, *April* 15, 1857.

At a meeting of the executive committee of the State Kansas Aid Committee of Massachusetts, held in Boston, April 11, 1857, it was

Voted, That Captain John Brown be authorized to dispose of one hundred rifles, belonging to this committee, to such free-State inhabitants of Kansas as he thinks to be reliable, at a price not less than fifteen dollars, and that he account for the same agreeably to his instructions, for the relief of Kansas.

GEORGE L. STEARNS,
Chairman Massachusetts State Kansas Committee.

At the same meeting it was

Voted, That Captain John Brown be authorized to draw on P. T.

Jackson, treasurer, for five hundred dollars, if, on his arrival in Kansas, he is satisfied that such sum is necessary for the relief of persons in Kansas.

GEORGE L. STEARNS,
Chairman Massachusetts State Kansas Committee.

The WITNESS. The first letter speaks of thirty one military caps. It should be thirty one thousand military caps, meaning percussion caps.

Question. The first letter directs Brown to take possession of the arms as your agent, and hold them subject to your order. Did I understand you to say that this was voluntarily proffered to him; and not at his request?

Answer. Yes, sir.

Question. Why did you desire to place these arms in his possession?

Answer. For safe-keeping.

Question. Were they not in safe-keeping where they were?

Answer. They were not substantially in our hands. We had passed them into the hands of the National Kansas Committee to be transported to Kansas, and they had an idea that they being called the National Kansas Committee, everything which was sent to them for transportation became their property the moment it passed into their hands, which we disputed; and after some letters had passed between us they gave them up to us again and we assumed the possession of them. That was a question which we had to settle with them—whether the property we sent to Kansas was theirs the moment it got into their possession. We denied it.

Question. Were the 100 Sharp's rifles, referred to in the letter of April 15, a different weapon from the Sharp's rifled carbine before spoken of?

Answer. The same weapon. A part of the same lot.

Question. Then the 100 rifles mentioned here were part of the 200 mentioned in that?

Answer. Yes, sir.

Question. Did he ever sell these rifles, as he was thus empowered?

Answer. I have no reason to suppose that he did. I never knew that he sold them. He never gave me any intimation that he sold them.

Question. Did he ever account with you for the proceeds?

Answer. No, sir.

Question. Did he ever advise you that he had sold them?

Answer. No, sir.

Question. Did he draw for the $500 that you authorized him to draw for?

Answer. Yes, sir.

Question. You saw Brown after that in 1857?

The WITNESS. After April, 1857?

Question. After April, 1857?

Answer. Yes, sir.

Question. Was there any conversation then between you as to those arms that were in his possession? Was any reference made to them?

Answer. I simply asked him if they were safe and in order. He told me they were.

Question. Did he tell you where they then were, in 1858?

Answer. I do not think he did. I do not recollect that he did.

Question. I find in the manuscript sent by Dr. Howe a copy of a letter written by you to John Brown, dated at Boston, May 14, 1858, addressed to him at Chatham, Canada West, which I will read, and ask you if it is a correct copy.

The letter was read to the witness, as follows:

BOSTON, *May* 14, 1858.

DEAR SIR: Inclosed please find a copy of a letter to Dr. Howe from Hon. Henry Wilson. You will recollect that you have the custody of the arms alluded to, to be used for the defense of Kansas, as agent of the Massachusetts State Kansas Committee. In consequence of the information thus communicated to me, it becomes my duty to warn you not to use them for any other purpose, and to hold them subject to my order as chairman of said committee. A member of our committee will be at Chatham early in the coming week, to confer with you as to the best mode of disposing of them.

Truly your friend,

GEORGE L. STEARNS,
Chairman Massachusetts State Kansas Committee.

Mr. JOHN BROWN,
Chatham, Canada West.

The following letter was also read to the witness:

BOSTON, *May* 15, 1858.

DEAR SIR: I wrote to you yesterday informing you that a member of the Massachusetts State Kansas Committee would visit Chatham, to confer about the delivery of the arms you hold.

As I can find no one who can spare the time, I have to request that you will meet me in New York city some time next week. A letter to me, directed to care of John Hopper, 110 Broadway, New York, will be in season. Come as early as you can. Our committee will pay your expenses.

Truly yours,

GEORGE L. STEARNS,
Chairman Massachusetts State Kansas Committee.

Mr. JOHN BROWN,
Chatham, Canada West.

Dr. Howe will go on as soon as he knows you are in New York.

Question. Will you give to the committee, as nearly as you can, the substance of Mr. Wilson's letter to Dr. Howe, which you inclosed in this letter to Brown?

Answer. It is so long since, that I do not recollect it. I cannot recall any expression of the letter.

Question. Do you remember what was the subject of it—what it referred to?

Answer. I think, as near as I can recollect, it must have referred to some communication of a Mr. Forbes, or, if the name was not mentioned, to some information that Mr. Wilson had received that those arms were to be used improperly; but further than that I have no recollection of it at all.

Question. Do you recollect what was the impropriety of the use that it was suspected Brown would make of them? What was the nature of the improper use that it was feared he would make of them?

Answer. I can only give you the general impression. It was that Brown had other designs than that for which the arms were put into his hands; that is, that he might invade Missouri, or, instead of defending Kansas, as we proposed, that he might carry his plans beyond that, and perform the same work in Missouri that had been performed by Missourians in Kansas. General Wilson, from the first, even immediately after the attack on Lawrence, always strongly opposed, I think, even any attempts to repel outrage, except in the way of immediate defense. He opposed any organized system. He always said, "Don't you interfere with the United States troops; if you do, the United States will crush you." I have heard him use that expression time and again.

Question. This letter is addressed to Brown, at Chatham, Canada West. How did you derive the information that he was at that time in Chatham, in Canada?

Answer. I presume I must have received a letter from him, which I have not got; because, among my letters which I did not think it was necessary to bring here, was one about a fortnight or three weeks before, simply inquiring of him where I could find him.

Question. Did you know the object of his visit to Chatham, in Canada?

Answer. No, sir.

Question. Were you aware that a convention was held there by Brown about that time?

Answer. No, sir.

Question. In this letter you requested him to meet you in New York city. Did he comply with that request to go to New York city?

Answer. No, sir; he did not. I was in New York, and he did not come there.

Question. Did you have any communication with him on the subject of these arms, after the date of these letters on the 14th and 15th of May?

Answer. Once only, when I asked him where they were, and he told me that they were stored in Ohio.

Question. When was it that you had that conversation with him?

Answer. That I cannot recollect. It was subsequent to these letters.

Question. Did you see John Brown in Boston, some time in May or June, 1858?

Answer. Yes, sir.

Mr. Collamer. Do you know the exact time you saw him in Boston, in 1858?

Answer. I think it must have been in June. At that time nearly all our regular operations had ceased.

By Mr. FITCH:

Question. Did the Massachusetts State Kansas Committee, through you or anybody else, ever withdraw those arms from Brown's charge?

Answer. No, sir.

Question. Were the pistols paid for out of your own means?

Answer. Yes, sir.

Question. Did the order addressed to Brown, at Chatham, to hold the arms subject to the future order of the committee, embrace the pistols as your private property?

Answer. They did not, perhaps, technically; but that was my understanding at the time.

Mr. COLLAMER. You supposed Brown would understand it so?

Answer. I presumed so.

By the CHAIRMAN:

Question. Did you see Brown in Boston, or that neighborhood, in the spring or summer of 1859?

Answer. Yes, sir; in the spring of 1859.

Question. Was he at your house?

Answer. No, sir.

Question. Did you see his son, John Brown, jr., there in the summer of 1859?

Answer. Yes, sir.

Question. Was he at your house?

Answer. Yes, sir.

Question. Can you tell the committee, so far as you know, what was the object of young Brown's visit to Boston at that time?

Answer. He came to me one day, at my store, and introduced himself as the son of John Brown. I asked him what he came for, after some conversation, and he said he came to see his father's friends. I was at that time very busily engaged in building; and as he wanted to go out on the same railroad that I was going, I invited him to come and dine with me. We dined together. During that time he seemed to be interested in what I had about my house; and I was particularly struck with the fact that he inquired about some bas-reliefs I had put into the walls. He criticised them in a most remarkable manner. He looked at the garden, and picked one or two flowers, and asked that he might take them home to his wife. I told him that he might take as many as he chose. In a few minutes, I found that he was holding them up and contrasting the colors—what not one man in five hundred would do. I was struck particularly with the natural love he showed not only for art but for nature. That was all that occurred at that time.

Question. Did he speak of his father, and say where he was, or what he was engaged in?

Answer. No, sir.

Question. Was there no reference to his father in his conversation at that time, so far as you can recollect now?

Answer. I think there must have been, but I do not now recollect what it was.

Question. Was that the only time you saw him during his visit to Boston?

Answer. Yes, sir.

Question. Was nothing said by young Brown of his desire to make collections in money for his father's use?

Answer. Nothing whatever.

Question. Did you give him any money?

Answer. No, sir.

Question. You saw John Brown the elder, in Boston, some time in the spring of 1859. Will you state under what circumstances you saw him there; what brought him there, so far as you know?

Answer. He came to Boston, as he told me, to get money for anti-slavery purposes.

Question. What were those anti-slavery purposes? Did he disclose them?

Answer. No, sir.

Question. Did you give him any money at that time?

Answer. Yes, sir.

Question. How much?

Answer. I do not recollect how much. I have no means of knowing—some hundreds of dollars.

Question. Will you state to the committee what were the anti-slavery purposes to which you intended that money to be devoted—what sort of purposes?

Answer. Well, my object, in giving him the money was because I considered that so long as Kansas was not a free State, John Brown might again be a useful man there. That was one object. Another was a very high personal respect for him. Knowing that the man had an idea that he was engaged in a work that I believed to be a righteous one, I gave him money to enable him to live or to do whatever he thought was right. When I first talked with John Brown in regard to Kansas affairs, he told me that it was the worst possible policy for a man to reveal his plans. I recollect his taking several scraps of newspapers from his pocket and saying, "The United States government immediately disclose their orders to their military officers. Before the orders leave Washington, they are published all through the papers; well, now, that is not the way; if a man is to do anything, he must keep his plans to himself." Respecting that, I never inquired of him afterwards about his plans, and he never revealed them to me.

Question. I understand you to say that in the month of May, 1858, in consequence of a letter from Henry Wilson, you thought it prudent and wise to endeavor to control the use of those arms in Brown's hands? Did you not think it necessary when you met him again, in 1859, to take further steps to control the use of those arms and prevent him putting them to what you have spoken of as an improper purpose? Was no attempt of that sort made?

Answer. No, sir; I did not suppose they would be put to any such purpose, as it has since appeared they were put to.

Question. There is a letter from John Brown, jr., dated at Syracuse, New York, on the 17th of August, 1859, and addressed to a man named Kagi, in which he says:

"While in Boston, I improved the time in making the acquaintance of these staunch friends of our friend Isaac. First called on Dr. H.—— He gave me a letter to the friend who does business on Milk street."

Is your place of business on Milk street?
Answer. Yes, sir.
Question. He goes on: "Went with him to his house, at Medford, and took dinner." Do you recollect whether he brought you a letter from Howe?
Answer. I think he did.
Question. He continues: "The last word he said to me was, 'Tell friend Isaac that we have the fullest confidence in his endeavor whatever may be the result.'" Do you remember that message sent to his father?
Answer. No, sir; I do not. I recollect sending a complimentary message to his father that I had confidence in him; but I have no recollection of that.
Mr. DAVIS. Was it his father who was called Isaac?
Answer. I do not know.
The CHAIRMAN. I was going to ask you whether you did or did not know that the father at that time passed by the name of Isaac, or Isaac Smith?
Answer. I did not.
Question. Did you never refer to old John Brown as Isaac or Isaac Smith?
Answer. No, sir.
Question. Were you aware that Brown had ordered a parcel of pikes to be made in the preceding year, in Connecticut?
Answer. I think I heard him say something about pikes, but whether it was that he had ordered them to be made, or what he said about them, I do not recollect. I think I heard him say something about pikes.
Mr. COLLAMER. When?
Answer. That must have been in May, 1857.
The CHAIRMAN. Do you know whether he told you that he had ordered any pikes to be made in that region of country?
Answer. No, sir, I do not.
Question. Do you remember in what connection he spoke of having pikes at all?
Answer. He might have spoken of them as being useful for military purposes.
Question. Did you know a young man named Francis J. Meriam of Boston?
Answer. No, sir.
Question. Did you ever hear of such a man?
Answer. I have heard of him in connection with this affair.
Question. Have you never met with him?
Answer. A man came and introduced himself to me as Mr. Lockwood who I supposed to be this Mr. Meriam, and he began to talk to me about the Harper's Ferry affair.
Question. Was that after the Harper's Ferry affair?
Answer. Yes, sir. I told him that I was very busy and could not

attend to him. He still continued talking, and at last I was obliged to tell him "sir, my time is so occupied that I cannot have anything to say to you; you must let me go."

Question. You did not know who he was?

Answer. No, sir; but I suspected.

Question. Did you know his family in Boston?

Answer. No, sir; I have no acquaintance with them.

Question. Have you any acqaintance with his parents or grand-parents?

Answer. I know Francis Jackson, who I believe is a connection of his. I know Wendell Phillips, who I think is related to him.

Question. Have you any information, derived from any proper source, as to whether Brown asked for authority or permission in any way to bring those arms to Harper's Ferry—the carbines and the pistols?

The WITNESS. Asked it of our committee?

The CHAIRMAN. Of you or anybody else who had control of them?

Answer. He did not.

Question. Were you aware of where Brown was in the summer of 1859, say July, August, or September, of that year?

Answer. No, sir.

Question. Had you any correspondence with him?

Answer. I had no correspondence with him. I knew that he was moving about, but I did not know where.

Question. As to those pistols, which were your private property, have you ever taken any measures to reclaim them?

Answer. No, sir.

Question. Have you never made any inquires as to where they were since Brown's death?

Answer. No, sir; I did not suppose it would be of any use to do so.

Question. Why did you suppose so?

Answer. I presumed that in the confusion at Harper's Ferry everything was distributed.

By Mr. FITCH:

Question. Was there any communication, either written or oral, between Brown and any member of your committee, to your knowledge, which enabled him to claim those arms as his property?

The WITNESS. Before I answer that, I wish to make this statement—that I have no knowledge or evidence that the arms at Harper's Ferry were the same arms. Still, I suppose they were, because they were about the same number, and of the same character and description. Now I am ready to answer the question.

Mr. FITCH. The question was, whether as Brown claimed those arms as his private property, (assuming that they were the same,) he had any right or authority, by virtue of any correspondence or communication with any member of your committee, to thus claim them?

Answer. I think he would have reason to consider the revolvers as his property, because I paid for them, not as something I intended to retain, but as something put into his hands for his use, expecting that they were to be used in Kansas. As to the other arms, the rifled carbines, if they were those that belonged to the Massachusetts State Kansas Committee, I see no reason why he should have claimed them, for they were never given to him as his property, but only intrusted to him.

Mr. FITCH. You think he had no other reason to claim them except the long silence of the committee on the subject?

Answer. That is all.

By the CHAIRMAN:

Question. Have any measures been taken to reclaim the 200 rifled carbines that were in his possession by your committee?

Answer. No, sir.

Question. Can you state why it is that no steps have been taken, by yourself or by the committee, to reclaim this property, which was thus left in Brown's possession, since his death?

Answer. Because we thought it would be of no use. Mr. Sennott came to me and asked me about them, and at the same time exhibited a letter from Mr. Brown to him authorizing him to take possession of this property as his, for the benefit of his family?

Question. What property?

Answer. The property at Harper's Ferry, whatever was there. I think the statement was a general one. I told him that as his agent he had better, in my opinion, go to Harper's Ferry and gather up whatever could be found; that the Adams' Express Company would bring them to Boston, and receive payment for their freight on delivery, and that then he could dispose of them; so far as I had any concern, I should not claim any of that property; that I was perfectly willing, if any of it could be saved, it should go to the benefit of the family.

The CHAIRMAN. Did the property referred to in his conversation, by Mr. Sennott, include the pistols and rifles Brown had brought to Harper's Ferry?

Answer. Yes, sir.

Mr. FITCH. You recognized, then, his right to the arms, in the conversation with Mr. Sennot?

Answer. So far as I was individually concerned, that is all.

The CHAIRMAN. Is that Kansas committee still in existence?

Answer. No; it cannot be called in existence, though I do not recollect that it was ever formally dissolved. We have had no meeting dissolving it.

Question. Were any steps taken by that committee or by you, as its chairman, after Brown's death, to inquire what became of those two hundred rifled carbines that you have referred to in your testimony?

Answer. None whatever.

Mr. FITCH. You have spoken of a conversation with John Brown, jr., at your house; did you converse or correspond with him subsequent to that time?

Answer. I feel very sure that I did not. If I did, it must have been since the Harper's Ferry affair. I may have written him a letter, but I think not. I think I never corresponded with John Brown, jr., at all.

The CHAIRMAN. You speak of seeing John Brown the elder in Boston, in 1859. Can you recollect whether it was as late as August, 1859?

Answer. Oh no, sir. I think it was in May.

Question. Did he then remain for some time in Boston?

Answer. I think he remained several days.

Question. Had you frequent intercourse and communication with him?

Answer. Yes, sir; I saw him often while he was there; not every day, but quite frequently.

Question. Did he tell you the object of his visit?

Answer. As I stated to you, his object was to obtain money for general anti-slavery purposes.

Question. Can you state the places in which you saw him, in what association, and whereabouts?

Answer. I saw him in his room at the United States Hotel several times. I saw him at Dr. Howe's room. I do not now recollect anywhere else.

Mr. DAVIS. Do I understand you to say that you do not recollect seeing him at any other place than at his room?

Answer. I saw him at his room at the United States Hotel, and I saw him at Dr. Howe's office. I recollect once seeing him at a meeting of a club that dined at the Parker House. I went in there late in the afternoon and saw him there.

Mr. DAVIS. Was it a dining party?

Answer. Yes, sir.

Question. Can you give any idea of who constituted it?

The WITNESS. At that time?

Mr. DAVIS. Certainly; at that dinner.

Answer. The only persons that I recollect, who were present at that time—they were mostly strangers to me—were F. W. Bird and, I think, Dr. Howe.

The CHAIRMAN. Did you say it was a club dining together?

Answer. Yes, sir.

Question. What was the name of the club?

Answer. There is no name to it. A number of gentlemen dine together; and, I believe, they are called Bird's club, from the name of Mr. Bird.

Question. Was it a party of gentlemen who meet periodically or occasionally to dine together?

Answer. Yes; they dine there every Saturday; a sort of half political club.

Question. Were you a member of that club?

Answer. No, sir; I was not at that time. Since that time I have dined with them.

Question. Was John Brown one of their guests on that occasion? Was he present at the dinner?

Answer. I cannot tell you whether he was or not.

The CHAIRMAN. I think you said you saw him there?

The WITNESS. I saw him there after the dinner. Whether he dined with them, or came in after the dinner, I cannot tell.

Mr. DAVIS. Had the company dispersed at all before you went there, or were all who had dined still there?

Answer. They were just about dispersing when I went in. Some of the gentlemen were standing and some were sitting at the table, and very soon they left.

The CHAIRMAN. The dinner was over?

Answer. Yes, sir.

Question. Did you go there to see Brown? What took you there?

Answer. I went there to meet him as I had met him before.

Question. You knew he was there?

Answer. Yes, sir; I went there because he was there.

Question. Did any conversation take place there in the presence of those gentlemen who were assembled as to the object of his visit to Boston; his desire to collect money for anti-slavery purposes, as you express it?

Answer. I presume there had been; I presume he went there for that purpose.

Question. Can you recollect none others who were there except the two whom you have mentioned?

Answer. No, sir; I think they were the only two that I knew personally at that time; the others I should not be likely to have remembered.

Question. But you might have known who they were, without knowing them personally?

Answer. Yes; but I do not recollect them, because I was not personally acquainted with them.

Question. How long did you remain there with Brown?

Answer. I should think it might be twenty minutes.

By Mr. DAVIS:

Question. What kind of a house is this Parker House?

Answer. It is one of the best eating houses in the town.

Question. Are select dinners given there?

Answer. Yes, sir; it is a place where everybody goes for a good dinner. If a literary club wish to dine, they go to the Parker House; if a political club wish to dine, they go to the Parker House.

Question. Is it a place where fine and expensive dinners are given?

Answer. A place where you can get the rarities of the season, and cooked in the best manner.

By Mr. FITCH:

Question. Did the Massachusetts State Kansas Committee keep a list of its contributors?

Answer. No, sir; the contributions were mostly in small sums. The way in which we were enabled to make our contributions so large was because we made them general through the towns.

Question. Was there any distinction in the funds contributed? Was one fund contributed specially for the purpose of purchasing arms, with a knowledge on the part of those who contributed, that the money was to be devoted to that purpose?

Answer. No, sir. There were two committees; first, the Boston

Relief Committee, for the relief of sufferers in Kansas; they collected in Boston chiefly in large sums, some $18,000 or $20,000. It was done under the spur of the moment, and I first worked with them; but very soon, when they had collected their large sums, they were not efficient, their operations stopped, and it was found necessary to make operations more extensive; that led to the establishment of the Massachusetts State Kansas Committee, who made general subscriptions, so far as they could, throughout the State. For instance, we could go into a town, appoint a lecture, organize a committee in that town for subscriptions, and in the course of one week, that committee would be sufficiently extensive to go to every house in the town; every individual would be approached and asked to give any sum—five or ten cents, a dollar, or whatever he chose. The result of that was a large subscription.

Question. But with no attempt to discriminate, as to the use to which the money of different contributors was to be put?

Answer. No, sir.

Question. Was it known that some of this was applied for the purchase of arms?

Answer. I do not think it was generally known; I do not think that the question was ever asked. I think, however, if it had been, the response would have been quite as large for arms as it would have been for other purposes.

Question. Do you remember the names of any prominent contributors? I do not mean prominent for the amount, but for their position, any men connected with the United States government in any capacity?

Answer. No, sir; I think that those men would not contribute at all.

Question. Not local officers, but members of Congress or any other body?

Answer. No, sir; you can see that in our operations we did not go in that way. Instead of getting money as you would in a political contest, in large sums from individuals, to distribute among the people, we went to the lower class of people. Our dependence was upon the laborers, the mechanics, the farmers, and such persons, much more than it was upon the professional men and merchants. As an instance of that, while we were collecting money freely in the country, the Boston merchants having made a heavy subscription in the spring, in May, I think, of 1856, an attempt being made in October or November, to get a further subscription was almost an entire failure. They said "we have given, and will not give any more." Those first subscriptions were given to the Boston committee; and I think the second subscription in Boston resulted, as near as I can recollect, in some two or three thousand dollars.

Question. There was no return of names then anywhere?

Answer. None, whatever.

By the CHAIRMAN:

Question. Did you ever know, or meet with Hugh Forbes?

Answer. No, sir.

Question. Did you ever have any correspondence with him?

Answer. No, sir.

By Mr. Collamer:

Question. Was any distinction made between you and Brown at any time, or any difference in his holding the pistols and the rifles?

Answer. None, whatever; they were contributed for the same purpose.

Question. Then the arms alluded to in your letter to Brown at Chatham included both?

Answer. I so understood at the time.

Question. You supposed it to include the whole?

Answer. Yes, sir.

Question. Did Mr. Brown to whom you gave the order to take those arms in Iowa, and also an order to procure the 200 revolvers, which you say you afterwards paid for yourself, know there was any difference as to their being paid for by you or by the committee?

Answer. He knew that the rifles were put into his hands by the committee, and he knew that I paid for the pistols.

Question. But they were both subject to your order?

Answer. Yes, sir.

Question. Did you at any time before the transaction at Harper's Ferry, in any way, directly or indirectly, understand that there was any purpose on the part of Brown to make any inroad upon the subject of slavery in any of the States?

Answer. No, sir; not except that Brown was opposed to slavery, and as he had in Kansas he would work again. I did not suppose that he had any organized plan.

Question. My idea is, making any forcible entry upon Virginia, or any other State?

Answer. No, sir.

Question. Had you ever any intimation of that kind, any idea of it?

Answer. No, sir. Perhaps I do not understand you. I did suppose he would go into Virginia or some other State and relieve slaves.

Question. In what way?

Answer. In any way he could give them liberty.

Question. Did you understand that he contemplated doing it by force?

Answer. Yes, sir; by force, if necessary.

Question. Will you explain in what manner, by force, you understood he contemplated doing it?

Answer. I cannot explain any manner, because, as I say to you, I never talked with him on the subject?

Question. Had you any idea that these arms were to be used for any such purpose as making an inroad into any State?

Answer. I think I do not understand you.

Question. John Brown has made an inroad into Virginia, with force and arms, to relieve slaves; you understand that?

Answer. Yes, sir.

Question. Now, did you ever, before that took place, have any intimation that that was contemplated to be done, intended to be done by him?

Answer. No, sir; I never supposed that he contemplated anything like what occurred at Harper's Ferry.

By the CHAIRMAN:

Question. What was your general information, then, if you did not know specifically what he intended to do?

Answer. I supposed that if he had an opportunity, and it came in his way to do what he did in Missouri, where he went in and took several slaves and ran them off, he would do that.

By Mr. DAVIS:

Question. And, if resisted, what then?

Answer. That is not for me to say?

Mr. DAVIS. He would have use for the arms that you furnished, if he were resisted; that was the idea, I presume. I intended to ask whether that was your idea.

Mr. FITCH. Was the supposition that Brown would resort to force a supposition of others as well as yourself?

The WITNESS. Let me explain what I mean by this?

The CHAIRMAN. Do so, fully; you have a right.

The WITNESS. I understood that John Brown ———

Mr. COLLAMER. State the time when you understood it?

The WITNESS. From first to last, I understood John Brown to be a man who was opposed to slavery, and, as such, that he would take every opportunity to free slaves where he could; I did not know in what way; I only know that from the fact of his having done it in Missouri in the instance referred to; I furnished him with money because I considered him as one who would be of use in case such troubles arose as had arisen previously in Kansas; that was my object in furnishing the money; I did not ask him what he was to do with it, nor did I suppose that he would do anything that I should disapprove of.

Mr. COLLAMER. Then I ask you, do you disapprove of such a transaction as that at Harper's Ferry?

Answer. I should have disapproved of it if I had known of it; but I have since changed my opinion; I believe John Brown to be the representative man of this century, as Washington was of the last—the Harper's Ferry affair, and the capacity shown by the Italians for self-government, the great events of this age. One will free Europe and the other America.

I wish to insert a copy of my letter to John Brown, dated Boston, November 7, 1857, as evidence of my intentions in sending the arms to Kansas:

"BOSTON, *November* 7, 1857.

"MY DEAR FRIEND: Your most welcome letter of the 16th ultimo came to hand on Saturday. I am very glad to learn that, after your hard pilgrimage, you are in more comfortable quarters, with the means to meet present expenses.

"Let me hear from you as often as you can, giving your impressions of passing events in Kansas.

"I have written Whitman, to whom I shall inclose this, that, in my opinion, the free-State party should wait for the border ruffian moves, and check-mate them as they are developed. Don't attack *them;* but if they attack you 'give them Jessie,' and Frémont besides. You know how to do it. But I think, both in Kansas and in Congress, if we let the Democratic party *try* to play their game, we shall find that they will do themselves more harm than we can do them.

"Mrs. Stearns joins me in the heartiest respect for you, and the hope that soon you will turn up in our neighborhood. We are all well, and have only our share of the trouble that now sweeps over the land.

"Truly your friend,

"GEO. L. STEARNS.

"John Brown, *Topeka.*"

By Mr. Davis:

Question. When you last furnished him money, was there any trouble in Kansas?

Answer. No, sir; there was not; but until Kansas is admitted as a free State I do not feel sure that there will not be. I do not consider that there is a guarantee yet against trouble until she is able to take care of herself.

By Mr. Fitch:

Question. The witness gave us his own idea as to the use to which Brown might put means and money—that he was to resort to force, if necessary. Now, I desire to know if that was the supposition of other members of the committee, as this gentleman has learned from conversation with them; whether they supposed, in placing arms and money at Brown's disposal, that he might use them to free negroes by force, if necessary?

Answer. I should answer that the committee did not place the arms in his possession for that purpose, and neither did I. They were placed in his hands for the defense of Kansas; they were continued in his hands for the defense of Kansas.

The Chairman. Do you know what the opinions, or views, or ideas of the other members of the committee were on the subject of Brown's using force to free negroes in the slave States?

Answer. No, sir; I do not.

Mr. Davis. At the time you wrote to Brown, withdrawing from him authority to use those arms as you understood he was about to use them, was it not the action of the committee?

Answer. The letter was signed by me as chairman of the Massachusetts State Kansas Committee, and of course carried their authority.

Mr. Davis. Were not the committee then aware of Brown's purpose to use the arms for some other end than that for which they were put in his possession?

Answer. I have answered that in my previous testimony, by stating that I wrote this letter at the request of Dr. Howe. It was not done by a regular committee meeting. We had no regular committee meetings in those days. He handed me this letter of Wilson, and requested me to write, or suggested the propriety of my writing to John Brown.

I wrote the letter to John Brown simply because Wilson had written a letter to Dr. Howe—not with any idea that there was any necessity for it.

The CHAIRMAN. Did you consult the committee about the propriety of writing such a letter?

Answer. No, sir.

Mr. DAVIS. Or afterwards, as to having written it?

Answer. No, sir; the committee did not know that such a letter was written, and I presume that most of them do not know it now. The committee was considered as virtually dissolved.

The CHAIRMAN. Did you communicate to any members of the committee, or to all of them, the substance of Wilson's letter to Howe, and Howe's request to you?

Answer. No, sir.

Mr. DAVIS. The committee, then, never did withdraw the authority from Brown?

Answer. Except so far as the letter of the chairman of the committee did it, in no other way. There was no action of the committee on the matter. The committee were not called together; no vote, no action of the committee, was taken upon it. It was an informal transaction.

Mr. COLLAMER. Had the committee before that time ceased to act?

Answer. Virtually.

GEORGE L. STEARNS.

NOTE.—I desire to add that John Brown sent word to his friends in Boston, by Mr. Hoyt, his counsel, not to make any attempt to rescue him, because his relations with his jailor (Mr. Avis) were such that he should not go out of the jail if he had an opportunity.

G. L. S.

In accordance with a request of Mr. Stearns, the following letters are appended to his testimony:

BOSTON, *March* 22, 1860.

SIR: By the testimony of Horace White, *Assistant* Secretary of the National Kansas Committee, as published in the newspapers, it would be inferred that John Brown, having been unsuccessful in his application to that committee for the Sharp's rifled carbines, proceeded to Boston to obtain them from the Massachusetts State Kansas Committee.

I have this morning received a letter from H. B. Hurd, Secretary of tho National Kansas Committee, which explains the transaction, and agrees with my testimony taken before your committee, on reference to which you will find that, on the 8th day of January, 1857, I gave John Brown an order to receive the arms; consequently he, on the twenty-fourth of the same month applied to the National Kansas Committe as our agent.

Please allow me to add these letters to my testimony, and oblige your obedient servant,

GEORGE L. STEARNS.

Hon. J. M. MASON,
Chairman of Select Committee of United States Senate.

CHICAGO, *March* 19, 1860.

There was only one meeting of the National Kansas Committee in the city of New York, and that was appointed for the 22d January, 1857, but on account of railroad accidents was delayed till 24th same month, when it commenced and continued in session for six days.

I wish to call your attention to one matter in connection with that meeting, and the application of John Brown for aid from that committee; it is this: When Mr. Brown was pressing his claim for the aid desired, I asked him this question: "If you get the arms and money you desire, will you invade Missouri or any slave territory?" To which he replied: "I am no adventurer; you all know me; you are acquainted with history; you know what I have done in Kansas; I do not expose my plans; no one knows them but myself, except, perhaps, one; I do not wish to be interrogated; if you wish to give me anything, I want you to give it freely; I have no other purpose but to serve the cause of liberty." This is the substance of what he said. I have not thought it over in some time, and could perhaps give more exactly his words.

Although it had been understood by the members of the committee that Mr. Brown intended to arm one hundred men to be scattered about in the Territory, and to be actual settlers and engaged in their several pursuits, only to be called out to repel invasion or defend the Kansas free-State settlers, yet this reply was not satisfactory to all, and the arms were voted back to your committee to be disposed of as you thought best.

It was thought by all present that, in making the inquiry above-mentioned, I was imagining a course of action that was out of the question.

I shall be happy to reply to any interrogations from you in regard to Kansas matters.

Yours, &c.,

H. B. HURD.

Mr. GEORGE L. STEARNS,
Boston, Massachusetts.

FEBRUARY 27, 1860.

HORACE WHITE sworn and examined.

By the CHAIRMAN:

Question. Will you state to the committee where you reside and what your occupation is?

Answer. I reside in the city of Chicago, in the State of Illinois. My occupation is that of an editor.

Question. Editor of what paper?

Answer. The Chicago "Press and Tribune."

Question. Were you acquainted with John Brown who was put to death recently in Virginia?

Answer. Yes, sir.

Question. How long had you known him?

Answer. Since August, 1856.

Question. Where did you make his acquaintance?

Answer. In the rooms of the National Kansas Committee, in Chicago, in that month and year.

Question. Will you state how that National Kansas Committee was composed?

Answer. It was composed of one person from each of the States of Vermont, Massachusetts, Connecticut, Rhode Island, New York, Pennsylvania, Ohio, Indiana, Michigan, three members from Illinois, one from Wisconsin, and one from Iowa.

Question. Where was their place of meeting?

Answer. Their executive committee or executive board met at Chicago. Their general meeting, the only one they ever held, was in New York city.

Question. How were the members of the committee appointed? What constituency appointed them?

Answer. They were appointed at a meeting held in Buffalo, in May, 1856; it was a sort of spontaneous gathering; it was called, I do not know that I can say by what authority, but there was a general meeting of the friends of free Kansas in the North, at Buffalo; it was held in pursuance of some sort of a call, and that meeting appointed these members of whom I speak, and they authorized the three members from Illinois to conduct the business of the committee as an executive committee.

Question. They were appointed by the national committee?

Answer. Yes, sir.

Question. Were they all residents of Chicago?

Answer. Yes, sir.

Question. Will you give their names?

Answer. Mr. J. D. Webster, chairman; Mr. George W. Dole, treasurer; Mr. H. B. Hurd, secretary.

Question. Did they possess all the powers of the committee proper, the large committee?

Answer. Yes, sir.

Question. Will you state what was the object and purpose of that committee? What were their functions?

Answer. Their purposes were to take charge of and distribute the contributions of the people of the North for free-State citizens of Kansas, whether of money, clothing, arms, or whatever else was contributed.

Question. In what way were those contributions to be used by the people of Kansas?

Answer. I can state in what way they *were* used.

Question. Was any mode pointed out by the committee for their use?

Answer. No, sir; they were to use them at their own discretion.

Question. The executive committee distributed them?

Answer. Yes, sir.

Question. Did they receive contributions in money?

Answer. Yes, sir.

Question. Were you officially connected with that committee?

Answer. Yes, sir.

Question. By what designation—in what office?

Answer. Assistant secretary.

Question. Who was the secretary?

Answer. H. B. Hurd. I would state here that the details of the business devolved upon me almost exclusively, but I had no vote in the committee.

Question. Can you state the amount of money that was received and distributed by that committee?

Answer. I can state very nearly the amount.

Question. I mean the gross amount?

Answer. Yes, sir; $120,000.

Question. Was any portion of that money expended by the committee in the purchase of arms?

Answer. Yes, sir.

Question. Can you tell how much.

Answer. Say $10,000 at the outside. It is proper to say that the large portion of our arms were received in contributions, as arms from first hands.

Question. Was any money expended by the committee in the purchase of arms?

Answer. Yes, sir.

Question. To the amount of $10,000?

Answer. I should say inside of $10,000, but about that. The transportation amounted to something.

Question. You said you formed the acquaintance of Brown in August, 1856. Was he at your committee room?

Answer. Yes, sir.

Question. More than once?

Answer. Yes, sir. He was there twice at Chicago, and once at the general meeting in New York city.

Question. Were you present in New York?

Answer. Yes, sir. Once I met him in Chicago, during the existence of this committee, when he did not appear at the rooms of the committee.

Question. Can you tell what the object of his visit to New York, during the session of the committee there was, so far as the committee was concerned?

Answer. Yes, sir.

Question. What was it?

Answer. It was in January, 1857, and he appeared there to petition the committee for two hundred Sharp's rifles and a field piece—I think a six-pounder—that were then stored at Tabor, Iowa; for a certain amount of money, as much as the committee were willing to give him; and for an amount of clothing, with which he hoped to fit out a company that he designed to use for military purpose in Kansas.

Question. Was the request granted?

Answer. No, sir. The request was granted in part. They voted him twelve boxes of selected clothing. There was a good deal of opposition to the policy of granting him those arms. Meanwhile, the Massachusetts State Kansas Committee had presented, through Mr.

F. B. Sanborn, a claim to those arms which were stored at Tabor, Iowa, and the committee finally restored the arms to the Massachusetts committee, and, therefore, Mr. Brown had no further petition for them, so far as we were concerned.

By Mr. COLLAMER:

Question. Were they arms that the Massachusetts committee had furnished?

Answer. Yes, sir.

By the CHAIRMAN:

Question. At that meeting in New York, in January, 1857, this claim of the Massachusetts society was proffered as a claim of property or control over those arms, and it was considered by the committee in New York. Did they pass from the control and possession of the New York committee to the Massachusetts committee?

Answer. They passed from the control of the National Kansas Committee to the Massachusetts committee.

Question. You say some clothing was voted to Brown; what else?

Answer. Nothing else at that time.

Question. Was anything voted to him at any other time by that committee? What was it, and when?

Answer. About November, 1856, they voted him $500 in money, and paid him $150. The balance, $350, was never paid to him.

Question. Why not?

Answer. Because the committee had not the money when he sent his draft to draw it. If the committee had had it, I have no doubt, it would have been paid.

Question. Was there any use specified, to which that money was to be put, which was thus voted?

Answer. I think not. There was a general understanding that Mr. Brown was to use it at his own discretion for the protection of the free-State men in Kansas.

Question. That was previous to the meeting in New York?

Answer. Yes, sir. At the New York meeting, Brown made his last appearance before the committee.

Question. Did you see Brown in Chicago after you had seen him in New York?

Answer. No, sir. I knew of his being there, but I did not see him.

Question. The last aid that was given by this committee, so far as we have gone, was in January, 1857, at New York.

Answer. Yes, sir; twelve boxes of clothing.

Question. Were any other contributions of any kind given to Brown afterwards by that committee?

Answer. No, sir.

Question. Did he receive any arms in any form from the committee, or any members of the committee?

Answer. The committee sent him twenty-five navy revolvers of Colt's manufacture, by Mr. Arny, but they never reached him. I think it was in August, 1856. They were sent to Lawrence, and were stored there a short time, subject to Brown's order. He did not appear to claim them, and they were loaned to a military company in Lawrence called the Stubbs, and Brown never appeared to claim them. He told me personally that the reason why he did not was, that he had had so much trouble and fuss and difficulty with the people of Lawrence, that he never would go there again to claim anything.

Question. Did you ever hold any conversation with Brown, or were you ever privy to any conversation he held with any person, in which he disclosed any plans that he had in reference to slavery outside of Kansas?

Answer. No, sir.

Question. Were you never aware that he contemplated any attempts to affect the condition of slavery in other States than Kansas?

Answer. No, sir.

Question. If you did not hear it from him, did you hear it from any others?

Answer. I have heard it since the Harper's Ferry affair took place.

Question. You have heard what since?

Answer. I have heard pretty much all that has appeared in the newspapers, and I have also seen Barclay Coppic since that affair.

Question. Did you know of any persons who were privy to the affair before it?

Answer. I knew Realf; I knew John E. Cook.

Question. Any except those who were with Brown at Harper's Ferry?

Answer. No, sir.

Question. Were any arms given to him by your committee except those navy revolvers, that you can recollect?

Answer. Not to Brown himself. I gave two rifles to two of his sons. After all the arms of the committee had been distributed in Kansas, or all but two or three, Mr. Brown made his appearance at the committee rooms with two of his sons.

Question. When was that?

Answer. October or November, 1856. One of them was Watson, and the other, I think, was Owen Brown. We had three or four rifles left, and I gave one to each of those sons, and, as they were very poorly clad, I went down to a fur store in Chicago and purchased each of them a pair of fur gloves and fur overshoes and caps, and I think that was the last dealings we had with Brown till the meeting in January, 1857, of which I have told you.

Question. At the time the arms were given to him, the navy revolvers, or at the time he applied in New York for a larger amount of arms, was any ammunition of any kind voted to him?

Answer. I think there was a considerable amount of ammunition in the form of cartridges accompanying the Massachusetts rifles at Tabor; but, as we voted those arms back to the Massachusetts committee, the ammunition accompanied them.

Question. Did you never have any conversations with Brown as to his general plans in reference to the condition of slavery in the country?

Answer. No, sir.

Question. Were you ever present at any meeting between him and the executive committee, when those plans were disclosed?

The WITNESS. You mean *our* executive committee?

The CHAIRMAN. Yes.

Answer. No, sir.

The CHAIRMAN. What I want to know is, whether you were present at any conversations between Brown and others, whether members of the committee or not, wherein he disclosed his plans and views in reference to slavery in the United States?

Answer. No, sir. I do not think he ever did before our committee develop his plans.

Question. Were you acquainted with James Redpath?

Answer. Yes, sir.

Question. Did you ever hear from him anything in reference to Brown's plans?

Answer. No, sir. Perhaps I ought, in justice to myself, to mention to the committee the way I happened to meet Mr. Barclay Coppic. A colored man in Chicago told me that he expected to see Barclay Coppic in a few days—

Mr. COLLAMER. When was that?

The WITNESS. That was about the first of this month—and I expressed to him a wish that he would introduce me to Mr. Coppic or bring him where I could see him, because I wished to learn what Brown's purposes were at Harper's Ferry. He said he would do so. Mr. Coppic arrived in town on the 8th day of this month, (February,) and this colored man procured an interview for me with Coppic.

The CHAIRMAN. It is not necessary to state anything as to that. You said you wished to explain how you came to make his acquaintance.

The WITNESS. That is all.

The CHAIRMAN. What did pass, it is unnecessary to state.

HORACE WHITE.

MARCH 5, 1860.

Hon. JOHN B. FLOYD, Secretary of War, sworn, and examined.

By the CHAIRMAN:

Question. Will you please to examine that letter, and give to the committee any information you may have concerning it? [Exhibiting the following letter:

CINCINNATI, *August* 20.

SIR: I have lately received information of a movement of so great importance that I feel it my duty to impart it to you without delay.

I have discovered the existence of a secret association, having for its object the liberation of the slaves at the South by a general insurrection. The leader of the movement is "*old John Brown,*" late of Kansas. He has been in Canada during the winter, drilling the negroes there, and they are only waiting his word to start for the South to assist the slaves. They have one of their leading men (a white man) in an armory in Maryland—where it is situated I have not been able to learn. As soon as everything is ready, those of their number who are in the Northern States and Canada are to come in small companies to their rendezvous, which is in the mountains in Virginia. They will pass down through Pennsylvania and Maryland, and enter Virginia at Harper's Ferry. Brown left the North about three or four weeks ago, and will arm the negroes and strike the blow in a few weeks; so that whatever is done must be done at once. They have a large quantity of arms at their rendezvous, and are probably distributing them already.

As I am not fully in their confidence, this is all the information I can give you. I dare not sign my name to this, but trust that you will not disregard the warnings on that account.

The envelope is directed:

"Hon. Mr. FLOYD,

Secretary of War, Washington."

Marked "private," and postmarked Cincinnati, August 23.]

Answer. I received this letter last summer, whilst I was at the Red Sweet Springs, in Virginia. My attention was a little more than usually attracted by it, and therefore I laid it away in my trunk. I receive so many anonymous letters that, of course, I pay no attention to them. I do not know but that I should have paid some little attention to this, notwithstanding it was anonymous, as the man seemed to be particular in the details, but he confused me a little by saying that these people were at work at an armory in Maryland; and I knew there was no armory in Maryland, and supposed, therefore, that it had gone into details for the purpose of exciting the alarms of the Secretary of War, and to have a parade about that for nothing; and that mistake in the statement satisfied me that there was nothing in it. Besides, I was satisfied in my own mind that a scheme of such wickedness and outrage could not be entertained by any citizens of the United States. I put the letter away, and thought no more of it until the raid broke out. Then I instantly remembered the letter, and believed the first intelligence that we received from Harper's Ferry to be true, because I recollected the contents of the letter. I showed the letter to nobody, I believe, except some member of my family, until the outbreak at Harper's Ferry. Then I remembered to have laid it away, and sat to work among my letters of the previous summer to find it. I have no means of knowing who wrote it, or what the object in writing it was. A gentleman from Cincinnati that I knew, wrote to me for the letter, stating that it was thought that the handwriting might be traced.

By Mr. COLLAMER:

Question. Had it been published?

Answer. Yes, sir. Immediately upon the intelligence of the outbreak at Harper's Ferry, the letter was hunted up and published. The object of publishing it was to show that the thing had a little more significance than a mere local outbreak, and that the country might be put on their guard against anything like a concerted movement.

The CHAIRMAN. The person referred to in Cincinnati wrote to you for the letter?

Answer. Yes, sir; and after the publication of the letter, I sent it to him, with injunctions to return it carefully to me, as it had become somewhat a link in the history of the transaction. I heard nothing of it until recently. It was miscarried in the mails, he told me, and when it was found he told me it was laid before the committee. That is all my knowledge about it.

Question. Your correspondent did not discover the writer?

Answer. He did not discover the writer, he told me; but said they had strong suspicions that certain persons somewhere in Kentucky had written it; but his statements amounted to nothing—mere suspicion.

Question. Have you received, from any source, any communication from any one apparently privy to the writing of that letter, since it was received by you, anonymous or otherwise?

Answer. I have not. I have received, since the outbreak, a good many anonymous letters speaking of stirring up insurrections in Virginia, and going into very minute and particular detail of how it was to be done, where the blows were to be struck, and all that; but I paid no attention to them at all. I knew it was evidently done to terrify the country; but none of those letters had any reference to this one.

Question. Will you state to the committee whether those anonymous letters you refer to as received since the outbreak, or any information of any kind has reached you since then, showing any purpose to interfere with the course of justice in Virginia, to rescue the prisoners or otherwise?

Answer. One single anonymous letter I did get, postmarked "Philadelphia," purporting to have been written at the seat of government of Pennsylvania, Harrisburg, saying that they would be there with a strong force, which was already mustered and prepared.

Question. Would be where?

Answer. At Charlestown; that the writer would be there, and that a rescue was certain. But of course that was extremely ridiculous, and I paid not the slightest attention to it. I remembered it, and that is all. It was a man, evidently, that would have been in the ranks, and no commander; but he professed to be one who had the thing in charge, and signed himself Cook. I remember that. That was the only one of that kind. There was one other that I recollect; and my attention was drawn to that afterwards, because there was an alarm down in Accomac. I got a letter from a man, saying there was an arrangement made on a very extensive scale; three clipper ships, he said, were loaded with arms and persons, who were to instigate insurrection upon the southern coast of Virginia. Of course, that would be understood at once to be a ridiculous thing to any man. But not very long after that intelligence came—it was printed, I believe, in all the newspapers—that there had been an alarm about insurrection in Accomac. I suppose it was just such a letter as that sent to me which excited the alarm down there. I paid no attention to it, and did not preserve any of those letters at all. They were very numerous.

JOHN B. FLOYD.

MAY 2, 1860.

Hon. WILLIAM H. SEWARD sworn and examined.

By the CHAIRMAN:

Question. Mr. Seward, will you be good enough to state to this committee whether a man named Hugh Forbes sought an interview with you at any time prior to the late attack on Harper's Ferry; where it was; and what passed between you?

Answer. I have seen a letter, purporting to be by a man named Forbes, in the newspapers, in which he states that he had an interview with me, and by the date of that letter I am able to identify the time. It was sometime in the spring of 1858, if that is the date of his letter. The person who called himself Forbes was a stranger to me, and came to my house in Washington, during the session of Congress, and asked to see me. I saw him alone. He began with a story of great personal distress, involving himself and family, and stated that he had come to me on that account. I supposed that the object of his visit was to solicit charity. I allowed him to go on with his story without interruption. I found it very incoherent, very erratic, and thought him a man of an unsound or very much disturbed mind. My object was to see whether he made any claim to the charity which I supposed he was going to solicit, and to be through with the visit as soon as I could, and therefore I did not interrupt him. What he said, in substance, was, that he was a revolutionist in Italy in the year 1848, or about that period. He was a foreigner, either an Englishman or a Scotchman, I do not know which. This fact appeared by a book that he showed me, of which, if I remember, he was the author. It was a strange and absurd book, as I thought, giving the art of exciting or getting up military revolutions, and it seemed to be the fruit of his revolutionary and military experience. Having shown me this book to show me how important a person he was, he then told me that some persons in New York, in the year previous, had induced him to go to Kansas to instruct, or under the idea that he could instruct, the free-State men of Kansas how to defend themselves against the armed invasion which was then there, or was expected there from the slave States, for the purpose of establishing slavery. He said that his family was at the time in Paris, and that he was in business in New York—what business I do not remember—which paid him, I think, a salary of $1,200; that he went to Kansas on the understanding that he should be indemnified for the loss of this business; that he met with delays in getting to Kansas and in reaching John Brown, with whom he was to coöperate; that when he reached Brown he (Forbes) was out of money and had suffered a great deal of privation, and applied to Brown for money, and Brown did not pay him any. Very soon a misunderstanding arose between him and Brown. He said Brown was a very bad man, and would not keep his word; was a reckless man, an unreliable man, a vicious man.

He said further, that in the course of their conversations as to the plan by which they should more effectually counteract this invasion—whether it was then there, or whether it was expected, I did not know—he (Forbes) suggested the getting up of a stampede of slaves secretly on the borders of Kansas, in Missouri, which Brown disapproved, and on his part suggested an attack upon the border States, with a view to induce the slaves to rise and so to keep the invaders at home to take care of themselves. He said that in their conversations Brown gave up and abandoned his own project as impracticable, and that soon after the disturbances in the Territory became quiet and ceased, and there was no longer anything for him to do there. He was penniless and Brown refused to pay him anything. He could not stay; he could not get back to New York. He appealed to persons in New England for relief and fulfillment of engagements which had been made to him, giving to them the information that his family in Paris were turned out of doors into the streets by reason of the non-fulfillment of those engagements and of his extreme poverty, and they not only refused to give him any relief, but disavowed any connection with him, or any knowledge of him, and said that John Brown was the man they knew and recognized, and the defense of the free-State men in Kansas was left to him. He replied to them by letter, as I understood, telling them that John Brown was worthless and unreliable, and fastening upon them an obligation to pay him, which they repudiated.

At that stage of the conversation, I interposed and arrested his statement, telling him (what was true) that from the beginning of the difficulties about Kansas until then, it seemed to me that everybody who had anything to do with the Territory came to me for instruction and advice how to proceed, and that from the beginning I had determined that I would have no consultations and confer with nobody, give no advice on the subject of what was done or to be done in Kansas, because it was inconsistent with the relations that I maintained here, where I was to act as a Senator upon what was done in Kansas and advise what ought to be done by Congress, that I had listened to him so far in violation of that rule, simply because he seemed to be in deep distress, and I was touched with the story he told me about his family, and was waiting to see whether he would state any grounds on which I could give him some money, that he had failed to do so, and that there the conversation must end, that he must leave me and see me no more. He went away.

Question. I understand then from your statement, that he made no reference to any purpose of Brown's to make an attack upon any of the slave States, except as you have given it in Kansas?

Answer. None whatever. On the contrary, until I read in Europe of John Brown's demonstration at Harper's Ferry, I had no more idea of an invasion by John Brown at that place, than I had of one by you or myself. The thing was one of those occurrences which, as I suppose happen often to all persons in our situation; certainly they do in mine. All kinds of erratic and strange persons call on me with all manner of strange communications and applications. This was one of them, and it passed out of my memory without leaving attached to it the least idea of any importance. Forbes told me nothing of any cotton speculation by any humanitarians or anybody else. I state this in answer to a statement contained in his published letter.

Question. Will you state whether Forbes was introduced to you, or whether he introduced himself?

Answer. He brought a simple letter of introduction from Doctor Bailey.

WILLIAM H. SEWARD.

INDEX TO TESTIMONY.

www.ingramcontent.com/pod-product-compliance
Lightning Source LLC
LaVergne TN
LVHW020219110826
845151LV00003B/759

9781425532642